100 Years of General Relativity – Vol. 3

AN INTRODUCTION TO COVARIANT QUANTUM GRAVITY AND ASYMPTOTIC SAFETY

100 Years of General Relativity

ISSN: 2424-8223

Series Editor: Abhay Ashtekar *(Pennsylvania State University, USA)*

This series is to publish about two dozen excellent monographs written by top-notch authors from the international gravitational community covering various aspects of the field, ranging from mathematical general relativity through observational ramifications in cosmology, relativistic astrophysics and gravitational waves, to quantum aspects.

100 Years of General Relativity – Vol. 3

AN INTRODUCTION TO COVARIANT QUANTUM GRAVITY AND ASYMPTOTIC SAFETY

Roberto Percacci

SISSA - International School for Advanced Studies, Trieste, Italy

 **World Scientific**

NEW JERSEY · LONDON · SINGAPORE · BEIJING · SHANGHAI · HONG KONG · TAIPEI · CHENNAI · TOKYO

Published by

World Scientific Publishing Co. Pte. Ltd.

5 Toh Tuck Link, Singapore 596224

USA office: 27 Warren Street, Suite 401-402, Hackensack, NJ 07601

UK office: 57 Shelton Street, Covent Garden, London WC2H 9HE

Library of Congress Cataloging-in-Publication Data

Names: Percacci, Roberto, author.

Title: An introduction to covariant quantum gravity and asymptotic safety /
    by Roberto Percacci (SISSA, Italy & Perimeter Institute for Theoretical Physics, Canada).

Other titles: 100 years of general relativity ; v. 3.

Description: Singapore ; Hackensack, NJ : World Scientific, [2017] |
    Series: 100 years of general relativity, ISSN 2424-8223 ; vol. 3

Identifiers: LCCN 2016051260| ISBN 9789813207172 (hardcover ; alk. paper) |
    ISBN 9813207175 (hardcover ; alk. paper)

Subjects: LCSH: Quantum gravity. | Perturbation (Mathematics) |
    Functions of complex variables. | Renormalization group.

Classification: LCC QC178 .P47 2017 | DDC 539.7/54--dc23

LC record available at https://lccn.loc.gov/2016051260

British Library Cataloguing-in-Publication Data

A catalogue record for this book is available from the British Library.

Typeset by Stallion Press

Email: enquiries@stallionpress.com

Printed in Singapore

Preface

The aim of this book is primarily pedagogical. It is intended as an introduction to the covariant formalism of quantum gravity and in particular to current research on asymptotic safety. The first four chapters, forming the first half of the book, are based on a short course entitled "Introduction to quantum gravity" that I have taught at SISSA in the last three academic years. They contain a concise review of well-established facts about the covariant approach to quantum gravity, the central result being the derivation of one loop divergences by 't Hooft and Veltman, described in chapter 3. Another classic result in quantum gravity is asymptotic freedom of higher derivative gravity, which is reported in detail later, together with other calculations of beta functions. This first part closes with a discussion of open options.

The second half of the book is an introduction to the current work on asymptotic safety. It begins with two technical chapters containing material that is useful also outside the domain of asymptotic safety, mainly heat kernel techniques and the functional renormalization group. Chapter 7 contains the general definition of asymptotic safety and then a number of explicit calculations in gravity. I have chosen to present in some detail the one loop calculation of beta functions in Einstein-Hilbert gravity, in higher derivative gravity and in three-dimensional topologically massive gravity, which are well understood, and some sample calculations that go beyond the one-loop approximation. The limitations of these examples are clear. Still one may hope, by virtue of their relative simplicity, that some aspects of these calculations will withstand the test of time.

These results are presented with very few references. There follows in chapter 8 a fairly comprehensive overview of the literature and a section describing the state of the art. All the current research lines and open problems are briefly discussed there with full references. This, together with the technical material, should enable the motivated reader to start working in this field.

Trieste, summer 2016

Acknowledgments

I will begin by mentioning some other books, both to acknowledge their influence and to recommend them for further reading.

• D. Oriti, editor "Approaches to Quantum Gravity", Cambridge University Press (2009),

is an up-to-date and complete collection of contributions, representing all the current research lines in quantum gravity.

• K. Kiefer, "Quantum gravity", Oxford University Press (2004),

is a more organic, single-author view of the topic.

There are several excellent books dedicated specifically to canonical quantum gravity. The covariant approach is less represented.

• B.S. DeWitt, "The Global Approach to Quantum Field Theory", Oxford University Press (2002)

is a compendium of the work of one of the creators of this subject.

• G.W. Gibbons and S.W. Hawking, editors "Euclidean Quantum Gravity", World Scientific (1993)

is a useful summary of classic results of the Cambridge school, quite close to the material in chapter 3.

• I.L. Buchbinder, S.D. Odintsov and I.L. Shapiro, "Effective action in quantum gravity", IOPP Publishing, Bristol (1992).

is in several ways closest to this book and provides a wealth of additional material. I wish to thank the authors for many informative discussions over the years and for an invitation to Tomsk in 2007.

The covariant background field method turns calculations in quantum gravity into calculations in quantum field theory in a fixed background. At a technical level there is therefore considerable overlap with

• N.D. Birrell and P.C.W. Davies "Quantum fields in curved space", Cambridge University Press (1984),

• V.F. Mukhanov and S. Winitzki "Introduction to quantum effects in gravity", Cambridge University Press (2007),

• L. Parker and D. Toms "Quantum field theory in curved spacetime", Cambridge University Press (2009).

These works discuss many examples of calculations of observable effects directly in Lorentzian signature and are a useful complement to the Euclidean calculations discussed here.

My understanding of the material in the second part of the book has been shaped through interactions with many collaborators, students and colleagues, starting with Roberto Floreanini and Luca Griguolo, with whom I tried in the 1990's to apply renormalization group ideas to quantum gravity. This game changed with Martin Reuter's seminal paper applying the functional renormalization group to gravity. His influence on this field cannot be overestimated and I am much indebted to him for sharing his insights with me on many occasions.

I have also learned much from collaborating with Astrid Eichhorn, Marco Fabbrichesi, Pedro Machado, Nobuyoshi Ohta, Ergin Sezgin, Alberto Tonero, and Gian Paolo Vacca, and from numerous discussions with Dario Benedetti, Alfio Bonanno, Holger Gies, Daniel Litim, Renate Loll, Tim Morris, Jan Pawlowski, Frank Saueressig, and Christof Wetterich. It is a pleasure to acknowledge the special role of my students Djamel Dou, Daniele Perini, Alessandro Codello, Christoph Rahmede, Omar Zanusso, Gaurav Narain, Adriano Contillo, Leslaw Rachwal, Pietro Donà, Giulio d'Odorico, Carlo Pagani, and Peter Labus who have shared with me the joys and frustrations of scientific research. Working with them has been a constant source of inspiration. Finally, special thanks go to Dario Benedetti, Alessandro Codello, Nobuyoshi Ohta, Vedran Skrinjar, and Gian Paolo Vacca for reading parts of an early draft of this book and making many useful comments, and to Peter Labus for help with the figures.

As a rule calculations in quantum gravity require a substantial amount of work. In all but the simplest cases discussed in this book algebraic manipulation software is indispensable. Extensive use has been made of commercial software, and for the material in sections 7.6 and 7.8 also of the free tensor manipulation package xAct.

Contents

Chapter 1

Quantum gravity:
A brief historical overview

1.1 Early years

The four known fundamental interactions can be divided in two groups. Gravity and electromagnetism have long range, manifest themselves macroscopically and have been known for centuries. The weak and strong nuclear forces act only at very short distances and were discovered only relatively recently. Gravity and electromagnetism are described by classical field theories. Both are examples of gauge theories, namely theories whose equations are covariant under an infinite dimensional group of transformations.

All this was well appreciated in 1916, after the discovery of General Relativity (GR). Because of these analogies, steps towards a quantum theory of electromagnetism would soon be followed, or sometimes even anticipated, by similar steps towards a quantum theory of gravity.[1] Significant early contributions were made by Rosenfeld [4] and Bronstein [5]. The notion of "graviton" as the quantum of the gravitational field was established already in the 1930's. In 1939 Fierz and Pauli derived the linearized Einstein equations as the relativistic field equation for a spin-2 field propagating in Minkowski space [6]. This was further developed by Gupta [7] using methods similar to Gupta-Bleuler quantization of the electromagnetic field. At this point there was a quantum theory of free gravitons and the next step was understanding their interactions.

1.2 DeWitt's era

Mechanical problems can be formulated in Lagrangian or in Hamiltonian formalism and even though one of the two formalisms may be more convenient in some particular circumstance, it is usually the case that when a problem can be understood in one way, it can also be understood in the other. In relativistic quantum field theories the use of Hamiltonian methods has the disadvantage of hiding Lorentz

---

[1] For an account of the early history see [1], and in particular [2] for the role of M. Bronstein. A concise general history of quantum gravity can also be found in [3].

invariance, so the Lagrangian formalism is often preferred. Attempts at quantum gravity proceeded in parallel but the difficulties encountered in the two schemes are seemingly very different and unrelated. Bryce DeWitt received his PhD in 1950 and in the subsequent decades played a central role in the development of both research lines.

Gauge invariance manifests itself at Hamiltonian level in the non-invertibility of the relation between velocities and momenta. This gives rise to constraints on the canonical variables and a proper way of dealing with these constraints was developed in the 1950's by Dirac and Bergmann. The application of this formalism to gravity culminated in a famous work by Arnowitt, Deser and Misner [8], where they presented the "ADM formalism" for the canonical treatment of gravity. Gravity can be seen as the motion of a point particle in an infinite dimensional "superspace" (not to be confused with the superspace of SUSY theories) consisting of three-dimensional metrics modulo spatial diffeomorphisms. In the mid-1960's the Wheeler-DeWitt equation was written down as a formal analogue of the Schrödinger equation on superspace [9]. It soon appeared that this nice formalism is fraught with mathematical and conceptual difficulties, making progress nigh impossible. At the mathematical level, one would have to make sense of the Wheeler-DeWitt operator as a sort of Laplacian on a Hilbert space of wave functionals over superspace; at the conceptual level one has to deal with the "problem of time": the fact that the Hamiltonian is a constraint that must vanish.

In the meantime, the parallel development of a quantum theory of the electromagnetic and gravitational fields had diverged. In 1947 a famous conference took place at Shelter Island, which led to great steps forward in the understanding of interacting quantum field theories, and the construction of QED was completed in the following years. The development of quantum gravity was obviously a much less urgent affair, since no experimental data existed, but when the issue was addressed, no comparable success could be claimed. The problem of computing transition amplitudes in quantum gravity had been addressed in earnest in the early 1960's by Feynman and DeWitt. Feynman showed that tree amplitudes reproduce the results of General Relativity[2] but then stumbled over apparent violations of unitarity. This was due to the presence of the unphysical gauge degrees of freedom and was fixed by DeWitt, who essentially introduced what we now call the Faddeev-Popov ghosts [10,11]. Thus the correct quantization of a non-abelian gauge theory was established for gravity *before* Yang-Mills theory. The issue of the renormalizability of the theory remained however open. The different dimension of the electromagnetic and gravitational couplings, and its potentially nefarious consequences, had already been remarked by Heisenberg in 1938 [12]. It was only in 1972, however, that the one-loop effective action of Einstein's theory was calculated by 't Hooft and Veltman [13], using the tools developed by Feynman and DeWitt. This milestone

---

[2]If GR had not been discovered by Einstein in 1915, it would have presumably been discovered by particle physicists at about this time.

paper established the non-renormalizability of gravity coupled to a scalar field, but left open the possibility of "miraculous" cancellations in pure gravity. The non-renormalizability of gravity coupled to various types of matter was established soon thereafter [14–19]. The remaining gap was only closed several years later by Goroff and Sagnotti [20] and van de Ven [21], who established the existence of a divergent term cubic in curvature in the effective action of pure gravity at two loops.

These papers led to a firm conclusion: the perturbative treatment of Einstein's theory as a quantum field theory, either on its own or coupled to generic matter fields, leads to the appearance of infinitely many divergences that spoil the predictivity of the theory. There were subsequently several attempts to wiggle out of this impasse, while remaining within the context of quantum field theory. They can be divided in three categories. First, one could change the gravitational equations. Stelle proved that a theory containing four-derivative terms in the Lagrangian (*i.e.* terms quadratic in curvature) is perturbatively renormalizable [22, 23]. Unfortunately it also appeared that those Lagrangians leading to a renormalizable theory contain propagating ghosts. These ghosts, unlike those introduced by DeWitt-Faddeev-Popov, would be physical particles and hence would lead to violation of unitarity. Conversely those Lagrangians that do not contain ghosts are non-renormalizable. At the perturbative level, one is thus led to conclude that there is incompatibility between unitarity and renormalizability.

The second attempt was based on the following idea: while a generic theory of gravity coupled to matter is non-renormalizable, perhaps there are some special combinations of matter fields that would lead to a renormalizable theory. By far the most important examples in this class are supergravity theories (SUGRA), pioneered by Freedman et al. [24]. They come in many varieties, depending on the dimension, the number N of superpartners of the graviton and the presence of additional multiplets or additional invariances (e.g. conformal SUGRA). Supersymmetric theories are very special because the balance of bosonic and fermionic degrees of freedom leads to cancellation of divergences in loop diagrams and indeed even the simplest SUGRAs do not have the two-loop divergence that is present in GR. Besides the improved quantum behavior, these theories have the aesthetic appeal of being less arbitrary than non-SUSY theories, and there were hopes that in this way one could arrive at a unique realistic unified theory of all interactions (a "Theory of Everything" or "TOE"). The difficulty of obtaining chiral fermions from dimensional reduction and the presence of anomalies largely thwarthed these hopes.

The third possibility is that the failure of renormalizability is a pathology of the perturbative approach, and not of gravity itself. There is more than one way of implementing this idea. The Hamiltonian approach to quantum gravity can be viewed as falling in this broad category, and will be discussed separately below. Within the covariant formalism, most work has been based, more or less explicitly, on the Feynman "sum over histories" approach. The earliest attempt goes back to 1957 and is due to Misner [25]. The Euclidean version of the gravitational functional integral was developed in the late 1970's especially by the Cambridge school [26, 27]. The non-perturbative nature of this approach is highlighted by the sum over gravitational instantons [28–30]. This approach led to important insights on the thermal nature of black holes [31] and of the creation of the universe *ex nihilo* [32]. Some of the issues arising in this context will be discussed later.

A less formal way of defining the gravitational path integral, parallel to similar work in QCD, is the lattice approach. Using a hypercubic lattice does not go well with our understanding of gravity, so two approaches have been mostly used: quantum Regge calculus (fixing a triangulation and allowing the edge lengths to fluctuate) [33], and Euclidean Dynamical Triangulations (EDT, building spacetime with identical, equilateral simplices) [34, 35]. Both programs ran into difficulties, in particular EDT was shown in the mid 1990's to lead only to pathological phases (a "crumpled" state in which all simplexes are directly connected or a "branched polymer" phase) separated by a first order phase transition [36, 37]. Work in this direction languished for a while.

The non-perturbative way out of the issue of the UV divergences is known as "nonperturbative renormalizability" and originates from the work of Wilson on the renormalization group. A theory with this property would be described by a renormalization group trajectory that tends to a non-free fixed point in the UV. This would define a continuum limit that is outside the perturbative domain. This idea was first floated by Weinberg at an Erice school in 1978 [38] then again, and in more detail, in his contribution to the Einstein centenary volume [39]. He used the term "asymptotic safety" for such a behavior, to emphasize the similarity to "asymptotic freedom". The original evidence for asymptotic safety of gravity came from calculations in $2+\epsilon$ dimensions, but it was not known how to take the physically interesting limit $\epsilon \to 2$ and so work on this line of research also subsided very quickly.

1.3   Loop quantum gravity

Let us now return to the Hamiltonian approach. Among the most problematic features of the ADM formalism is the non-polynomiality of the constraints, and in particular of the Wheeler-DeWitt equation, giving rise to particularly severe factor-ordering issues. In the 1980's, Ashtekar discovered a canonical transformation that makes the constraints polynomial [40]. One of the most interesting features of this formulation is that prior to imposing the diffeomorphism constraints the phase space of the theory is the same as that of a Yang-Mills theory. The main issue with this formulation is that one has to use complex variables and impose reality constraints in the end. A few years later Rovelli and Smolin reformulated the theory in such a way that its Hilbert space (again prior to imposing diffeomorphism constraints) consists of spin networks [41]. This was the beginning of Loop Quantum Gravity (LQG), one of the main contenders for a quantum formulation of gravity. In the course of the years, LQG went through several changes of emphasis. One was the replacement of Ashtekar's complex variables by a set of real ones. Another goes under the name "spin foams", which are four-dimensional structures interpolating between spin network states. The transition amplitude between two spin networks is then given by a sum over spin foams having the given networks as boundaries and governed by a purely combinatorial action. Note that in this way spin foams can also be seen as a definition of the gravitational sum over histories. Further developments along these lines are Group Field Theory [42] and tensor models [43], which can also be seen as more refined versions of dynamical triangulations. With these developments the original canonical approach has morphed from a direct attempt at "quantizing GR" into something resembling more closely the Feynman path integral of a statistical model. The action of these models is only vaguely inspired by GR, with the main issue now being whether the models reproduce GR in some suitable continuum limit.

1.4 The standard model

Let us return to the non-renormalizability results of the late 1970's and early 1980's. To understand the subsequent developments, one has to recall the main developments in particle physics after the Shelter Island conference. After the success of QED, attention turned towards the weak and strong interactions. Although one can try to model them by classical fields with exponentially decaying potentials, this is not particularly useful. It is the nature of the nuclear forces that they always act between elementary particles, so they must necessarily be described in terms of quantum fields. The first useful theory of the weak interactions was written by Fermi in 1933. Unlike gravity and electromagnetism, in Fermi's theory the weak interactions do not proceed through a field propagating *in vacuo* but are rather described as a contact interaction between four fermions. More precisely, Fermi postulated an interaction term of the form

$$\frac{G_F}{\sqrt{2}} J^\mu J^\dagger_\mu \,, \tag{1.1}$$

where J_μ is a current bilinear in the spinor fields. Initially Fermi used the vector current $J_\mu = V_\mu$. Gamow and Teller later introduced the possibility of parity violation through an admixture of axial currents A_μ and still later Marshak and Sudarshan and independently Feynman and Gell-Mann determined that the correct combination of currents is of the form $J = V - A$. These four-fermion interactions are non-renormalizable. The idea that emerged was that looking at this seemingly point-like interaction at distances of order $\sqrt{G_F}$ or shorter, one would see that the fermions actually exchange a vector boson, just as in QED two electrons scatter by exchanging a photon. The only difference is that to account for the short range of the interaction the vector boson of the weak interactions would have to be massive, with a mass proportional to $G_F^{-1/2}$. The technical difficulty then lay in understanding the properties of massive vector bosons.

In 1953 Pauli generalized the Kaluza-Klein construction from five to six dimensions, using 2-dimensional spheres as fibers. In this way he arrived at the notion of non-abelian gauge fields (specifically, for the group $SO(3)$), but he did not publish this work because he was not able to write a gauge-invariant mass term (see [44]). Yang and Mills did not have such qualms and one year later proposed non-abelian gauge fields, initially as possible carriers of the strong interactions [45]. Their work was criticized by Pauli, but this issue was eventually resolved in 1960 with the notion of the Higgs mechanism.[3] As a side remark, it is worth mentioning that shortly after Yang and Mills' work, Utiyama recognized the similarity between gravity and Yang-Mills fields, starting the research line known as "gauge theories of gravity" [47]. In the subsequent decade, the Yang-Mills theory for the group $SU(2) \times U(1)$ was proposed by Weinberg and Salam, based on earlier work by Glashow, as a model for electromagnetic and weak interactions. The proof of renormalizability of this theory was provided by 't Hooft in 1972, as already mentioned, opening the way to wide acceptance of the model. Crucial experimental confirmations were the discovery of the neutral currents at CERN in 1973 and eventually direct detection of the W and Z bosons, again at CERN in 1983.

The story of the strong interactions is quite different. The original field-theoretic model for the binding of protons and neutrons in the nuclei had been proposed by Yukawa in 1934: it consisted of the exchange of a scalar particle called "meson", whose mass he could approximately predict, leading to an exponentially falling interaction potential. These particles were indeed discovered in 1947 (they are now known as the π mesons). Unfortunately, the fact that the coupling constant is of order one precluded the use of this field theory beyond tree level. Accelerator experiments in the 1950's produced a plethora of new strongly interacting particles (collectively called hadrons) and it proved impossible to describe them by a perturbative QFT. This led many physicists, including Heisenberg, to doubt that QFT could ever describe the strong interactions, and led them pragmatically to an alternative approach called "S-matrix theory". Proponents of this approach were ready to give up the notion of spacetime at nuclear distances and tried to directly determine the S matrix from general properties such as unitarity and analyticity. The observation that hadrons can be arranged on Regge trajectories led to the conclusion that not all hadrons can be elementary particles and to the development of a string theory of the strong interactions. This kind of approach was popular in the 1960's.

In the meantime, hadrons had been classified according to their charge and isospin (Wigner, Heisenberg) and later also strangeness (Nishijima, Gell-Mann). Neglecting their mass differences, they turned out to fall into multiplets of either $SU(2)$ or $SU(3)$. Gell-Mann observed that this structure could be explained if the

---

[3]This story is told in detail in O'Raifeartaigh's book [46].

hadrons were made of more fundamental constituents called quarks, carrying the fundamental representation of $SU(3)$, but given that they cannot be isolated it was not clear in what sense they could be thought of as particles. There was another difficulty with this idea, namely some hadrons would require combinations of quantum numbers that are forbidden by Pauli's exclusion principle. This led Han, Nambu and Greenberg to postulate in 1965 that the quarks carry an additional quantum number of another $SU(3)$, now called "color", and that they interact exchanging vector bosons (nowadays called gluons) in an octet of this group. Feynman had a somewhat similar model of hadrons as being composed of elementary constituents he called "partons", but unlike Gell-Mann he was more inclined to think of them as ordinary particles. Based on Feynman's ideas, Bjorken predicted that certain scaling properties should hold in the deep inelastic scattering of electrons and protons, and this was spectacularly verified in experiments at SLAC in 1969. The final theoretical development leading to the success of QCD was the discovery of asymptotic freedom by Gross and Wilczek [48] and Politzer [49] in 1973. This explained why the quarks/partons behave like free particles inside the hadrons at high energy, in spite of the theory being strongly interacting at large distances. These developments led physicists to abandon the *S*-matrix approach in favor of QCD, so also the strong interactions could be successfully described, at least at high energy, by a perturbative Yang-Mills theory.

1.5 GUTs, supergravity and superstrings

By 1973 the Standard Model (SM) of the electroweak and strong interactions was therefore in place, essentially in the same form that it maintains today. The Weinberg-Salam model gives a coherent description of the electromagnetic and weak interactions, but is not technically a unified theory, since the group $SU(2) \times U(1)$ is not simple and thus there are two gauge couplings. There followed years in which many tried to extend the gauge group of the SM in such a way as to have a truly unified theory. This led to so-called grand-unified theories (GUTs) which are quite successful at collecting all known elementary particles in a few multiplets but become quite cumbersome when one has to explain the way in which the group breaks down to the SM group. Above all, persistent failure to observe proton decay, the main theoretical prediction of GUTs, has led to disillusion and loss of interest.

Attempts to bring gravity into the picture rekindled interest in old approaches to unification and in the late 1970's and early 1980's there was a revival of the Kaluza-Klein approach. The original idea of obtaining the gauge fields from the mixed (internal/spacetime) components of the metric proved too restrictive and usually all matter fields (including the gauge fields) were assumed to be present in the higher dimensional theory. Rather than being simply postulated, the local product structure of the higher dimensional space was obtained by a dynamical mechanism of "spontaneous compactification". Harmonic expansion of the matter fields in the compact dimensions gave rise to an effective dynamics in spacetime, including both light modes, that can be identified with the known fields, and an infinite tower of yet unobserved very massive modes. All this was often in supersymmetric form. In particular, the realization that there exists a unique $N = 1$ supergravity (SUGRA) in $d = 11$, enjoying very special properties and related by dimensional reduction to $N = 8$ SUGRA in $d = 4$, raised hopes that a TOE could be found along these lines. Some of these SUGRAs turned out to be related to superstring theories in $d = 10$. One of the major difficulties with some of these models were gauge anomalies, which imply that these are not consistent as quantum theories. It was therefore a major breakthrough when Green and Schwartz discovered that certain very special superstring theories in $d = 10$ are free of anomalies. This came to be known as the first superstring revolution.

The preceding review of the history of particle physics should help the reader understand the subsequent massive re-orientation of the particle physics community. The first reason is that after QED, the understanding of the fundamental interactions as quantum phenomena had progressed through phases of unification. By the mid-1980's there had been great theoretical steps forward in this direction and it seemed that gravity would be the last piece in a puzzle that had already been largely built. There was therefore little interest, or even little faith, in the idea of constructing a quantum theory of gravity as a standalone interaction.

The second reason, which was important for a subset of particle physicists, was the existing work on string theory as a model for strong interactions. Closed string theory predicts the existence of a massless spin-2 state. As long as strings were used as models for hadrons, there was no natural interpretation for this state, and the reinterpretation of this state as the graviton was the first step of the first superstring revolution.

The third reason is the importance of the perturbative approach both in establishing the existence of the theory and as a tool to extract from it quantitative predictions. Essentially all the calculations leading to quantitative predictions in the SM are based on this method. The main area where perturbation theory fails are the strong interactions at low energy, where QCD becomes strongly coupled, and here the situation is less brilliant. In principle one could try to derive all the parameters governing the physics of hadrons out of QCD. Lattice gauge theory is now able to predict the masses of these particles (or more precisely their mass ratios): this is a great success, but it came after thirty years of hard work and we are still far from understanding the dynamics. This explains why the majority of particle physicists would put more faith in a perturbative approach such as superstrings rather than a non-perturbative one such as Weinberg's asymptotic safety.

Superstring theory went through enormous development in the 1980's and 1990's but has not fulfilled some of the initial hopes, such as being able to calculate some of the free parameters of the SM. The reason is that while superstring theories in $d = 10$ come only in a handful of varieties, and are perhaps subsumed by a unique structure in $d = 11$, they admit a huge number of dimensional reductions, removing much of the predictive power. On the other hand superstring theory has spawned many new theoretical ideas and tools that are useful in other contexts, perhaps the most important one being the AdS/CFT correspondence. It has also led to major progress in understanding field theories at nonperturbative level. For this reason superstring theory has come to be viewed by some as a more general and more powerful theoretical framework in which to discuss QFT.

1.6 Recent developments and future prospects

Around the turn of the century some developments have taken place that somewhat changed the perspective on the problem of quantum gravity. The first is the realization, within the particle physics community, that non-renormalizable theories can be useful and predictive in some limited energy domain, if one uses a set of techniques that go under the name of "Effective Field Theory" (EFT). In this way the notions of renormalizability and UV completeness lost some importance. These ideas percolated to gravity mainly through the effort of J. Donoghue, who provided a concrete example of a loop calculation that can be done in perturbative quantum GR and is not affected by the uncertainties about its UV behavior: it is the calculation of the leading quantum correction to the Newtonian potential, that will be described in section 4.5.4. It is based on methods that have been tested in a wide variety of other phenomena, it is free of ambiguities and it is hard to imagine that one of its premises could be invalidated in the future. It is probably not an exaggeration to say that this is the most reliable result we have in quantum gravity.

Quite generally, the modern EFT approach implies that the old Feynman-DeWitt perturbative treatment of gravitons in a background spacetime, with interactions dictated by the Hilbert action of GR, is a useful and consistent QFT as long as one only asks questions about "low energy" physics, where by "low energy" one means energies lower than the Planck mass m_{Planck}. In fact the expansion parameter of the EFT of gravity is the ratio E/m_{Planck}, where E is the characteristic mass scale of the phenomenon under study. Even at the highest energies available in accelerators, or in cosmic rays, this number is extremely small, in fact much smaller than the couplings of the other interactions at the same energies. In this sense one could paradoxically say that the EFT of gravity is *the best* perturbative QFT.

The predictions of this QFT deviate from those of classical GR only by amounts that are way too small to be detected. This is unfortunate, because it means that it is probably going to be very hard to obtain proof of genuine quantum effects in gravity, but it can also be viewed positively, because every test of GR is also a test of the EFT of gravity. If we keep in mind that the SM itself is incomplete (at least because the abelian gauge interactions are not asymptotically free), then one should conclude that we do have a QFT of gravity that is no worse than the QFT of the electroweak interactions, with a domain of applicability ranging from cosmological scales up to scales that are way higher than anything that can be currently achieved in accelerators.

While this may be viewed as a spectacularly successful theory, there are several reasons not to be completely satisfied. The first is conceptual. The EFT is definitely a quantum theory of gravity, as long as we understand "gravity" in the Newtonian sense of "the force that makes apples fall". But Einstein taught us to think of gravity as the geometry of spacetime rather than a force, and in the EFT, spacetime is still Minkowski space, or at most some fixed curved background, so this is definitely not a "quantum theory of spacetime". Understandably, this motivation is felt more strongly among General Relativists, whereas old-school particle physicists tended to be more comfortable with the notion that the geometric structures of GR may be merely an illusion emerging in some limit from the dynamics of gravitons.[4] With the recognition that also Yang-Mills theories have a geometrical interpretation, and the related work on instantons, this attitude has become unpopular even among particle physicists, so that there is now a wide consensus that "quantum gravity" cannot be merely a theory of gravitons propagating on a fixed background.

One may argue about how much of the structure of GR is going to be needed, but in any case by "quantum gravity" most people nowadays mean a "quantum theory of spacetime", and there is no generally accepted theory of this type.

A second and more practical reason has to do with the existence of singularities in GR. John Wheeler justly called this "the greatest crisis in physics of all time". It is generally expected that quantum gravity would provide an answer to this issue, but the EFT of gravity is of no help here. This is because the quantum effects are generally expected to come to the rescue only when the curvature reaches the Planck scale, and this is the regime where the EFT breaks down too. So, the quantum theory of gravity that solves the problem of the singularities will be a quantum theory of spacetime in some sense.

The sense in which the EFT breaks down at the Planck scale is that all the operators in the effective action, containing any number of curvatures and derivatives, become equally important. The coefficients of all these terms are not calculable in the EFT, leading to a breakdown of predictivity. In principle, some notion of UV completeness can be used as a criterion to select theories and may restore some form of predictivity. This is the same logic that led to the formulation of the Standard Model, and it may be appropriate to use it again in the context of gravity. Here too there has been progress.

Returning to our historical review, in addition to developments in superstring theory and in LQG, the turn of the century has seen a resurgence of covariant non-perturbative approaches. Work on asymptotic safety, which had languished since the 1980's, has been restarted by M. Reuter's application of "functional renormalization group" techniques to gravity. This has led to new evidence for the existence of a nontrivial fixed point directly in four dimensions. This notion of non-perturbative

---

[4]This point of view has been expressed for example in the preface and section 6.9 in [50].

renormalizability makes the theory predictive also in the trans-Planckian regime. It will be the main topic of the last chapters of this book.

The dynamical triangulation approach has been revamped in the form of "Causal Dynamical Triangulations" (CDT), where the sum is restricted to Euclidean configurations that derive from a Lorentzian spacetime. This has led to a different and much more promising phase structure, with a phase that looks like an extended, de Sitter space. In principle these Monte-Carlo calculations could provide numerical tests of the asymptotic safety idea.

A different idea that can also be tested by CDT has been proposed by Hořava. It consists in giving different scaling dimensions to time and space and writing an action that contains four space derivatives but only two time derivatives. In this way higher derivative gravity can be made renormalizable and free of ghosts at the expense of local Lorentz invariance.

At the same time a dedicated group of researchers led by Bern and Dixon, making use of ideas and techniques that originate from superstring theory, has made great progress in the calculation of SUGRA amplitudes at previously unthinkable loop order, finding unexpected cancellations and prompting them to conjecture that the theory may even be finite. The divergence structure of $N = 8$ SUGRA is still being actively investigated.

All the above research lines, and to some extent also spin foams, group field theory and tensor models, are direct extensions of the covariant QFT approach to quantum gravity. In spite of the perturbative non-renormalizability of GR, this research line initiated by Feynman and DeWitt more than 50 years ago, is therefore still very active.

In particle physics, many new ideas for physics beyond the SM (BSM) have been put forward over the years and the success of the SM has become a source of frustration. The results of the first LHC run leave open the possibility of a "great desert" between the Fermi and the Planck scale. This would be a completely new situation: up to now, every time a new energy scale has been opened up for exploration, new phenomena have always appeared. There is much hope that the second LHC run will reveal something new. On the other hand, the desert scenario may be positive for quantum gravity, since it may give an unimpeded view of some Planck scale phenomena. Either way, it is conceivable that a deeper understanding of some outstanding issues in particle physics may also lead to new insights into quantum gravity.

Chapter 2

Gravitons

2.1 The linear field equations

There are in principle two ways to approach General Relativity. One may call them the "top down" and the "bottom up" approach. Historically the top-down approach came first: it is the route followed originally by Einstein. Armed with some physical intuition and with the notions of Riemannian geometry, in 1915 he arrived by pure thought at a unique set of nonlinear second order differential equations for the gravitational field. For weak fields these equations can be linearized and have solution describing the propagation of gravitational waves in a background Minkowski space.

The bottom up approach is more laborious: it consists of starting from Lorentz-covariant linear field equations and trying to reconstruct the full nonlinear theory from there. The appropriate linear equation was written in 1939 by Fierz and Pauli, who were searching for the relativistic wave equation for a spin-2 particle [6]. The question of reconstructing the interactions of such a theory was addressed and only partly answered by Feynman around 1962 [51]. The reconstruction of the full nonlinear theory can be achieved with a clever trick found later by Deser [52].

The two approaches are complementary and equally instructive. Due to its elegance and in part perhaps also to Einstein's charisma, the top-down approach is far better known. Given that in this chapter we will be dealing with the linear theory, we shall use both approaches and start from the bottom-up one.

2.1.1 *The relativistic spin-2 field equation*

In particle physics, all forces are thought of as due to the exchange of some mediator. For example, in the Yukawa model the nuclear forces between nucleons are due to the exchange of scalar particles (the mesons). In the standard model, electroweak and strong interactions are due to the exchange of spin-1 particles. It is then natural to think also of gravity as the effect of the exchange of some particle that we may call "graviton".

Gravity is a universal interaction that is proportional to the masses of the interacting bodies and decays like the inverse of the square of the distance. The corresponding potential energy is

$$V(r) = -G\frac{m_1 m_2}{r} \tag{2.1}$$

where $G = 6.674 \times 10^{11} m^3/(s^2 kg)$. Most of the time we shall use natural units $\hbar = 1$, $c = 1$, where Newton's constant corresponds to $2.612 \times 10^{-70} m^2$ (see Appendix A.1 for a table of units). The fundamental properties of the graviton can be deduced from these basic facts.

Since gravity is a long range force the exchanged particles must be massless. What is their spin? It cannot be half-integer, because half-integer particles have only bilinear interactions and therefore cannot act at tree level as mediators of exchange interactions: force fields correspond to particles with integer spin. In order to further constrain the spin, let us calculate the interaction energy between two heavy particles, mediated by a particle of spin s. From the representation theory of the Lorentz group, a spin s field is contained in a symmetric tensor with s indices $\phi_{\mu_1,\ldots,\mu_s}$, and it interacts with a conserved current $J^{\mu_1,\ldots,\mu_s}$ that is also a symmetric tensor. We choose source particles at rest with charges Q_1, Q_2 and four-velocities $u = (1,0,0,0)$, so that the sources are $J_{1,2}^{\mu_1,\ldots,\mu_s} = Q_{1,2} u^{\mu_1} \ldots u^{\mu_s}$. The Fourier transform of the interaction energy will be proportional to

$$J^{\mu_1,\ldots,\mu_s} P_{\mu_1,\ldots,\mu_s,\nu_1,\ldots,\nu_s} J^{\nu_1,\ldots,\nu_s} , \tag{2.2}$$

where $P_{\mu_1,\ldots,\mu_s,\nu_1,\ldots,\nu_s}$ is the propagator of the mediator field. By Lorentz covariance it can depend only on the metric and on q_μ, the four-momentum of the exchanged particle. Any term containing the momentum will drop out of (2.2) due to current conservation, so for the present purposes we can just assume

$$P_{\mu_1,\ldots,\mu_s,\nu_1,\ldots,\nu_s} = \frac{1}{-q^2}\eta_{\mu_1\nu_1} \cdots \eta_{\mu_s\nu_s} , \tag{2.3}$$

suitably symmetrized. In this kinematic configuration the momentum q is spacelike, so $q^2 > 0$ (we use signature $-+++$). The overall factor -1 in the propagator is essential in the following: it is dictated by the positivity of the energy of a freely propagating quantum. When this form is inserted in (2.2), it produces s powers of $g(u,u) = -1$. The sign of the interaction potential is thus given by $(-1)^{s+1}Q_1Q_2$. As a result, like charges attract and opposite charges repel when s is even, whereas like charges repel and opposite charges attract when s is odd. In the case of gravity, the charges are proportional to the masses, which for all physically realizable sources are positive. Given that the charges of all bodies have the same sign and that the resulting interactions are universally attractive, we conclude that the mediator of the gravitational interaction must have even spin.

The simplest possibility would be spin-0. In this case the interaction term would be of the form $\phi T^\lambda{}_\lambda$, where $T^\mu{}_\nu$ is the energy-momentum tensor. Since the energy-momentum tensor of electromagnetism is traceless, there could be no gravitational deflection of light, in contrast to observation. Thus the simplest remaining option is that the graviton has spin-2.

In order to describe a spin-2 particle we must start from a symmetric tensor $\phi_{\mu\nu}$. The free Lagrangian for such a field contains $\partial_\alpha\phi_{\mu\nu}\partial_\beta\phi_{\rho\sigma}$ with the six indices contracted in all possible ways. Terms without derivatives are not allowed because the mass is zero.[1] Using the symmetry of $\phi_{\mu\nu}$ and the freedom of performing integrations by parts, the free action can be reduced to the general form

$$\int d^4x \left(a_1\partial_\alpha\phi_{\mu\nu}\partial^\alpha\phi_{\mu\nu} + a_2\partial_\alpha\phi_{\mu\alpha}\partial^\beta\phi_{\mu\beta} + a_3\partial_\alpha\phi_{\mu\alpha}\partial^\mu\phi + a_4\partial_\alpha\phi\partial^\alpha\phi\right) , \qquad (2.4)$$

where we write $\phi = \phi^\lambda{}_\lambda$. Note that the field $\phi_{\mu\nu}$ has the canonical dimension of mass, as any bosonic field. The energy-momentum tensor of matter acts as linear source for the graviton

$$S_S = \kappa \int d^4x\, \phi_{\mu\nu}T^{\mu\nu} , \qquad (2.5)$$

where κ is a constant with the dimension of length. The resulting field equation is of the form

$$f_{\mu\nu} = -\kappa T_{\mu\nu} ,$$

with

$$f_{\mu\nu} = -2a_1\,\partial^2\phi_{\mu\nu} - a_2\left(\partial_\mu\partial_\alpha\phi_\nu{}^\alpha + \partial_\nu\partial_\alpha\phi_\mu{}^\alpha\right) - a_3\left(\partial_\mu\partial_\nu\phi + \eta_{\mu\nu}\partial_\alpha\partial_\beta\phi_{\alpha\beta}\right) - 2a_4\,\eta_{\mu\nu}\partial^2\phi .$$

Given that the r.h.s. of the field equation is conserved, the conservation of the l.h.s. must hold as an identity. Requiring that $\partial_\mu f^\mu{}_\nu = 0$ results in the conditions $2a_1 + a_2 = 0$, $a_2 + a_3 = 0$ and $a_3 + 2a_4 = 0$. We fix the overall normalization by choosing $a_1 = -1/2$, which leads to $a_2 = 1$, $a_3 = -1$, $a_4 = 1/2$. The sign has been chosen such that transverse plane waves have positive energy, as we shall see below.

The final form of the Fierz-Pauli equation is

$$\partial^2\phi_{\mu\nu} - \partial_\nu\partial_\alpha\phi_\nu{}^\alpha - \partial_\nu\partial_\alpha\phi_\mu{}^\alpha + \partial_\mu\partial_\nu\phi + \eta_{\mu\nu}\partial_\alpha\partial_\beta\phi^{\alpha\beta} - \eta_{\mu\nu}\partial^2\phi = -\kappa T_{\mu\nu} . \qquad (2.6)$$

For later reference we also write it in the form

$$\mathcal{O}_{\mu\nu}{}^{\rho\sigma}\phi_{\rho\sigma} = -\kappa T_{\mu\nu} . \qquad (2.7)$$

where

$$\mathcal{O}_{\mu\nu}{}^{\rho\sigma} = \frac{1}{2}(\delta^\rho_\mu\delta^\sigma_\nu + \delta^\rho_\nu\delta^\sigma_\mu)\partial^2 - \frac{1}{2}\left(\delta^\rho_\mu\partial_\nu\partial^\sigma + \delta^\sigma_\mu\partial_\nu\partial^\rho + \delta^\rho_\nu\partial_\mu\partial^\sigma + \delta^\sigma_\nu\partial_\mu\partial^\rho\right)$$
$$+ \eta^{\rho\sigma}\partial_\mu\partial_\nu + \eta_{\mu\nu}\partial^\rho\partial^\sigma - \eta_{\mu\nu}\eta^{\rho\sigma}\partial^2 . \qquad (2.8)$$

It can be derived from the action $S_{FP} + S_S$, where

$$S_{FP} = \int d^4x \left(-\frac{1}{2}\partial_\alpha\phi_{\mu\nu}\partial^\alpha\phi^{\mu\nu} + \partial_\alpha\phi_{\mu\alpha}\partial^\beta\phi^{\mu\beta} - \partial_\alpha\phi_{\mu\alpha}\partial^\mu\phi + \frac{1}{2}\partial_\alpha\phi\partial^\alpha\phi\right) . \qquad (2.9)$$

---

[1]Fierz and Pauli also considered a possible mass term. Massive gravitons have been investigated recently in view of possible infrared modifications of gravity, see [53, 54].

Notice that if the field $\phi_{\mu\nu}$ is transverse ($\partial_\alpha \phi^\alpha{}_\mu = 0$) and traceless ($\phi = 0$), its action consists only of the first term. Recalling that we use signature $-+++$, its Hamiltonian would be

$$\int d^3x \left(\frac{1}{2} \partial_0 \phi_{\mu\nu} \partial_0 \phi^{\mu\nu} + \frac{1}{2} \partial_i \phi_{\mu\nu} \partial_i \phi^{\mu\nu} \right) ,$$

where $i = 1, 2, 3$ labels the space coordinates. Thus the sign of the action is the right one for a transverse traceless wave to have positive energy. On the other hand for a pure trace field $\phi_{\mu\nu} = \frac{1}{4}\eta_{\mu\nu}\phi$, the Fierz-Pauli action is

$$\frac{3}{16} \int d^4x \, \partial_\mu \phi \partial^\mu \phi . \tag{2.10}$$

This has the wrong sign: it gives a negative Hamiltonian. In the classical theory this does not matter because the trace field does not propagate. This issue is more serious in the quantum theory and we shall encounter it repeatedly later.

2.1.2   *Linearizing Einstein's equations*

The equations for the gravitational field written by Einstein were very nonlinear:

$$R_{\mu\nu} - \frac{1}{2}g_{\mu\nu}R = 8\pi G\, T_{\mu\nu} . \tag{2.11}$$

In the weak field approximation we can expand

$$g_{\mu\nu} = \eta_{\mu\nu} + h_{\mu\nu} \tag{2.12}$$

with $|h_{\mu\nu}| \ll 1$. The linearized Christoffel symbols are

$$\Gamma_\mu{}^\lambda{}_\nu = \frac{1}{2}\eta^{\lambda\tau}\left(\partial_\mu h_{\tau\nu} + \partial_\nu h_{\tau\mu} - \partial_\tau h_{\mu\nu}\right) ; \tag{2.13}$$

and the linearized Riemann and Ricci tensor and Ricci scalar are

$$R_{\mu\nu\rho\sigma} = \frac{1}{2}\left(\partial_\mu\partial_\sigma h_{\rho\nu} - \partial_\mu\partial_\rho h_{\sigma\nu} - \partial_\nu\partial_\sigma h_{\rho\mu} + \partial_\nu\partial_\rho h_{\sigma\mu}\right) , \tag{2.14}$$

$$R_{\mu\nu} = \frac{1}{2}\left(-\partial^2 h_{\mu\nu} + \partial_\mu\partial_\rho h^\rho{}_\nu + \partial_\nu\partial_\rho h^\rho{}_\mu - \partial_\mu\partial_\nu h\right) , \tag{2.15}$$

$$R = -\partial^2 h + \partial_\alpha\partial_\beta h^{\alpha\beta} . \tag{2.16}$$

The linearized Einstein equations are then

$$\partial^2 h_{\mu\nu} - (\partial_\mu\partial^\rho h_{\rho\nu} + \partial_\nu\partial^\rho h_{\rho\mu}) + \partial_\mu\partial_\nu h + \eta_{\mu\nu}\partial_\rho\partial_\sigma h^{\rho\sigma} - \eta_{\mu\nu}\partial^2 h = -16\pi G\, T_{\mu\nu} . \tag{2.17}$$

When we compare this to the Fierz-Pauli equation (2.6) one has to pay attention to the fact that the field $h_{\mu\nu}$ used here is dimensionless whereas the field $\phi_{\mu\nu}$ in the Fierz-Pauli equation has dimension of mass, as is clear from the form of the action (2.9). The two equations agree if the fields are related by a rescaling

$$h_{\mu\nu} = 2\kappa\,\phi_{\mu\nu} \qquad \text{with} \qquad \kappa = \frac{1}{m_P} = \sqrt{8\pi G} . \tag{2.18}$$

The mass m_P is called the reduced Planck mass. In standard units it is given by

$$m_P = \sqrt{\frac{\hbar c}{8\pi G}} = 4.34 \times 10^{-6}\text{g} = 2.43 \times 10^{18}\text{GeV}/c^2 . \tag{2.19}$$

2.1.3 *Plane waves*

It is convenient to define a bar operation on symmetric tensors:

$$\bar{t}_{\mu\nu} = t_{\mu\nu} - \frac{1}{2}g_{\mu\nu}t \, ,$$

where by t we denote the trace $g^{\rho\sigma}t_{\rho\sigma}$. In dimensions $d \neq 2$ it has an inverse

$$\underline{t}_{\mu\nu} = t_{\mu\nu} - \frac{1}{d-2}g_{\mu\nu}t$$

denoted by an underbar (note that in four dimensions $\underline{t}_{\mu\nu} = \bar{t}_{\mu\nu}$). In terms of "barred" variables, the linearized Einstein equations (2.17) can be rewritten more compactly as

$$\partial^2 \bar{h}_{\mu\nu} - (\partial_\mu \partial^\rho \bar{h}_{\rho\nu} + \partial_\nu \partial^\rho \bar{h}_{\rho\mu}) + \eta_{\mu\nu}\partial_\rho\partial_\sigma \bar{h}^{\rho\sigma} = -16\pi G \, T_{\mu\nu} \, . \tag{2.20}$$

These equations have an infinite dimensional kernel consisting of fields of the form

$$h_{\mu\nu} = \partial_\mu \epsilon_\nu + \partial_\nu \epsilon_\mu \, , \tag{2.21}$$

or equivalently

$$\bar{h}_{\mu\nu} = \partial_\mu \epsilon_\nu + \partial_\nu \epsilon_\mu - \eta_{\mu\nu}\partial_\lambda \epsilon^\lambda \, . \tag{2.22}$$

The fluctuations of this form are simply infinitesimal coordinate transformations of the flat metric and the existence of the kernel is a consequence of the diffeomorphism invariance of the gravitational action. This redundancy has to be eliminated by imposing a gauge condition. When this is done, the operator on the l.h.s. of the equation is then invertible on the subspace of fluctuations that satisfy the gauge condition. In the discussion of gravitational waves it is convenient to use the so-called de Donder condition

$$\partial_\mu \bar{h}^{\mu\nu} = 0 \, . \tag{2.23}$$

Given a fluctuation $h_{\mu\nu}$ which does not satisfy this condition, one looks for an infinitesimal coordinate transformation ϵ_μ such that $h_{\mu\nu} + \partial_\mu \epsilon_\nu + \partial_\nu \epsilon_\mu$ satisfies it. For this, ϵ must satisfy the equation

$$\partial^2 \epsilon_\nu = -\partial^\mu \bar{h}_{\mu\nu} \, . \tag{2.24}$$

This equation always admits a solution. In fact, the solution is determined only up to a solution of the homogeneous equation

$$\partial^2 \epsilon_\nu = 0 \, , \tag{2.25}$$

indicating that the gauge condition (2.23) leaves some residual gauge freedom that has to be fixed separately.

To summarize, the linearized fluctuations of the metric around flat space, in the de Donder gauge, satisfy the simple equation

$$\partial^2 \bar{h}_{\mu\nu} = -16\pi G \, T_{\mu\nu} \, . \tag{2.26}$$

We will now discuss the vacuum solutions of this equation. Since the bar is invertible in $d \neq 2$, the equation $\partial^2 \bar{h}_{\mu\nu} = 0$ is equivalent to

$$\partial^2 h_{\mu\nu} = 0 . \tag{2.27}$$

The general solution can be written as a Fourier superposition of plane waves. Let us concentrate on one particular Fourier mode with momentum p^μ:

$$h_{\mu\nu}(x) = \Pi_{\mu\nu} e^{i p_\mu x^\mu} + \Pi_{\mu\nu}^* e^{-i p_\mu x^\mu} . \tag{2.28}$$

The complex constant tensor $\Pi_{\mu\nu}$ is called polarization tensor. The wave equation and the gauge condition require that

$$p^2 = 0 \qquad \text{and} \qquad p^\mu \bar{\Pi}_{\mu\nu} = 0 . \tag{2.29}$$

We can exploit the residual gauge freedom (2.25) to impose four additional conditions on $\Pi_{\mu\nu}$. To this end we observe that also the solution of (2.25) can be written as a Fourier superposition of plane waves, and we pick the one with the same momentum p^μ of (2.28):

$$\epsilon_\mu(x) = \epsilon_\mu e^{i p_\mu x^\mu} + \epsilon_\mu^* e^{-i p_\mu x^\mu} . \tag{2.30}$$

Under this transformation, $\bar{\Pi}_{\mu\nu}$ changes into

$$\bar{\Pi}_{\mu\nu}' = \bar{\Pi}_{\mu\nu} + i(p_\mu \epsilon_\nu + p_\nu \epsilon_\mu - \eta_{\mu\nu} p^\lambda \epsilon_\lambda) . \tag{2.31}$$

We can choose the constant vector ϵ_μ such that

$$U^\mu \bar{\Pi}_{\mu\nu}' = 0 \tag{2.32}$$

for some constant vector U. These look like d conditions, but in reality only $d-1$ are independent since $p^\mu \bar{\Pi}_{\mu\nu} U^\nu$ is identically zero. We can thus impose one additional condition. Taking the trace of (2.31) we see that the trace of $\bar{\Pi}_{\mu\nu}$ changes by $-2ip \cdot \epsilon$. We can therefore choose ϵ_μ so as to make $\bar{\Pi}_{\mu\nu}$ traceless. When $\bar{\Pi}_{\mu\nu}$ is traceless, $\bar{\Pi}_{\mu\nu} = \Pi_{\mu\nu}$, so we can summarize the conditions on the polarization as follows

$$\Pi^\lambda{}_\lambda = 0 ; \quad p^\mu \Pi_{\mu\nu} = 0 ; \quad U^\mu \Pi_{\mu\nu} = 0 . \tag{2.33}$$

The polarization tensor has $d(d+1)/2$ free parameters on which there are altogether $2d$ conditions, thus leaving $d(d-3)/2$ physically distinct polarization states. This number is zero (or negative) for $d \leq 3$. This is related to the fact that in $d \leq 3$ the Riemann tensor is zero when the Ricci tensor is zero. In four dimensions a gravitational wave has two polarization states, that we now describe in more detail.

Let us consider a plane wave propagating in the z direction. The wave vector has components $p^\mu = (p, 0, 0, p)$; we choose $U^\mu = (1, 0, 0, 0)$. Then the conditions (2.33) imply that the polarization tensor has only two degrees of freedom that we call e_+ and $e_\times$:

$$\Pi_{\mu\nu} = \begin{pmatrix} 0 & 0 & 0 & 0 \\ 0 & e_+ & e_\times & 0 \\ 0 & e_\times & -e_+ & 0 \\ 0 & 0 & 0 & 0 \end{pmatrix} \tag{2.34}$$

The two free real parameters e_+ and $e_\times$ are the amplitudes of the two polarization states of the gravitational wave with four-momentum p^μ. A rotation by an angle θ in the (x, y) plane, given by the Lorentz transformation

$$\Lambda^\mu{}_\nu = \begin{pmatrix} 1 & 0 & 0 & 0 \\ 0 & \cos\theta & \sin\theta & 0 \\ 0 & -\sin\theta & \cos\theta & 0 \\ 0 & 0 & 0 & 1 \end{pmatrix}$$

transforms the polarization tensor into another with the same form, but

$$e'_+ = e_+ \cos 2\theta - e_\times \sin 2\theta , \tag{2.35}$$

$$e'_\times = e_+ \sin 2\theta + e_\times \cos 2\theta , \tag{2.36}$$

so that the combinations $e_L = \frac{1}{\sqrt{2}}(e_+ - ie_\times)$ and $e_R = \frac{1}{\sqrt{2}}(e_+ + ie_\times)$ transform by a phase:

$$e_L \mapsto e'_L = e_L e^{-2i\theta} , \qquad e_R \mapsto e'_R = e_R e^{2i\theta} . \tag{2.37}$$

The coefficient in the exponent means that e_L and e_R represent gravitational waves with helicity ± 2.

The line element of the plane wave propagating in the z direction, with $+$ polarization, amplitude e_+ and frequency q is

$$ds^2 = -dt^2 + (1 + 2e_+ \sin(p(z - t)))dx^2 + (1 - 2e_+ \sin(p(z - t)))dy^2 + dz^2 , \tag{2.38}$$

thus the proper distance between two points with fixed coordinates (x, z) and the distance between two points with fixed coordinates (y, z) oscillate in time with a phase shift of π. Using (2.35) we see that the polarization states e_+ and $e_\times$ are equivalent up to a rotation by $\pi/4$. A ring of freely falling particles is deformed under the effect of such waves as shown in the following figure:

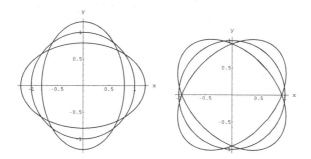

As is well-known, in GR it does not make sense to talk of the energy and momentum of the gravitational field. In the linearized theory on Minkowski space, there is no conceptual difficulty of this kind and we can define the energy-momentum tensor of the Fierz-Pauli field. For a transverse traceless wave of the type described above, it is

$$t_{\mu\nu}^{(TT)} = \frac{1}{32\pi G} \partial_\mu h_{\rho\sigma} \partial_\nu h^{\rho\sigma} . \tag{2.39}$$

Averaging out the oscillations of the field, one gets

$$t_{\mu\nu}^{(TT)} = \frac{k_\mu k_\nu}{16\pi G}\Pi^{\rho\sigma*}\Pi_{\rho\sigma} \ . \tag{2.40}$$

2.1.4 *Quantization*

At the level of free fields, there is no difficulty in defining gravitons as the elementary quanta of the gravitational field and imagining a gravitational wave as being composed of a large number of gravitons.

Starting from the plane wave solutions described in the preceding section, we can Fourier expand

$$h_{\mu\nu}(x) = \sum_{\sigma=L,R}\int d\mathbf{p}\left(a_{\mathbf{p},\sigma}\Pi_{\mu\nu}(\sigma)e^{ip_\mu x^\mu} + a_{\mathbf{p},\sigma}^*\Pi_{\mu\nu}(\sigma)^*e^{-ip_\mu x^\mu}\right) , \tag{2.41}$$

where $\mathbf{p}$ denotes the space components of the momentum. We then promote the Fourier coefficients $a_{\mathbf{p}}(\sigma)$ and their conjugates to quantum operators satisfying the canonical commutation relations

$$[a_{\mathbf{p},\sigma}, a_{\mathbf{p}',\sigma'}] = 0$$
$$\left[a_{\mathbf{p},\sigma}^\dagger, a_{\mathbf{p}',\sigma'}^\dagger\right] = 0 \tag{2.42}$$
$$\left[a_{\mathbf{p},\sigma}, a_{\mathbf{p}',\sigma'}^\dagger\right] = i\delta_{\sigma\sigma'}\delta(\mathbf{p} - \mathbf{p}') \ .$$

We can define a Fock space whose vacuum state is the unique Poincaré-invariant state, satisfying

$$a_{\mathbf{p},\sigma}|0\rangle = 0 \tag{2.43}$$

and where $a_{\mathbf{p},\sigma}$ and $a_{\mathbf{p},\sigma}^\dagger$ act as raising and lowering operators. For example, the single-graviton states are defined by

$$a_{\mathbf{p},\sigma}^\dagger|0\rangle = |p,\sigma\rangle \ . \tag{2.44}$$

Before proceeding with a discussion of graviton interactions, it will be a useful (and sobering) exercise to estimate what it would take to detect such a particle. We momentarily stop using the natural units $\hbar = c = 1$ for the rest of this section. Classically the energy-momentum tensor of a particle at position $\vec{x}(t)$ and with four-momentum p_μ is

$$t_{\mu\nu}(\vec{x},t) = c^2\frac{p_\mu p_\nu}{E}\delta(\vec{x} - \vec{x}(t))$$

Thus for a large number of gravitons all with wave-vectors k_μ contained in a volume, the averaged-out energy-momentum tensor is

$$t_{\mu\nu} = c^2\hbar\frac{k_\mu k_\nu}{\omega}N$$

where N is the number density of gravitons. Comparing with the formula (2.40) for the energy-momentum tensor of a gravitational wave we deduce that the number density of gravitons is

$$N = \frac{\omega c^2}{16\pi\hbar G}\Pi^{\rho\sigma *}\Pi_{\rho\sigma} \ . \tag{2.45}$$

Typical values of the frequency and amplitude for a gravitational wave such as the one that has been recently observed [55] would be of the order of $\omega = 10^3 \text{Hz}$ and $\Pi \approx 10^{-21}$, leading to $N \approx 10^{14} \text{cm}^{-3}$.

Considering how difficult it is to detect classical gravitational waves, this should make it clear that the detection of a single graviton is way beyond our experimental capabilities. This has led Freeman Dyson to actually question the usefulness of the notion of graviton. Continuing along the line of the previous reasoning, the energy density of the kind of wave discussed above is $t_{00} \approx 10^{-10} \text{erg/cm}^3$. An individual graviton cannot be contained in a volume smaller than $(c/\omega)^3$, so its energy density would be smaller than $\hbar\omega^4/c^3 \approx 10^{-47} \text{erg/cm}^3$. So to detect a graviton, a detector working on the physical principles as LIGO would have to have a sensitivity that is 10^{37} times higher.

Dyson and others [56] have considered several sources of gravitons and several possible designs of graviton detectors and concluded that no design leads to anything that looks remotely feasible, even in principle. At the moment, the most convincing evidence could come from the polarization of the cosmic microwave background due to gravitational waves produced during inflation [57, 58], which is a quantum mechanical phenomenon and would establish the existence of gravitons. This has not been observed yet, but is not beyond the reach of the next generation of experiments.

2.1.5 *Spin projectors*

In Minkowski space it is often convenient to split a field into irreducible representations of the Lorentz group, which correspond to degrees of freedom of different spin J and parity P. For example a vector field A_μ can be split into a $(d-1)$-dimensional representation with $J^P = 1^-$ and a one-dimensional representation with $J^P = 0^+$, corresponding to the transverse and longitudinal components. In Fourier space, they are obtained by acting with the projectors

$$L^\mu_{\ \nu} = \frac{p^\mu p_\nu}{p^2} \ ; \qquad T^\mu_{\ \nu} = \delta^\mu_\nu - \frac{p^\mu p_\nu}{p^2} \ , \tag{2.46}$$

where p_μ is the momentum carried by the wave.

Similarly, a symmetric rank-2 field $\phi_{\mu\nu}$ can be decomposed into four irreducible representations of the Lorentz group. If we choose coordinates such that x_L is in the direction of the momentum and x_i, $i = 1 \ldots d$ are transverse, then the irreducible representations can be listed as follows:

- $\frac{(d+1)(d-2)}{2}$-dimensional representation ϕ_{ij}^{TT} with $J^P = 2^+$

- a $(d-1)$-dimensional representation $\xi_i = \phi_{iL}$ with $J^P = 1^-$
- a one-dimensional representation $w = \phi_{LL}$ with $J^P = 0^+$
- a one-dimensional representation $s = \phi_{ii}$ with $J^P = 0^+$

where $\phi_{\mu\nu}^{TT}$ is transverse on both indices ($L_\rho{}^\mu \phi_{\mu\nu}^{TT} = 0$) and traceless, and the vector ξ_μ is transverse ($L_\rho{}^\mu \xi_\mu = 0$).

The analogs of the L and T projectors are now four-index tensors denoted $P^{(J,aa)}{}_{\mu\nu}{}^{\alpha\beta}$ where J is the spin label and a serves to distinguish between different representations with the same spin [59–61]. It has only one value for $J = 1, 2$ and so need not be written in those cases, while for $J = 0$, a runs over the values s, w. In addition to the projectors, there are two operators $P^{(0,sw)}$ and $P^{(0,ws)}$ intertwining between the spin-0 representations s and w.

Explicitly we have

$$P^{(2)}{}_{\mu\nu}{}^{\rho\sigma} = \frac{1}{2}(T_\mu^\rho T_\nu^\sigma + T_\mu^\sigma T_\nu^\rho) - \frac{1}{d-1}T_{\mu\nu}T^{\rho\sigma} \tag{2.47}$$

$$P^{(1)}{}_{\mu\nu}{}^{\rho\sigma} = \frac{1}{2}(T_\mu^\rho L_\nu^\sigma + T_\mu^\sigma L_\nu^\rho + T_\nu^\rho L_\mu^\sigma + T_\nu^\sigma L_\mu^\rho) \tag{2.48}$$

$$P^{(0,ss)}{}_{\mu\nu}{}^{\rho\sigma} = \frac{1}{d-1}T_{\mu\nu}T^{\rho\sigma} \tag{2.49}$$

$$P^{(0,ww)}{}_{\mu\nu}{}^{\rho\sigma} = L_{\mu\nu}L^{\rho\sigma} \tag{2.50}$$

$$P^{(0,sw)}{}_{\mu\nu}{}^{\rho\sigma} = \frac{1}{\sqrt{d-1}}T_{\mu\nu}L^{\rho\sigma} \tag{2.51}$$

$$P^{(0,ws)}{}_{\mu\nu}{}^{\rho\sigma} = \frac{1}{\sqrt{d-1}}L_{\mu\nu}T^{\rho\sigma} \tag{2.52}$$

These operators satisfy the orthogonality relation

$$P^{(J,ab)}{}_{\mu\nu}{}^{\alpha\beta} P^{(K,cd)}{}_{\alpha\beta}{}^{\rho\sigma} = \delta^{JK}\delta^{bc} P^{(J,ad)}{}_{\mu\nu}{}^{\rho\sigma} \ , \tag{2.53}$$

and the completeness

$$P^{(2)}{}_{\mu\nu}{}^{\rho\sigma} + P^{(1)}{}_{\mu\nu}{}^{\rho\sigma} + P^{(0,ss)}{}_{\mu\nu}{}^{\rho\sigma} + P^{(0,ww)}{}_{\mu\nu}{}^{\rho\sigma} = \mathbf{1}_{\mu\nu}{}^{\rho\sigma} \ , \tag{2.54}$$

where

$$\mathbf{1}_{\mu\nu}{}^{\rho\sigma} = \frac{1}{2}(\delta_\mu^\rho \delta_\nu^\sigma + \delta_\mu^\sigma \delta_\nu^\rho) \tag{2.55}$$

is the identity in the space of symmetric tensors. Acting with these projectors we can decompose

$$\phi_{\mu\nu} = \phi_{\mu\nu}^{TT} + i(p_\mu \xi_\nu + p_\nu \xi_\mu) + \frac{1}{d}T_{\mu\nu}s + \frac{1}{d}L_{\mu\nu}w \ , \tag{2.56}$$

where the first term is transverse and traceless, the second is traceless but not transverse, the third is transverse but not traceless and the last is neither transverse nor traceless.

Note that the projectors/intertwiners belonging to the same spin are conveniently arranged into a matrix, so for spin-0 we have

$$P^{(0)} = \begin{bmatrix} P^{(0,ss)} & P^{(0,sw)} \\ P^{(0,ws)} & P^{(0,ww)} \end{bmatrix} \tag{2.57}$$

Any quadratic action for a symmetric rank-2 tensor can be written as

$$S^{(2)} = \frac{1}{2} \int \frac{d^d q}{(2\pi)^d} \, \phi_{\mu\nu}(-q) \mathcal{O}^{\mu\nu\rho\sigma}(q) \phi_{\rho\sigma}(q) \, , \qquad (2.58)$$

where $\mathcal{O}$ is a differential operator.[2] Since a symmetric rank-2 tensor can be decomposed in its irreducible components as described above, it follows that this action can also be written as

$$S^{(2)} = \frac{1}{2} \sum_{J,a,b} \int \frac{d^d q}{(2\pi)^d} \, \phi^{Ja}(-q) \cdot a_{(J,ab)}(q) \, P^{(J,ab)} \cdot \phi^{Jb}(q) \, , \qquad (2.59)$$

where the dot stands for contraction of pairs of indices and $a_{J,ab}$ are matrices of coefficient depending on momentum. These coefficients can be computed by acting with $\mathcal{O}_{AB}$ on the (diagonal) spin projectors $P^{(Jaa)}$ and subsequently reexpressing the result in terms of spin projectors and intertwiners.

This way of writing has several virtues. If the operator $\mathcal{O}$ is Lorentz covariant, it does not mix irreducible representations of the Lorentz group with different values of spin and parity. Therefore one achieves at least a partial diagonalization of the kinetic operator, where mixing can only occur within degrees of freedom with the same spin and parity (in our case, only between the spin-0 fields s and w). Second, one can immediately see what are the propagating degrees of freedom and their masses. Third, having partially diagonalized the problem, it is now easy to invert the operator to obtain the propagators. In fact, the propagator is

$$\sum_{J,a,b} a_{(J,ab)}^{-1}(q) \, P^{(J,ab)} \, , \qquad (2.60)$$

where $a_{(J,ab)}^{-1}(q)$ are the inverses of the coefficient matrices (for $J = 1, 2$ these are one-by-one matrices, *i.e.* simple functions of q).

Finally, the spin projectors can be used to disentangle the gauge degrees of freedom from the physical ones, at least at linear level. Before doing this for gravity, let us first recall how it works for electromagnetism. The analog of the decomposition (2.56) is (for a Fourier component with wave-number p_μ)

$$A_\mu = A_\mu^T + i p_\mu \phi \, , \qquad (2.61)$$

where $p^\nu A_\nu^T = 0$. A generic gauge transformation with parameter $\epsilon(x) = \tilde{\epsilon} e^{i p_\mu x^\mu}$ only changes the longitudinal part of A_μ, shifting $\phi \to \phi + \tilde{\epsilon}$. However, $\partial_\mu \epsilon$ can also be transverse: this happens if ϵ is a solution of the equation $\partial^2 \epsilon = 0$. In Fourier space, this means $p^2 \tilde{\epsilon} = 0$. This freedom can be used to set one component of A_μ equal to zero, for example $U^\mu A_\mu = 0$, for some vector U^μ. Then, the physical states of the electromagnetic field are parametrized by the polarization vectors Π_μ satisfying $k^\mu \Pi_\mu = 0$ and $U^\mu \Pi_\mu = 0$, *i.e.* $d - 2$ degrees of freedom per Fourier mode.

---

[2]As we shall discuss in section 3.4, an operator $\mathcal{O}$ acting on covariant symmetric tensors carries indices $\mathcal{O}_{\mu\nu}{}^{\rho\sigma}$ and in Eq. (2.58) there is an implicit choice of a metric in the space of symmetric tensors. Here and throughout this chapter we simply assume that indices are raised and lowered with the Minkowski metric $\eta_{\mu\nu}$.

Also in the case of gravity, in order to understand the effect of gauge tranformations on the degrees of freedom $h_{\mu\nu}^{TT}$, ξ_μ, w and s we need to distinguish gauge transformation for which $i(p_\mu\epsilon_\nu + p_\nu\epsilon_\mu)$ is longitudinal, from those for which it is transverse. In the former case

$$P_{\mu\nu\rho\sigma}^{(2)}i(p^\rho\epsilon^\sigma + p^\sigma\epsilon^\rho) = 0 \ ,$$
$$P_{\mu\nu\rho\sigma}^{(1)}i(p^\rho\epsilon^\sigma + p^\sigma\epsilon^\rho) = i(p_\mu\epsilon_\nu + p_\nu\epsilon_\mu) - 2iL_{\mu\nu}p_\rho\epsilon^\rho \ , \qquad (2.62)$$
$$P_{\mu\nu\rho\sigma}^{(ss)}i(p^\rho\epsilon^\sigma + p^\sigma\epsilon^\rho) = 0 \ ,$$
$$P_{\mu\nu\rho\sigma}^{(ww)}i(p^\rho\epsilon^\sigma + p^\sigma\epsilon^\rho) = 2iL_{\mu\nu}p_\rho\epsilon^\rho \ ,$$

showing that ξ_μ and w are gauge degrees of freedom. The spin-2 and s are invariant under this class of transformations. In the latter case

$$p^2\epsilon_\nu + p_\nu(p_\mu\epsilon^\mu) = 0 \ , \qquad (2.63)$$

and one finds

$$P_{\mu\nu\rho\sigma}^{(2)}i(p^\rho\epsilon^\sigma + p^\sigma\epsilon^\rho) = i(p_\mu\epsilon_\nu + p_\nu\epsilon_\mu) - \frac{2i}{d-1}T_{\mu\nu}p_\rho\epsilon^\rho \ ,$$
$$P_{\mu\nu\rho\sigma}^{(1)}i(p^\rho\epsilon^\sigma + p^\sigma\epsilon^\rho) = 0, \qquad (2.64)$$
$$P_{\mu\nu\rho\sigma}^{(ss)}i(p^\rho\epsilon^\sigma + p^\sigma\epsilon^\rho) = \frac{2i}{d-1}T_{\mu\nu}p_\rho\epsilon^\rho \ ,$$
$$P_{\mu\nu\rho\sigma}^{(ww)}i(p^\rho\epsilon^\sigma + p^\sigma\epsilon^\rho) = 0 \ .$$

These transformations can be used to set $s = 0$ and to impose $d - 1$ additional conditions $U^\mu\phi_{\mu\nu}^{TT} = 0$, bringing it to the standard form described in section 2.1.3.

This formalism is particularly useful when one deals with complicated kinetic operators, such as the ones we will discuss in the next section. As a preliminary exercise, let us see here how this works for the Fierz-Pauli operator (2.8). Going to the momentum representation by the standard rule $\partial_\mu \to ip^\mu$ it can be rewritten in the form (round bracket around indices denoting symmetrization):

$$\mathcal{O}^{\mu\nu\rho\sigma} = (-p^2)\left(1^{\mu\nu\rho\sigma} - \eta^{\mu(\rho|}L^{\nu|\sigma)} - \eta^{\nu(\rho|}L^{\mu|\sigma)} + \eta^{\mu\nu}L^{\rho\sigma} + \eta^{\rho\sigma}L^{\mu\nu} - \eta^{\mu\nu}\eta^{\rho\sigma}\right)$$

then expanding each occurrence of the metric as $\eta^{\mu\nu} = L^{\mu\nu} + T^{\mu\nu}$ this becomes

$$\mathcal{O}^{\mu\nu\rho\sigma} = (-p^2)\left(1^{\mu\nu\rho\sigma} - P^{(1)\mu\nu\rho\sigma} - P^{(ww)\mu\nu\rho\sigma} - (d-1)P^{(ss)\mu\nu\rho\sigma}\right)$$

and finally using the completeness relation (2.54) one gets

$$\mathcal{O}^{\mu\nu\rho\sigma} = (-p^2)\left(P^{(2)\mu\nu\rho\sigma} - (d-2)P^{(ss)\mu\nu\rho\sigma}\right) \ . \qquad (2.65)$$

This implies that the Pauli-Fierz action can be written in the form (2.59) with coefficient matrices

$$a_2 = -p^2 \ , \qquad a_1 = 0 \ , \qquad a_0 = \begin{bmatrix} (d-2)p^2 & 0 \\ 0 & 0 \end{bmatrix} \ . \qquad (2.66)$$

A naive reading of this formula seems to imply that GR contains a spin-0 degree of freedom. However, as explained above, s can be eliminated by a residual gauge transformation and is not a physical degree of freedom. Still, we note that the s degree of freedom has a kinetic term with the wrong sign. This is the same sign that had been noticed in the end of section 2.1.1 for the trace of $\phi_{\mu\nu}$. Insofar as the corresponding degree of freedom does not propagate, it does not cause any instability, but we shall encounter this issue in a different form when we discuss the Euclidean version of the theory.

The coefficient matrices (2.66) make it manifest that the kinetic operator of GR is not invertible in the spin-1 and in the w sectors, this being a consequence of the gauge invariance of the theory. To make the operator invertible one can add to the Pauli-Fierz action

$$S_{FP} = \frac{1}{2} \int d^d x \, \phi_{\mu\nu} \mathcal{O}^{\mu\nu\rho\sigma} \phi_{\rho\sigma} \tag{2.67}$$

a gauge fixing term

$$S_{GF} = -\frac{1}{\alpha} \int d^d x \, \eta^{\mu\nu} F_\mu F_\nu \tag{2.68}$$

with

$$F^\mu = \partial_\nu \phi^{\mu\nu} - \frac{1}{2} \partial^\nu \phi \ . \tag{2.69}$$

Proceeding as before, this contributes to $\mathcal{O}^{\mu\nu\rho\sigma}$ the terms

$$\frac{-p^2}{2\alpha} \left[2P^{(1)} + (d-1)P^{(ss)} + P^{(ww)} - \sqrt{d-1}(P^{(sw)} + P^{(ws)}) \right] \ . \tag{2.70}$$

Putting together (2.65) and (2.70), the kinetic operator in the gauge-fixed Pauli-Fierz action is

$$-p^2 \left[P^{(2)} + \frac{P^{(1)}}{\alpha} - \frac{2\alpha(d-2)-d+1}{2\alpha} P^{(ss)} - \frac{\sqrt{d-1}}{2\alpha}(P^{(sw)} + P^{(ws)}) + \frac{P^{(ww)}}{2\alpha} \right] , \tag{2.71}$$

whence we read off the coefficient matrices

$$a_2 = -p^2 \ , \qquad a_1 = \frac{1}{\alpha}(-p^2) \ , \qquad a_0 = \begin{bmatrix} \frac{2\alpha(d-2)-d+1}{2\alpha}p^2 & \frac{\sqrt{d-1}}{2\alpha}p^2 \\ \frac{\sqrt{d-1}}{2\alpha}p^2 & \frac{1}{2\alpha}(-p^2) \end{bmatrix} \ . \tag{2.72}$$

The propagator is then given by the sum of the inverses of the coefficient matrices, multiplied by the respective projectors:

$$\frac{1}{-p^2} \left[P^{(2)} + \alpha P^{(1)} \right.$$

$$\left. - \frac{P^{(ss)}}{d-2} - \frac{\sqrt{d-1}}{d-2}(P^{(sw)} + P^{(ws)}) + \frac{2\alpha(d-2)-d+1}{d-2}P^{(ww)} \right] \ . \tag{2.73}$$

2.2 Four derivative theories

2.2.1 *Actions*

The Fierz-Pauli equations are the relativistic field equation for a spin-2 particle involving two derivatives. If one allows also more derivatives, other equations are possible. Here we shall discuss the equations containing four derivatives. Instead of trying to construct them from the bottom up in Minkowski space, we will obtain them from the linearization of fully nonlinear equations. We start from the action principle, which will be needed anyway in later chapters.

The Einstein field equations (possibly with a cosmological constant Λ), can be derived from the Hilbert action

$$S_H(g) = \frac{1}{2\kappa^2} \int d^d x \sqrt{|g|}(-2\Lambda + R) \ , \qquad \kappa = \sqrt{8\pi G} \ . \qquad (2.74)$$

(Here we are assuming $d > 2$, since in $d = 2$ this action is a topological invariant.) Fourth order equations will be obtained from actions that contain four derivatives of the metric. The most general diffeomorphism invariant action of this type is:

$$\int d^d x \sqrt{|g|} \left[\alpha R^2 + \beta R_{\mu\nu} R^{\mu\nu} + \gamma R_{\mu\nu\rho\sigma} R^{\mu\nu\rho\sigma} + \tau \nabla^2 R \right] \ , \qquad (2.75)$$

where ∇ is the Levi-Civita connection, $R_{\mu\nu\rho\sigma}$ the Riemann tensor and α, β, γ, τ are arbitrary couplings. The last term is a total derivative, and we shall mostly ignore it.

There is some arbitrariness in the choice of the basis of invariants entering the action. The basis of invariants used in (2.75) will be referred to as the "Riemann basis". Let us discuss two alternative choices.

The Weyl tensor is the tracefree part of the Riemann tensor:

$$C_{\mu\nu\rho\sigma} = R_{\mu\nu\rho\sigma} - \frac{1}{d-2}(g_{\mu\rho}R_{\nu\sigma} - g_{\mu\sigma}R_{\nu\rho} - g_{\nu\rho}R_{\mu\sigma} + g_{\nu\sigma}R_{\mu\rho})$$
$$+ \frac{1}{(d-1)(d-2)}R(g_{\mu\rho}g_{\nu\sigma} - g_{\mu\sigma}g_{\nu\rho}) \ . \qquad (2.76)$$

and one has

$$C_{\mu\nu\rho\sigma} C^{\mu\nu\rho\sigma} = R_{\mu\nu\rho\sigma} R^{\mu\nu\rho\sigma} - \frac{4}{d-2}R_{\mu\nu}R^{\mu\nu} + \frac{2}{(d-1)(d-2)}R^2 \ . \qquad (2.77)$$

Another significant combination of curvature terms is

$$E = R_{\mu\nu\rho\sigma} R^{\mu\nu\rho\sigma} - 4R_{\mu\nu}R^{\mu\nu} + R^2 \ . \qquad (2.78)$$

In three dimensions the Weyl tensor is identically zero and then from Eq. (2.77) one deduces that also $E = 0$. In four dimensions E is locally a total derivative. This fact is proven in section 2.4. Thus in four dimensions only two linear combinations of the terms in (2.75) have local effects. It is then obviously useful to have E as one of the independent combinations of curvatures. There are two particularly useful choices for the remaining two invariants. The first choice is to use (2.78) to eliminate the square of the Riemann tensor in favor of the squares of the Ricci tensor and Ricci scalar:

$$\int d^d x \sqrt{|g|} \left[a_1 R_{\mu\nu} R^{\mu\nu} + a_2 R^2 + a_3 E \right] . \tag{2.79}$$

We will call this the "Ricci basis". The relation between the couplings is

$$\alpha = a_2 + a_3 ; \qquad \beta = a_1 - 4a_3 ; \qquad \gamma = a_3 , \tag{2.80}$$

or conversely

$$a_1 = \beta + 4\gamma ; \qquad a_2 = \alpha - \gamma ; \qquad a_3 = \gamma . \tag{2.81}$$

The second choice is to further replace the square of the Ricci tensor by the square of the Weyl tensor. This can be achieved by use of the identity

$$C_{\mu\nu\rho\sigma} C^{\mu\nu\rho\sigma} = E + \frac{d-3}{d-2} \left(4 R_{\mu\nu} R^{\mu\nu} - \frac{d}{d-1} R^2 \right) , \tag{2.82}$$

which can be obtained substituting (2.78) in (2.77). Then we can rewrite (2.75) in the physically more significant form

$$\int d^d x \sqrt{|g|} \left[\frac{1}{2\lambda} C_{\mu\nu\rho\sigma} C^{\mu\nu\rho\sigma} + \frac{1}{\xi} R^2 - \frac{1}{\rho} E \right] . \tag{2.83}$$

We call this the "Weyl basis". In four dimensions only the first two terms in (2.79) and (2.83) affect the equations of motion. Also we note that in four dimensions the Weyl squared term has the property of being invariant under Weyl transformations:

$$g_{\mu\nu}(x) \to \Omega^2(x) g_{\mu\nu}(x) . \tag{2.84}$$

Thus in four dimensions the only term in the Weyl basis that is not Weyl invariant is the R^2 term. From (2.82) we see that in four dimensions the Weyl term is equal, modulo total derivatives, to the combination $2 R_{\mu\nu} R^{\mu\nu} + \frac{2}{3} R^2$, so the action (2.79) is Weyl-invariant when $a_1 = 3a_2$.

The relations between the couplings in (2.75) and (2.83) are

$$\lambda = \frac{2(d-3)}{(d-2)(\beta+4\gamma)}, \quad \rho = \frac{4(d-3)}{(d-2)\beta+4\gamma}, \quad \xi = \frac{4(d-1)}{4(d-1)\alpha+d\beta+4\gamma} . \tag{2.85}$$

or conversely

$$\alpha = -\frac{1}{\rho} + \frac{1}{\xi} + \frac{1}{(d-1)(d-2)\lambda}, \quad \beta = \frac{4}{\rho} - \frac{2}{(d-2)\lambda}, \quad \gamma = -\frac{1}{\rho} + \frac{1}{2\lambda}. \tag{2.86}$$

Note that in $d = 3$, C^2 and E both vanish identically and the form (2.83) is not appropriate. The couplings λ, ρ and ξ have mass dimension $4 - d$. In dimensions higher than three, it is customary to define the dimensionless combinations

$$\omega \equiv -\frac{(d-1)\lambda}{\xi}, \quad \theta \equiv \frac{\lambda}{\rho} . \tag{2.87}$$

2.2.2 *Linearized equations*

In this section we shall assume that the cosmological constant is zero. The action consisting of the Hilbert term (2.74) and the four-derivative terms (2.75) gives rise to the following equations of motion:

$$\frac{1}{2\kappa^2}\left(R_{\mu\nu} - \frac{1}{2}g_{\mu\nu}R\right) + \alpha E_{\mu\nu}^{(1)} + \beta E_{\mu\nu}^{(2)} + \gamma E_{\mu\nu}^{(3)} = 0, \qquad (2.88)$$

where

$$E_{\mu\nu}^{(1)} = 2RR_{\mu\nu} - 2\nabla_\mu\nabla_\nu R + g_{\mu\nu}\left(2\Box R - \frac{1}{2}R^2\right),$$

$$E_{\mu\nu}^{(2)} = 2R_{\mu\lambda}R_\nu^\lambda - 2\nabla^\lambda\nabla_{(\mu}R_{\nu)\lambda} + \Box R_{\mu\nu} + \frac{1}{2}(\Box R - R_{\rho\sigma}R^{\rho\sigma})g_{\mu\nu}, \qquad (2.89)$$

$$E_{\mu\nu}^{(3)} = 2R_{\mu\rho\lambda\sigma}R_\nu^{\rho\lambda\sigma} + 4\nabla_{(\rho}\nabla_{\lambda)}R_\mu{}^\rho{}_\nu{}^\lambda - \frac{1}{2}g_{\mu\nu}R_{\rho\sigma\lambda\tau}R^{\rho\sigma\lambda\tau}.$$

and $\Box = \partial^2$.

When linearized[3] they have the form $2\mathcal{K}_{\mu\nu\alpha\beta}h^{\alpha\beta} = 0$, where

$$\mathcal{K}_{\mu\nu\alpha\beta} = \left(\frac{1}{8\kappa^2} + \left(\frac{\beta}{4} + \gamma\right)\Box\right)(\eta_{\mu\alpha}\eta_{\nu\beta}\Box - \eta_{\nu\beta}\partial_\mu\partial_\alpha - \eta_{\mu\beta}\partial_\nu\partial_\alpha)$$

$$+ \left(\frac{1}{8\kappa^2} - \left(\alpha + \frac{\beta}{4}\right)\Box\right)(\eta_{\alpha\beta}\partial_\mu\partial_\nu + \eta_{\mu\nu}\partial_\alpha\partial_\beta - \eta_{\mu\nu}\eta_{\alpha\beta}\Box)$$

$$+ \left(\alpha + \frac{\beta}{2} + \gamma\right)\partial_\mu\partial_\nu\partial_\alpha\partial_\beta. \qquad (2.90)$$

It is more useful to rewrite this operator using the spin projectors. Using the identities (2.54) and

$$\eta_{\mu\nu}\eta_{\alpha\beta} = (d-1)P_{\mu\nu\alpha\beta}^{(ss)} + \sqrt{d-1}\left(P_{\mu\nu\alpha\beta}^{(ws)} + P_{\mu\nu\alpha\beta}^{(sw)}\right) + P_{\mu\nu\alpha\beta}^{(ww)}$$

$$\eta_{\mu\nu}\partial_\alpha\partial_\beta + \eta_{\alpha\beta}\partial_\mu\partial_\nu = \left[2P_{\mu\nu\alpha\beta}^{(ww)} + \sqrt{d-1}\left(P_{\mu\nu\alpha\beta}^{(ws)} + P_{\mu\nu\alpha\beta}^{(sw)}\right)\right]\Box$$

$$\eta_{(\mu|\alpha}\partial_{|\nu)}\partial_\beta + \eta_{(\mu|\beta}\partial_{|\nu)}\partial_\alpha = \left[P_{\mu\nu\alpha\beta}^{(1)} + 2P_{\mu\nu\alpha\beta}^{(ww)}\right]\Box \qquad (2.91)$$

$$\partial_\mu\partial_\nu\partial_\alpha\partial_\beta = P_{\mu\nu\alpha\beta}^{(ww)}\Box^2$$

we can rewrite it in the form

$$P_{\mu\nu\alpha\beta}^{(2)}\left(\frac{1}{8\kappa^2} + \left(\frac{\beta}{4} + \gamma\right)\Box\right)\Box + P_{\mu\nu\alpha\beta}^{(ss)}\left(-\frac{d-2}{8\kappa^2} + \left((d-1)\alpha + \frac{d\beta}{4} + \gamma\right)\Box\right)\Box$$

$$= P_{\mu\nu\alpha\beta}^{(2)}\left(\frac{1}{8\kappa^2} + \frac{d-3}{2(d-2)\lambda}\Box\right)\Box + P_{\mu\nu\alpha\beta}^{(ss)}\left(-\frac{d-2}{8\kappa^2} + \frac{d-1}{\cdot\,\xi}\Box\right)\Box. \qquad (2.92)$$

In the last step we passed to the Weyl basis. These forms reveal several important facts. First we notice that, just like in the case of GR, only the projectors $P^{(2)}$ and $P^{(ss)}$ appear. Thus clearly the field operator is not invertible in the $P^{(1)}$ and $P^{(ww)}$ sectors. This is again the effect of the gauge invariance of the action. Furthermore,

---

[3] for this calculation one can use the general formulae given in section 3.4, see also section 7.4.1.

the field components that appear are the gauge-invariant ones (modulo harmonic gauge transformations, as we have seen in the preceding section). This neat separation between the gauge-invariant and the gauge degrees of freedom happens because we are on-shell: We are expanding around Minkowski space, which is a solution of the field equations only when the cosmological constant is zero. Using the Weyl basis reveals that only the Weyl-squared term contributes to the propagation of the spin-2 mode, and only the R^2 term contributes to the propagation of the spin-0 mode.

Let us now restrict ourselves to four dimensions, where the higher-derivative couplings are dimensionless. In this case we see from the first line of (2.92) that for $\beta = -4\alpha = -4\gamma$ the four-derivative terms do not contribute to the propagator at all. This was to be expected, since this particular combination corresponds to the Euler topological invariant. Using (2.18) we write $\frac{1}{\kappa^2} = m_P^2$. Then (2.92) becomes

$$P^{(2)}_{\mu\nu\alpha\beta}\frac{1}{4\lambda}\left(\Box + \frac{1}{2}\lambda m_P^2\right)\Box + P^{(ss)}_{\mu\nu\alpha\beta}\frac{3}{\xi}\left(\Box - \frac{1}{12}\xi m_P^2\right)\Box. \qquad (2.93)$$

Writing a propagator for the theory requires that we fix the gauge. However, the propagator of the spin-2 degrees of freedom and of the scalar s can be studied without gauge fixing. In momentum space the spin-2 propagator can be decomposed in two fractions

$$\frac{4\lambda}{p^4 - \frac{1}{2}\lambda m_P^2 p^2} = \frac{8}{m_P^2}\left(\frac{1}{-p^2} - \frac{1}{-p^2 + \frac{1}{2}\lambda m_P^2}\right). \qquad (2.94)$$

We see that the theory contains two spin-2 degrees of freedom: a massless spin-2 particle that can be identified with the ordinary graviton, but also a massive spin-2 particle, which is either a tachyon (when $\lambda > 0$) or a massive ghost (when $\lambda < 0$). In both cases this is a pathology. We will return to this issue in section 4.3.

Likewise in the spin-0 sector the propagator can be written

$$\frac{\xi/3}{p^4 + \frac{1}{12}\xi m_P^2 p^2} = \frac{4}{m_P^2}\left(-\frac{1}{-p^2} + \frac{1}{-p^2 - \frac{1}{12}\xi m_P^2}\right). \qquad (2.95)$$

The first term is a massless particle with negative residue at the pole: a ghost. This, however, is the same as the spin-0 particle in GR and, as we have discussed in section 2.1.5, does not propagate. There remains a physical massive particle with the correct sign for the propagator. To avoid tachyonic propagation, one must have $\xi > 0$.

From this type of reasoning one can also deduce that the four-derivative terms are completely negligible at ordinary scales. This can be seen by asking what experimental bound there is on the coefficients α and β. We discuss β but obviously the same holds for α. From the fact that general relativity works well at large scales, comparing the different terms in the action, written in momentum space, one gets that $\beta p^4 \ll m_P^2 p^2$ or equivalently $\beta \ll m_P^2/p^2$. The strongest bound comes from the highest momenta. Newton's law has been tested down to distances of the order of a millimeter, which corresponds to momenta p of the order of the milli-eV. Using the fact that the Planck mass is of the order of 10^{28}eV, we get $\beta \ll 10^{62}$, which is not much of a bound. Said differently, it would take an enormous coefficient for the higher derivative terms to become relevant at the macroscopic scales at which we have some experience of gravity. (Note that if β was so large, the mass of the ghost would also be correspondingly lowered and the issue would become much more urgent.)

2.3 Power counting

The interactions of gravitons are obtained by expanding the action in powers of the graviton field h. For the Hilbert action we can write schematically

$$S = \frac{1}{4\kappa^2} \int d^d x \left[-\frac{1}{2}(\partial h)^2 + h(\partial h)^2 + h^2(\partial h)^2 + \ldots \right], \qquad (2.96)$$

and we recall that $\kappa = \sqrt{8\pi G}$. In this section we will systematically ignore all index structures and numerical factors of order unity.

For the sake of a perturbative treatment it is desirable to canonically normalize the field. To this end we absorb a factor 2κ in the definition of h, as in (2.18). Then the action becomes

$$S = \int d^d x \left[-\frac{1}{2}(\partial\phi)^2 + \kappa\,\phi(\partial\phi)^2 + \kappa^2\phi^2(\partial\phi)^2 + \ldots \right]. \qquad (2.97)$$

We see that there are infinitely many interaction terms. This is due to the non-polynomial nature of the Hilbert action. All interactions contain exactly two derivatives and are proportional to powers of κ according to the number of legs on the vertex.

Consider a one-loop diagram with E external legs and let us begin by assuming for simplicity that all vertices are three-point vertices. See for example Fig. 2.1, where $E = 5$. Then the diagram also has E vertices and E internal propagators. Let q be the momentum in the loop. Each propagator contributes a power $1/(-q^2)$ and each vertex contributes a factor κ times two powers of momenta. These momenta could be either external momenta p^μ, or the loop momentum q^μ. The highest divergence occurs when all momenta in the vertices belong to the internal lines, because in this case each vertex contributes a factor $(q + p)^2$ to the numerator of the integrand, where p is some combination of external momenta. Altogether the powers of q from vertices and propagators cancel out, and we conclude that the diagram diverges at worst as Λ_{UV}^d, where Λ_{UV} is a momentum cutoff.

Now consider a diagram with the same number of external legs, but suppose that two three-point vertices are replaced by a four-point vertex, as in Fig. 2.2. Relative to the previous diagram, there is one less vertex and one less internal propagator. The overall power of κ will be the same, however, because two powers of κ from the three-vertices will be replaced by one power of κ^2 from the new four-vertex. Also the powers of momenta in the integral will still cancel out so that the degree of divergence will be the same. By considering more general cases one can easily see that the superficial degree of divergence of a one-loop diagram is Λ_{UV}^d independent of the type of vertices that enter in the diagram.

Now suppose we add an internal line ending at two three-point vertices, as in Fig. 2.3. We have a new momentum integration with a momentum q', two new vertices and three new internal propagators. This changes the divergence by a factor Λ_{UV}^{d-2}. If one of the ends of the new internal line ended at one of the pre-existing vertices, there would be only one new vertex and two new internal propagators, and if both ends of the new line ended on pre-existing vertices, there would be only one new internal line. In each case the degree of divergence changes by the same amount.

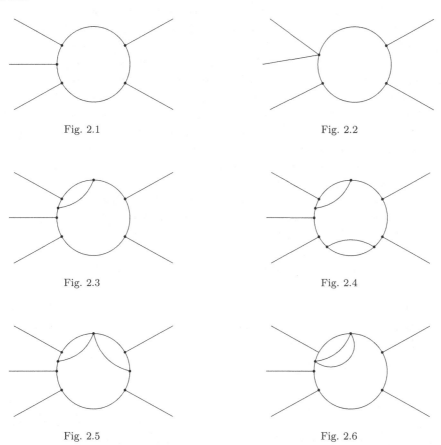

Fig. 2.1   Fig. 2.2

Fig. 2.3   Fig. 2.4

Fig. 2.5   Fig. 2.6

Each further loop produces a factor Λ_{UV}^{d-2} (see e.g. the diagrams in Figs. 2.4, 2.5, 2.6), so that at L loops the degree of divergence is

$$\Lambda_{UV}^{d+(L-1)(d-2)} \ . \tag{2.98}$$

These are all special cases of the following general rule: the degree of divergence of a diagram with L loops, I internal lines and V vertices is $dL + 2V - 2I$, so using the topological relation $L = 1 + I - V$ one finds (2.98).

Equation (2.98) shows that the degree of divergence in Einstein's theory does not depend on L when $d = 2$, so in this case the theory is power-counting renormalizable. In fact in two dimensions the Hilbert action is topological. In four dimensions the degree of divergence increases with the number of loops. The highest divergence is always proportional to the operator of lowest dimension, which is the cosmological term, but at a given loop order, there can be a finite number of lower divergences proportional to higher-dimension operators. For example, from dimensional analysis we see that at one loop the cosmological term has a quartic divergence, the Hilbert term has a quadratic divergence and terms of the form (2.75) can be logarithmically divergent. At two loops the cosmological term has a sixth-power divergence, the Hilbert term has a quartic divergence, terms of the form (2.75) can be quadratically divergent and the logarithmic divergence involves new terms with six derivatives, *e.g.* three powers of curvature. The increase in the degree of divergence with the number of loops implies that there will also be divergences proportional to operators of higher dimension, signalling that the theory is non-renormalizable.

Of course, this argument only lists the divergences that *can* occur at a certain loop order. Some other reason may prevent the divergence from actually appearing and the theory may be better behaved than expected. For this reason, actual calculations are needed to verify that the expected divergences actually occur. We will discuss these calculations for gravity in chapter 3.

Let us consider instead the theories of gravity that we discussed in section 2.2. As in (2.83) it is convenient to write the coupling in the form $1/\lambda$. This time one rescales the field with $\sqrt{\lambda}$, and the action can be expanded as

$$S = \int d^d x \left[(\Box h)^2 + \sqrt{\lambda} h (\Box h)^2 + \lambda h^2 (\Box h)^2 + \ldots \right] \qquad (2.99)$$

which shows that the perturbative coupling is $\sqrt{\lambda}$. The propagator is then of order $1/q^4$ and the vertices are also of order q^4. Repeating the preceding reasonings, one finds that the superficial degree of divergence of a diagram with L loops is

$$\Lambda_{UV}^{d+(L-1)(d-4)} . \qquad (2.100)$$

Note that if the Hilbert term was also present, it would only contribute subleading terms both to propagator and vertices, so the counting would not change. In four dimensions the degree of divergence does not increase with loop order, so that the theory is renormalizable.

One could proceed further and consider theories with still higher derivatives. In four dimensions, a theory with six or more derivatives is power-counting superrenormalizable [62]. Theories with infinitely many derivatives have a chance of being renormalizable and ghost-free. We shall mention this briefly again in section 4.3.

2.4 Appendix: Topological invariants

In this section we restrict ourselves to four dimensions. Let $*R$ and $R*$ be the dual of the Riemann tensor as a two-form or in the Lie algebra:

$$*R_{\mu\nu\rho\sigma} = \frac{1}{2}\eta_{\mu\nu\alpha\beta}R^{\alpha\beta}{}_{\rho\sigma} \; ; \qquad R^*_{\mu\nu\rho\sigma} = \frac{1}{2}\eta_{\rho\sigma\alpha\beta}R_{\mu\nu}{}^{\alpha\beta} \tag{2.101}$$

where $\eta_{\mu\nu\rho\sigma} = \sqrt{|g|}\epsilon_{\mu\nu\rho\sigma}$, $\eta^{\mu\nu\rho\sigma} = \frac{1}{\sqrt{|g|}}\epsilon_{\mu\nu\rho\sigma}$ and $\epsilon_{\mu\nu\rho\sigma}$ is the Kronecker symbol (a tensor density) with numerical values ± 1 or 0 according to the parity of the permutation $\mu\nu\rho\sigma$. (For the Riemann tensor these two operations are really the same due to the identity $R_{\mu\nu\rho\sigma} = R_{\rho\sigma\mu\nu}$, but for a generic metric connection they are conceptually distinct.) The quantity $*R^*$ is the "double dual":

$$*R^*_{\mu\nu\rho\sigma} = \frac{1}{4}\eta_{\mu\nu\alpha\beta}\eta_{\rho\sigma\gamma\delta}R^{\alpha\beta\gamma\delta} \; . \tag{2.102}$$

If the manifold is compact and without boundary the two quantities

$$\tau = \frac{1}{16\pi^2}\int d^4x \sqrt{|g|}R_{\mu\nu\rho\sigma}*R^{\mu\nu\rho\sigma} \tag{2.103}$$

$$\chi = \frac{1}{32\pi^2}\int d^4x \sqrt{|g|}R_{\mu\nu\rho\sigma}*R^{*\mu\nu\rho\sigma} \tag{2.104}$$

are integral topological invariants, called the Hirzebruch signature and the Euler number. (In particular, the Euler number is equal to the alternating sign of the Betti numbers: $\chi = b_0 - b_1 + b_2 - b_3 + b_4$.) Locally, the integrands in these expressions can be written as total derivatives.

We will prove here a weaker consequence of these statements, namely we will show by a direct calculation that χ is invariant under infinitesimal variations of the metric, up to surface terms. The proof requires that all dependence on the metric be made explicit. Thus we rewrite the tensors η in terms of ϵ and further recall that the curvature tensor is originally defined with three covariant and one contravariant index:

$$\chi = \frac{1}{128\pi^2}\int d^4x \frac{1}{\sqrt{|g|}}\epsilon^{\mu\nu\rho\sigma}\epsilon^{\alpha\beta\gamma\delta}g_{\alpha\lambda}g_{\gamma\tau}R_{\mu\nu}{}^\lambda{}_\beta R_{\rho\sigma}{}^\tau{}_\delta$$

There are thus three explicit dependences on the metric, plus the dependence through the curvature of the Levi-Civita connection. The variation gives:

$$\begin{aligned}
\delta\chi = \frac{1}{128\pi^2}\int d^4x \frac{1}{\sqrt{|g|}}\Big[&-\frac{1}{2}g^{\lambda\tau}\delta g_{\lambda\tau}\epsilon^{\mu\nu\rho\sigma}\epsilon^{\alpha\beta\gamma\delta}R_{\mu\nu\alpha\beta}R_{\rho\sigma\gamma\delta} \\
&+2\epsilon^{\mu\nu\rho\sigma}\epsilon^{\alpha\beta\gamma\delta}g_{\alpha\lambda}g_{\gamma\tau}(\nabla_\mu\delta\Gamma_\nu{}^\lambda{}_\beta - \nabla_\nu\delta\Gamma_\mu{}^\lambda{}_\beta)R_{\rho\sigma}{}^\tau{}_\delta \\
&+2\epsilon^{\mu\nu\rho\sigma}\epsilon^{\alpha\beta\gamma\delta}\delta g_{\alpha\lambda}g_{\gamma\tau}R_{\mu\nu}{}^\lambda{}_\beta R_{\rho\sigma}{}^\tau{}_\delta \Big]
\end{aligned} \tag{2.105}$$

We can now rewrite all terms as contractions of tensors:

$$\delta\chi = \frac{1}{128\pi^2} \int d^4x \sqrt{|g|} \left[-\frac{1}{2}\delta g^\lambda{}_\lambda \eta^{\mu\nu\rho\sigma} \eta^{\alpha\beta\gamma\delta} R_{\mu\nu\alpha\beta} R_{\rho\sigma\gamma\delta} \right.$$

$$\left. +4\eta^{\mu\nu\rho\sigma} \eta^{\alpha\beta\gamma\delta} \nabla_\mu \delta\Gamma_{\nu\alpha\beta} R_{\rho\sigma\gamma\delta} + 2\eta^{\mu\nu\rho\sigma} \eta^{\alpha\beta\gamma\delta} \delta g_\alpha{}^\epsilon R_{\mu\nu\epsilon\beta} R_{\rho\sigma\gamma\delta} \right] \quad (2.106)$$

Using the fact that a totally antisymmetric tensor with five indices in four dimensions is zero, we have

$$\eta^{\alpha\beta\gamma\delta} \delta g_\alpha{}^\epsilon = \eta^{\epsilon\beta\gamma\delta} \delta g_\alpha{}^\alpha + \eta^{\alpha\epsilon\gamma\delta} \delta g_\alpha{}^\beta + \eta^{\alpha\beta\epsilon\delta} \delta g_\alpha{}^\gamma + \eta^{\alpha\beta\gamma\epsilon} \delta g_\alpha{}^\delta.$$

Each of the last three terms on the r.h.s. gives the same contribution as the l.h.s., so altogether

$$2\eta^{\mu\nu\rho\sigma} \eta^{\alpha\beta\gamma\delta} \delta g_\alpha{}^\epsilon R_{\mu\nu\epsilon\beta} R_{\rho\sigma\gamma\delta} = 2\frac{1}{4}\delta g_\alpha{}^\alpha \eta^{\mu\nu\rho\sigma} \eta^{\epsilon\beta\gamma\delta} R_{\mu\nu\epsilon\beta} R_{\rho\sigma\gamma\delta}.$$

So the first and the last terms in (2.106) exactly cancel. The second term can be integrated by parts. Using $\nabla\eta = 0$, and the Bianchi identity, it reduces to a total derivative, Q.E.D.

Now let us write the Euler invariant as

$$\chi = \frac{1}{128\pi^2} \int d^4x \sqrt{|g|} \, \epsilon^{\mu\nu\rho\sigma} \epsilon_{\alpha\beta\gamma\delta} R_{\mu\nu}{}^{\alpha\beta} R_{\rho\sigma}{}^{\gamma\delta} \ .$$

We can now use the identity

$$\epsilon^{\mu\nu\rho\sigma} \epsilon_{\alpha\beta\gamma\delta} = \delta^{[\mu}_{[\alpha} \delta^\nu_\beta \delta^\rho_\gamma \delta^{\sigma]}_{\delta]}$$

and contract all the indices in the two curvature tensors with the Kronecker tensors. The contractions can give terms proportional to the square of the Riemann tensor, the square of the Ricci tensor and the square of the Ricci scalar. Counting each type of term gives

$$\chi = \frac{1}{32\pi^2} \int d^4x \sqrt{|g|} \left[R_{\mu\nu\rho\sigma} R^{\mu\nu\rho\sigma} - 4R_{\mu\nu} R^{\mu\nu} + R^2 \right] = \frac{1}{32\pi^2} \int d^4x \sqrt{|g|} E \ .$$

$$(2.107)$$

showing that E is the integrand of the Euler invariant.

It is worth stressing the difference between a total derivative term such as $\nabla^2 R$, and E. Both are of the form $\nabla_\mu \Omega^\mu$ but whereas in the former $\Omega^\mu = \nabla^\mu R$ is a globally defined vectorfield, in the case of the latter Ω^μ may not be globally defined. This happens when the tangent bundle is nontrivial and has to be described by different charts. The vectorfield Ω^μ is well-defined in each bundle chart, but if one tries to extend it over the whole manifold one will encounter singularities, analogous to the Dirac string in the case of the magnetic monopole.

Chapter 3

Failure of renormalizability

The main goal of this chapter is to calculate the one-loop divergences in Einstein's theory. This requires a number of technical steps. By using the background field method, the calculation can be reduced to the case of a spin-2 quantum field propagating in an external non-dynamical gravitational field. We review a very general method to compute the divergent part of the one-loop effective action in any Euclidean QFT, based on the early time asymptotic expansion of the heat kernel. In Minkowskian signature, this is known as the Schwinger-DeWitt method [63–65]. As a warmup we apply this method to scalar and gauge theories, then we come to gravity. In order not to interrupt the flow of the arguments, some necessary results on the heat kernel are left to Section 3.7.

3.1 Divergences in curved spacetime: Scalar field

Quantum field theories in Minkowski space generally exhibit divergences. For example, in the case of ϕ^4 theory, the effective potential has a quartic divergence that is field-independent and represents the vacuum energy density of the field, a quadratic divergence that renormalizes the mass and a logarithmic divergence that renormalizes the quartic self-coupling. Suppose we replace the flat metric by some fixed curved metric. At distance scales much smaller than the typical curvature radius, spacetime will look approximately flat. Since the ultraviolet divergences are manifestations of quantum fluctuations with very short wavelenghts, they are only sensitive to the local structure of spacetime and therefore we expect all the divergences that occur in a quantum field theory in Minkowski space also to occur, essentially unaltered, in a curved space. On the other hand, the curvature tensor defines a new mass scale and one expects new divergent terms proportional to powers of curvature to appear in the effective action. For example, dimensional analysis would permit a term $R\Lambda^2$ along with the term $\phi^2\Lambda^2$, and a term $R^2 \log\left(\frac{\Lambda^2}{\mu^2}\right)$ along with the term $\phi^4 \log\left(\frac{\Lambda^2}{\mu^2}\right)$. We will analyze here the origin of these new types of divergences. Since they are due to the interaction of the field with the background metric, it will be sufficient to consider initially fields that are minimally coupled to the metric and to neglect their self-interactions. The effective action of such fields is given by Gaussian functional integrals.

3.1.1 *The Euclidean functional integral*

We begin by considering a real scalar field ϕ propagating on a manifold M with a fixed external metric $g_{\mu\nu}$. We keep the spacetime dimension d arbitrary for now. The action is

$$S(\phi; g) = -\frac{1}{2} \int d^d x \sqrt{|g|} g^{\mu\nu} \partial_\mu \phi \partial_\nu \phi \tag{3.1}$$

and the functional integral, in the presence of a source coupled linearly to ϕ, is

$$\int (d\phi) e^{i(S(\phi;g) + \int d^4 x \sqrt{|g|} j\phi)} . \tag{3.2}$$

To make the functional integral better defined one performs a Wick rotation. We shall follow here the standard flat space procedure of continuing time in the complex plane. For this we will assume that spacetime is globally static with a metric of the form

$$g_{\mu\nu} = \begin{pmatrix} g_{00} & 0 \\ 0 & g_{ij} \end{pmatrix} \tag{3.3}$$

where $g_{00} < 0$. We shall discuss in section 5.2 a more general procedure. We define $t = -it_E$, then

$$iS(\phi; \bar{g}) = i\frac{1}{2} \int dt d^{d-1} x \sqrt{|g|} \left[-g^{00} (\partial_t \phi)^2 - g^{ij} \partial_i \phi \partial_j \phi \right]$$

$$= \frac{1}{2} \int dt_E d^{d-1} x \sqrt{|g|} \left[g^{00} (\partial_{t_E} \phi)^2 - g^{ij} \partial_i \phi \partial_j \phi \right] = -S_E(\phi; g_E) \tag{3.4}$$

where

$$S_E(\phi; g_E) = \frac{1}{2} \int d^d x_E \sqrt{g_E} \, g_E^{\mu\nu} \partial_\mu \phi \partial_\nu \phi \tag{3.5}$$

and

$$(g_E)_{\mu\nu} = \begin{pmatrix} -g_{00} & 0 \\ 0 & g_{ij} \end{pmatrix} \tag{3.6}$$

is a positive definite metric on an analytically continued manifold M_E. Similarly

$$i \int dt d^{d-1} x \sqrt{|g|} \, j\phi = \int d^d x_E \sqrt{g_E} \, j\phi . \tag{3.7}$$

We shall henceforth always work on the Euclidean manifold and drop the subscripts E. We will assume that the boundary conditions on the fields are such that it is possible to perform integration by parts without having any boundary terms left. This is possible *e.g.* if M is compact and without boundary. With an integration by parts the action can be rewritten in the form

$$S(\phi; g) = \frac{1}{2} \int d^d x \sqrt{g} \, \phi \Delta \phi , \tag{3.8}$$

where $\Delta = -\nabla_\mu \nabla^\mu$ is the covariant Laplacian and the partition function is

$$Z(j; g) = e^{W(j;g)} = \int (d\phi) e^{-S(\phi;g) + \int d^4 x \sqrt{g} \, j\phi} . \tag{3.9}$$

The functional W is the generating functional of connected Green's functions.

3.1.2 *Functional determinant*

We begin by analyzing the functional $W(0, g)$. It is not possible to use Fourier analysis in curved space. Instead, one can use the spectral decomposition of the operator Δ:

$$\Delta \phi_n = \lambda_n \phi_n \,, \tag{3.10}$$

with eigenvalues λ_n and eigenfunctions ϕ_n. It is convenient to assume that M is compact and without boundary, in which case the spectrum is discrete. Then n is an integer; each value of n labels a single eigenfunction and different values of n could correspond to the same eigenvalue. The operator $-\nabla^2$ acting on scalars on a compact manifold has an eigenvalue zero with eigenfunction ϕ_0 =constant. As will become clear shortly, this mode would give rise to an undamped integral and consequently to a divergent partition function. For this reason it has to be treated separately. One can avoid this issue by adding a small mass term.

We take the eigenfunctions to be dimensionless and orthonormal with respect to the natural inner product on $C^\infty(M)$:

$$(\phi_n, \phi_m) = \mu^d \int_M d^d x \sqrt{g}\, \phi_n(x)\phi_m(x) = \delta_{nm} \,,$$

where μ is a constant with the dimension of mass. The eigenfunctions form a basis in the space of functions on M, so we can decompose the field ϕ as

$$\phi(x) = \sum_n a_n \phi_n(x) \,.$$

This is the analog of the Fourier decomposition of the field in flat spacetime. Here the coefficients a_n have the same dimension as ϕ. The Euclidean action then becomes

$$S(\phi; g) = \frac{1}{2\mu^d} \sum_n \lambda_n a_n^2 = \frac{1}{2} \sum_n \tilde{\lambda}_n \tilde{a}_n^2 \,,$$

where we defined dimensionless quantities $\tilde{a}_n = a_n/\mu^{d/2-1}$, $\tilde{\lambda}_n = \lambda_n/\mu^2$. The path integral measure can be written formally

$$(d\phi) = N \, \Pi_n \frac{da_n}{\mu^{d/2-1}} = \mathcal{N} \, \Pi_n d\tilde{a}_n \,,$$

where $\mathcal{N}$ is an infinite, field-independent, dimensionless normalization factor such that the Gaussian normalization condition holds:

$$1 = \int (d\phi) e^{-\frac{\mu^2}{2} \int d^d x \sqrt{g}\, \phi^2} \,. \tag{3.11}$$

Explicitly this condition gives

$$1 = \mathcal{N} \Pi_n \left(\int d\tilde{a}_n e^{-\frac{1}{2}\tilde{a}_n^2} \right) \equiv \mathcal{N} \Pi_n \sqrt{2\pi} \,. \tag{3.12}$$

Then the path integral at zero source is seen to formally correspond to a functional determinant:

$$Z(0;g) = e^{W(0;g)} = \mathcal{N}\Pi_n \left(\int d\tilde{a}_n e^{-\frac{1}{2}\tilde{\lambda}_n \tilde{a}_n^2} \right) = \mathcal{N}\Pi_n \left(\sqrt{\frac{2\pi}{\tilde{\lambda}_n}} \right) = (\det\tilde{\Delta})^{-1/2} ,$$

(3.13)

where $\tilde{\Delta} = \Delta/\mu^2$.

If we take into account the nontrivial source term as in (3.9), we find

$$Z(j;g) = (\det\tilde{\Delta})^{-1/2} e^{\frac{1}{2}\int d^d x \sqrt{g}\, j\Delta^{-1}j} .$$

(3.14)

Now we define the expectation value of the field in the presence of the source j as

$$\varphi_j(x) \equiv \langle\phi\rangle_j = \frac{\delta W}{\delta j(x)} ,$$

(3.15)

where φ is a c-number field, and the connected two point function

$$G(x,y)_j = \langle\phi(x)\phi(y)\rangle - \langle\phi(x)\rangle\langle\phi(y)\rangle = \frac{\delta^2 W}{\delta j(x)\delta j(y)} .$$

(3.16)

The relation between the source j and the classical field φ in (3.15) can be inverted. We denote j_φ the source that is required to produce an expectation value φ. The effective action is a functional of φ that is defined via a Legendre transform:

$$\Gamma(\varphi;g) = -W(j_\varphi;g) + \int d^d x \sqrt{g}\, j_\varphi \varphi .$$

(3.17)

For the Gaussian integral (3.14), $j_\varphi = \Delta\varphi$, so we arrive at

$$\Gamma(\varphi;g) = \frac{1}{2} \int d^d x \sqrt{g}\, \varphi\Delta\varphi + \frac{1}{2} \operatorname{Tr} \log \tilde{\Delta} .$$

(3.18)

The first term is the classical action for φ. Note that for $\varphi = 0$ the effective action can be defined more directly without going through the Legendre transform

$$\Gamma(g) \equiv \Gamma(0;g) = -\log Z = \frac{1}{2} \operatorname{Tr} \log \tilde{\Delta} .$$

(3.19)

This functional is obviously ill-defined. We will now try to understand its properties.

3.1.3 *Zeta function regularization*

One way of making sense of the trace

$$\frac{1}{2} \operatorname{Tr} \log \tilde{\Delta} = \frac{1}{2} \sum_n \log \tilde{\lambda}_n$$

(3.20)

is zeta function regularization [66–68]. In analogy to the definition of Riemann's zeta function $\zeta_R(s) = \sum_{n=1}^{\infty} n^{-s}$, we define a zeta function of the operator Δ by

$$\zeta_\Delta(s) = \sum_n \tilde{\lambda}_n^{-s} ,$$

(3.21)

where $\tilde{\lambda}_n$ are the dimensionless eigenvalues. (In this notation, degenerate eigenvalues have to be counted separately; otherwise, each different term in the sum would have to be weighted by the multiplicity of the eigenvalue.) Using $\log z = -\frac{d}{ds}z^{-s}\big|_{s=0}$ we have

$$\frac{1}{2}\mathrm{Tr}\log\tilde{\Delta} = -\frac{1}{2}\sum_n \frac{d}{ds}\tilde{\lambda}_n^{-s} = -\frac{1}{2}\frac{d}{ds}\zeta_\Delta(s)\Big|_{s=0} . \tag{3.22}$$

A theorem of Weyl says that the number of eigenvalues that are less than x grows like $x^{d/2}$ for large x, so for large n the sum in the zeta function can be replaced by $\int d\lambda\lambda^{\frac{d}{2}-1-s}$.[1] Therefore, the zeta function is convergent for $\mathrm{Re}(s) > d/2$ and can be defined on the whole complex plane by analytic continuation. In this way one obtains a finite value for the effective action.

As an example of such a procedure we consider the partition function of the scalar field ϕ at temperature T in a large cubic box of side L and volume $V = L^3$. The goal is to establish the dependence on the temperature and volume. This is obtained by imposing on the field periodicity in Euclidean time with period $\beta = 1/T$ and Dirichlet conditions on the boundary of the box. With these boundary conditions the eigenvalues of the Laplacian are

$$\lambda_{n,\vec{k}} = \left(\frac{2\pi n}{\beta}\right)^2 + \vec{k}^2 , \quad \text{for } n = -\infty,\dots,0,\dots,\infty .$$

In the limit $L \gg \beta$, the Dirichlet condition can be replaced by periodicity in space with period L. Then the density of eigenvalues is $\rho = \frac{2V}{(2\pi)^3}\int d\vec{k}$ for $n > 0$ and half that for $n = 0$. The zeta function is

$$\zeta_\Delta(s) = \frac{4\pi V}{(2\pi)^3}\left[\int_0^\infty dk\, k^{2-2s} + 2\sum_{n=1}^\infty \int_0^\infty dk\, k^2\left(\left(\frac{2\pi n}{\beta}\right)^2 + k^2\right)^{-s}\right] . \tag{3.23}$$

The first term is independent of β and therefore uninteresting. Integrating by parts, the second integral becomes

$$-\frac{1}{2s+2}\int_0^\infty dk\,\left(\left(\frac{2\pi n}{\beta}\right)^2 + k^2\right)^{-s+1} .$$

When $\mathrm{Re}(s) > 2$, this is infrared divergent. It can be regulated by putting an IR cutoff ϵ at the lower end of integration. When analytically continued to $s \to 0$ this contribution can be neglected. Defining $k = \frac{2\pi n}{\beta}\sinh y$, the zeta function becomes

$$\zeta_\Delta(s) \approx -\frac{8\pi V}{(2\pi)^3}\sum_{n=1}^\infty\left(\frac{2\pi n}{\beta}\right)^{3-2s}\frac{1}{2s+2}\int_0^\infty dy(\cosh y)^{3-2s} .$$

---

[1]One can also arrive at this heuristically by arguing that at short distances the spectrum of the Laplacian is the same as the spectrum of the flat Laplacian, so that the spectral sum can be approximated by $\int dp\,p^{d-1}$, and then making the change of variables $\lambda = p^2$.

The integral over y is equal to $\frac{\sqrt{\pi}}{2}\frac{\Gamma\left(s-\frac{3}{2}\right)}{\Gamma(s-1)}$ and the sum over n yields a Riemann zeta function:

$$\zeta_\Delta(s) \approx -\frac{8\pi V}{(2\pi)^3}\left(\frac{2\pi}{\beta}\right)^{3-2s}\frac{1}{2s+2}\zeta_R(2s-3)\frac{\sqrt{\pi}}{2}\frac{\Gamma\left(s-\frac{3}{2}\right)}{\Gamma(s-1)} \ .$$

The logarithm of the partition function at zero field is

$$W = -\Gamma = \frac{1}{2}\zeta'_\Delta(0) = \frac{\pi^2 V}{90}T^3 \ . \tag{3.24}$$

From this, using standard thermodynamic relations (with Boltzmann's constant set equal to one), one can obtain the energy, pressure and entropy of the radiation:

$$E = -\frac{dW}{d\beta} = \frac{\pi^2 V}{30}T^4 \tag{3.25}$$

$$P = \frac{1}{\beta}\frac{dW}{dV} = \frac{\pi^2}{90}T^4 \tag{3.26}$$

$$S = \beta E + W = \frac{2\pi^2 V}{45}T^3 \ . \tag{3.27}$$

This procedure has reproduced the correct dependence of the partition function on the volume and on the temperature. Note that zeta function regularization is actually a renormalization procedure: it automatically discards infinities and yields finite results. For us, it is more interesting to understand what divergences were present, and to study their dependence on the metric. For this we need to make a little digression to discuss the heat kernel and some of its properties.

3.1.4   *The heat kernel*

Consider the "heat equation" for the covariant Laplacian Δ

$$\frac{d\Psi}{dt} + \Delta\Psi = 0 \ . \tag{3.28}$$

It describes a diffusion process on the manifold M with metric g, occurring in an external "time" t.[2] The heat kernel $K_\Delta(x,y;t)$ for the operator Δ is a function on $M \times M \times R$ satisfying the heat equation with the initial condition

$$K_\Delta(x,y;0) = \delta(x,y) \ . \tag{3.29}$$

Note that t has dimension of length squared and K_Δ has dimension of inverse volume. Given Ψ at the initial time $t = 0$, the solution of the heat equation is given at any later time by $\Psi(x,t) = \int d^d y \sqrt{g(y)}K_\Delta(x,y;t)\Psi(y,0)$. The heat kernel can be written formally as $K_\Delta(.,.;t) = e^{-t\Delta}$. Using (3.10) it has the spectral decomposition

$$K_\Delta(x,y;t) = \sum_n \phi_n(x)\phi_n(y)e^{-t\lambda_n} \ . \tag{3.30}$$

---

[2]The original heat equation describes the diffusion of heat in a conducting medium. In this case M is three-dimensional Euclidean space, Δ is the ordinary Laplacian and it has a prefactor $k/(c\rho)$, where k is thermal conductivity, c is specific heat, ρ is density, and t is ordinary time.

We shall be particularly interested in the trace of the heat kernel, which is the dimensionless function

$$\text{Tr}K_\Delta(t) = \int d^dx \sqrt{g} K_\Delta(x, x; t) = \sum_n e^{-t\lambda_n} . \tag{3.31}$$

In flat d-dimensional space the heat kernel can be easily calculated using Fourier analysis. In this case we denote the coordinates by d-dimensional vectors, *e.g.* $\vec{x}$. The Fourier transform of K_Δ on the first coordinate is a function $\tilde{K}_\Delta(\vec{q}, \vec{y}; t)$ satisfying the equation

$$\frac{d}{dt}\tilde{K}_\Delta(\vec{q}, \vec{y}; t) + q^2 \tilde{K}_\Delta(\vec{q}, \vec{y}; t) = 0 ,$$

with the initial condition

$$\tilde{K}_\Delta(\vec{q}, \vec{y}; 0) = e^{-i\vec{q}\cdot\vec{y}} .$$

The solution is $\tilde{K}_\Delta(\vec{q}, \vec{y}; t) = e^{-q^2 t - i\vec{q}\cdot\vec{y}}$. The inverse Fourier transform is a Gaussian integral:

$$K_\Delta(\vec{x}, \vec{y}; t) = \int \frac{d\vec{q}}{(2\pi)^d} e^{-q^2 t + i\vec{q}\cdot(\vec{x}-\vec{y})} = \frac{1}{(4\pi t)^{d/2}} e^{-\frac{|\vec{x}-\vec{y}|^2}{4t}} . \tag{3.32}$$

Therefore, in flat space the trace of the heat kernel is

$$\text{Tr}K_\Delta(t) = \frac{V}{(4\pi t)^{d/2}} , \tag{3.33}$$

where V is the (infinite) volume.

Let us return to a general curved manifold. Since every manifold looks locally like Euclidean space, in the limit $t \to 0$ the trace of the heat kernel must reduce to the form it has in flat space, Eq. (3.33). The deviations from this form must be proportional to the deviation of the metric from flatness, which is measured by curvature invariants. One is led to expect that the trace of the heat kernel has, for $t \to 0$, an asymptotic expansion of the form

$$\text{Tr}K_\Delta(t) \approx \frac{1}{(4\pi t)^{d/2}} \left[B_0(\Delta) + tB_2(\Delta) + t^2 B_4(\Delta) + \ldots\right] , \tag{3.34}$$

where

$$B_n(\Delta) = \int d^4x \sqrt{g} b_n(\Delta) \tag{3.35}$$

and $b_n(\Delta)$ are scalars constructed from the curvature and its covariant derivatives. For dimensional reasons, $b_n(\Delta)$ must contain n derivatives of the metric. Thus the coefficient b_2 must be proportional to R, b_4 must be a combination of the invariants $R_{\mu\nu\rho\sigma}R_{\mu\nu\rho\sigma}$, $R_{\mu\nu}R_{\mu\nu}$, R^2 and $\nabla^2 R$, and so on. There remains to determine the numerical coefficients. The calculation of the heat kernel expansion coefficients is a well-developed field of mathematics and many results can be found in the literature.

In order not to break the line of reasoning some results will be derived in section 3.7. Here we just report the result for the first three coefficients:

$$b_0 = 1 \; ; \qquad b_2 = \frac{1}{6} R \; ; \tag{3.36}$$

$$b_4 = \frac{1}{180} \left(R_{\mu\nu\rho\sigma} R^{\mu\nu\rho\sigma} - R_{\mu\nu} R^{\mu\nu} + \frac{5}{2} R^2 + 6 \nabla_\mu \nabla^\mu R \right) \; .$$

The zeta function is related to the heat kernel by an integral transform. Using the integral representation of the gamma function $\Gamma(s) = \int_0^\infty d\bar{t}\, \bar{t}^{s-1} e^{-\bar{t}}$, changing variable $\bar{t} = \lambda t$ we find

$$\lambda^{-s} = \frac{1}{\Gamma(s)} \int dt\, t^{s-1} e^{-\lambda t} \; .$$

Inserting into (15.2.1) and using (15.2.6) we find that the zeta function is related to the trace of the heat kernel by a Mellin transform:

$$\zeta_\Delta(s) = \frac{1}{\Gamma(s)} \int_0^\infty dt\, t^{s-1} \mathrm{Tr} K_\Delta(t) \; . \tag{3.37}$$

For $s \to 0$ we have $\frac{1}{\Gamma(s)} \approx s + \gamma s^2 + \ldots$ where γ is the Euler-Mascheroni constant. Then, inserting (3.37) in (3.22) and ignoring that the integrals are divergent we obtain the formal expression for the effective action

$$\Gamma(g) = -\frac{1}{2} \int_0^\infty dt\, t^{-1} \mathrm{Tr} K_\Delta(t) \; . \tag{3.38}$$

As mentioned earlier, small t corresponds to short distances, so in the integral (3.38) the lower end of the integration range corresponds to the UV, while the upper end corresponds to the IR. (This is also clear from the fact that the dimension of t is inverse squared length.) To exhibit the ultraviolet divergences we choose an ultraviolet cutoff Λ_{UV} and a finite reference mass $\mu < \Lambda_{UV}$, and we split the ultraviolet-regulated integral into $\int_{1/\Lambda_{UV}^2}^\infty = \int_{1/\Lambda_{UV}^2}^{1/\mu^2} + \int_{1/\mu^2}^\infty$. For $t \to \infty$ the trace of the heat kernel is dominated by the smallest eigenvalue; if Δ does not have negative or zero eigenvalues, $\mathrm{Tr} K(t) \approx e^{-t\lambda_1}$, so the second piece is convergent. In the first piece we can use the asymptotic expansion (3.34); the first three terms of the asymptotic expansion will give rise to divergences at the low end of the integration. The divergent part of the effective action is then

$$\Gamma(g) = -\frac{1}{2} \frac{1}{(4\pi)^{d/2}} \int d^d x \sqrt{g} \int_{1/\Lambda_{UV}^2}^{1/\mu^2} dt \left[t^{-\frac{d}{2}-1} b_0 + t^{-\frac{d}{2}} b_2 + \ldots + t^{-1} b_d + \ldots \right] \tag{3.39}$$

$$= -\frac{1}{2} \frac{1}{(4\pi)^{d/2}} \int d^d x \sqrt{g} \left[\frac{\Lambda_{UV}^d}{d/2} b_0 + \frac{\Lambda_{UV}^{d-2}}{\frac{d}{2}-1} b_2 + \ldots + \log \frac{\Lambda_{UV}^2}{\mu^2} b_d + \text{finite terms} \right] \; .$$

As anticipated, we find divergent terms proportional to integrals of curvature. The first term is proportional to the vacuum energy and can be absorbed in a renormalization of the cosmological constant. The second term is proportional to the Einstein-Hilbert action and can be absorbed in a renormalization of Newton's constant. The third term contains four derivatives and leads to renormalizations of the coefficients of the terms in the action that are of second order in the curvature. In the zeta function method all these divergences are automatically eliminated in the analytic continuation procedure by the addition of counterterms.

3.2 Generalizations

We have used the heat kernel to calculate the divergences of a quantum scalar field in an external metric. The same method can be applied to calculate the one-loop divergences in any QFT, not just in a gravitational context. Before applying it to quantum gravity, we shall further acquaint ourselves with this method by calculating the one-loop divergences in the effective potential of a scalar field in flat space and the one-loop divergences of a Yang-Mills theory in flat space.

Besides their didactical value, these calculations will serve the purpose of introducing other pieces in the heat kernel coefficients, all of which will be needed in the application to quantum gravity. Let us state first the general result.

3.2.1 *The master formula*

Let ψ be a quantum field carrying both spacetime and internal indices. Geometrically, it should be thought of in the following terms. There is a metric g on spacetime of signature (p, q) that defines a bundle of orthonormal frames OM; there is also a principal G-bundle P over M with a connection A. At this level g and A can be thought of as fixed external fields, or as background fields. The field ψ has a given spin, *i.e.* it carries a representation σ of the "Lorentz" group $SO(p, q)$. It also transforms under G in some representation ρ. Then ψ should be thought of as a section of the vectorbundle $S \otimes V$, where S is associated to OM by the representation σ and V is associated to P by the representation ρ. All the fields that are of interest in particle physics fall in this very broad class.

The metric g defines a unique torsion-free connection Γ in OM called the Levi-Civita connection. It can be used to define the covariant derivative of sections of S. The connection A can be used to define the covariant derivative of sections of V. We will denote ∇ the tensor product of these connections in $S \otimes V$. To make this more explicit, let $\theta_a = \theta_a{}^\mu \partial_\mu$ be a local field of orthonormal frames on M and e_i a local field of frames in V. We use $A, B, \ldots$ for the indices in the space carrying the representation σ. For example if σ is a spinor representation, A are spinor indices; if σ is a vector (fundamental) representation, $A = a$; if σ is a spin-2 representation, $A = (a, b)$, symmetric in (a, b) and so on. The field ψ has components ψ_A^i and its covariant derivative is

$$D_\mu \psi_A^i = \partial_\mu \psi_A^i + \Gamma_\mu^{ab} \sigma_{ab\,A}{}^B \psi_B^i + A_\mu^m \rho_m{}^i{}_j \psi_A^j \, , \qquad (3.40)$$

where σ_{ab} (antisymmetric in (a, b)) are the generators of $SO(p, q)$ in the representation σ and $\rho_m{}^i{}_j$ are the generators of G in the representation ρ (m is an index in the Lie algebra of G). One can then define the covariant Laplacian, mapping sections of $S \otimes V$ to sections of the same bundle, by

$$- g^{\mu\nu} D_\mu D_\nu \qquad (3.41)$$

We assume that the action of the field ψ is

$$S(\psi; g, A) = \frac{1}{2} \mathcal{G}(\psi, \Delta \psi) \qquad (3.42)$$

where $\mathcal{G}(\psi, \psi') = \int dx \sqrt{g} G_{ij}^{AB} \psi_A^i \psi_B'^j$ is a local inner product of sections of $S \otimes V$ and Δ is a general Laplace-type operators of the form

$$\Delta = -g^{\mu\nu} D_\mu D_\nu + \mathbf{E} \tag{3.43}$$

where $\mathbf{E}$ is an endomorphism of $S \otimes V$, with index structure $E_A{}^{Bi}{}_j$.

Following step by step the arguments of section 3.1, the divergences of the effective action $\Gamma(0; g, A)$ coming from the functional integral over ψ are given by

$$-\frac{1}{2} \frac{1}{(4\pi)^{d/2}} \int d^d x \sqrt{g} \left[\frac{\Lambda_{UV}^d}{d/2} b_0(\Delta) + \frac{\Lambda_{UV}^{d-2}}{\frac{d}{2}-1} b_2(\Delta) + \ldots + \log \frac{\Lambda_{UV}^2}{\mu^2} b_d(\Delta) \right]. \tag{3.44}$$

The heat kernel coefficients of the operator Δ are given by [69–72]

$$
\begin{aligned}
b_0 &= \operatorname{tr} \mathbf{1} \ ; \\
b_2 &= \frac{1}{6} R \operatorname{tr} \mathbf{1} - \operatorname{tr} \mathbf{E} \ ; \\
b_4 &= \frac{1}{180} \left(R_{\mu\nu\rho\sigma} R^{\mu\nu\rho\sigma} - R_{\mu\nu} R^{\mu\nu} + \frac{5}{2} R^2 + 6 \nabla^2 R \right) \operatorname{tr} \mathbf{1} \\
&\quad + \frac{1}{2} \operatorname{tr} \mathbf{E}^2 - \frac{1}{6} R \operatorname{tr} \mathbf{E} + \frac{1}{12} \operatorname{tr} \Omega^{\mu\nu} \Omega_{\mu\nu} - \frac{1}{6} \nabla^2 \operatorname{tr} \mathbf{E} \ ,
\end{aligned}
\tag{3.45}
$$

In these formulae $\mathbf{1}$ is the identity in $S \otimes V$, so $\operatorname{tr} \mathbf{1}$ is the dimension of the fiber of $S \otimes V$ (the product of the dimensions of the spaces carrying the representations σ and ρ). Ω is the curvature of the connection in $S \otimes V$, defined by

$$[D_\mu, D_\nu] \psi = \Omega_{\mu\nu} \psi \ . \tag{3.46}$$

Note that the last terms of both lines of b_4 are total derivatives; we will neglect such terms in the following. These heat kernel coefficients are independent of the dimension; the dependence of the heat kernel on the dimension is entirely in the prefactor in (3.34).

The terms proportional to $\mathbf{1}$ were already present in (3.36). They are due to the background metric g. Equations (3.45) contain new terms: terms due to the endomorphism $\mathbf{E}$, and a term proportional to Ω^2 due to the covariant derivative D. We will understand the origin and structure of these new terms by studying the following two examples.

3.2.2 *Effective potential*

In the examples considered so far, the quantum field had no self-interactions: its functional integral was purely Gaussian. In this and in the following example we shall see how to use the same technique to compute the one-loop effective action of a self-interacting theory.

We begin with the one-loop divergences in the effective potential of a real self-interacting scalar field in four dimensions, with Euclidean action

$$\int d^4 x \left[\frac{1}{2} (\partial\phi)^2 + V(\phi^2) \right] \ . \tag{3.47}$$

The trick is to expand $\phi = \bar{\phi} + \varphi$, where $\bar{\phi}$ is a constant background field and φ a shifted quantum field. The action is highly non-Gaussian, but at one loop only the part of the action quadratic in the fluctuation is needed. In this case it is

$$\frac{1}{2} \int d^4x \, \varphi(-\partial^2 + E)\varphi \,, \qquad \text{with} \qquad E = 2V'(\bar{\phi}^2) + 4\bar{\phi}^2 V''(\bar{\phi}^2) \,. \qquad (3.48)$$

The effective action $\Gamma(\varphi; \bar{\phi})$ is now a functional of two arguments: the "classical field" (the Legendre conjugate of the source that couples linearly to φ), which, by a slight abuse of terminology, we will still denote φ, and the background. The one-loop effective action evaluated at $\varphi = 0$ is given by

$$\Gamma(0; \bar{\phi}) = S(\bar{\phi}) + \frac{1}{2} \text{Tr} \log \tilde{\Delta} \,.$$

The divergent part of the effective action is given by the master formula (3.44), where b_n are the heat kernel coefficients of the operator $\Delta = -\partial^2 + E$. These are given by the **E**-terms in (3.45). The effective potential V_{eff} is defined by $\Gamma(0; \bar{\phi}) = \int d^4x V_{eff}(\bar{\phi}^2)$, for constant $\bar{\phi}$. Its divergent part is

$$-\frac{1}{2} \frac{1}{(4\pi)^2} \left[\frac{1}{2} \Lambda_{UV}^4 - \Lambda_{UV}^2 E + \frac{1}{2} \log \frac{\Lambda_{UV}^2}{\mu^2} E^2 \right] \,.$$

For example if

$$V = \frac{1}{2} m^2 \phi^2 + \frac{1}{4!} \lambda \phi^4 \,,$$

we have

$$E = m^2 + \frac{1}{2} \lambda \bar{\phi}^2 \,.$$

In this case, neglecting field-independent terms,

$$V_{eff}(\bar{\phi}^2) = V(\bar{\phi}^2) - \frac{1}{2} \frac{1}{(4\pi)^2} \left[-\frac{1}{2} \Lambda_{UV}^2 \bar{\phi}^2 + \frac{1}{2} \log \frac{\Lambda_{UV}^2}{\mu^2} \left(\frac{1}{4} \lambda^2 \bar{\phi}^4 + \lambda m^2 \bar{\phi}^2 \right) \right] \,, \qquad (3.49)$$

up to finite terms. We can define the renormalized coupling to be the coefficient of $\phi^4/4!$ in V_{eff}:

$$\lambda_R(\mu) = \lambda - \frac{3}{16\pi^2} \lambda^2 \log \frac{\Lambda_{UV}}{\mu} \,, \qquad (3.50)$$

then, as a by-product of the preceding calculation, one obtains the familiar beta function

$$\mu \frac{\partial \lambda_R}{\partial \mu} = \frac{3}{16\pi^2} \lambda^2 \,. \qquad (3.51)$$

Note that, up to a sign, the coefficient of the beta function is just the coefficient of the logarithmic divergence.

3.2.3 *Yang-Mills theory*

Next we consider the one-loop divergences in a non-abelian gauge theory in four dimensions. This will also be a useful preliminary example of background gauge fixing, in preparation for use in gravity.[3] The Euclidean action is

$$S_{YM}(A) = \frac{1}{4g^2} \int d^4x F_{\mu\nu}^a F^{\mu\nu a} , \tag{3.52}$$

where $F_{\mu\nu}^a = \partial_\mu A_\nu^a - \partial_\nu A_\mu^a + f^a{}_{bc} A_\mu^b A_\nu^c$ and g is the gauge coupling. As in the previous example, we use the background field method and split $A_\mu^a = \bar{A}_\mu^a + a_\mu^a$. Denoting $\bar{F}_{\mu\nu}^a$ the curvature of the background field, we have

$$F_{\mu\nu}^a = \bar{F}_{\mu\nu}^a + \bar{D}_\mu a_\nu^a - \bar{D}_\nu a_\mu^a + f^a{}_{bc} a_\mu^b a_\nu^c , \tag{3.53}$$

where $\bar{D}_\mu a_\nu^a = \partial_\mu a_\nu^a + \bar{A}_\mu^b f^a{}_{bc} a_\nu^c$ is the covariant derivative with respect to the background field.

An infinitesimal gauge transformation with parameter ϵ^a gives

$$\delta_\epsilon A_\mu^a = D_\mu \epsilon^a = \partial_\mu \epsilon^a + f^a{}_{bc} A_\mu^b \epsilon^c . \tag{3.54}$$

This can be split in different ways between background and fluctuation. One is to keep the background fixed and attribute all the variation to the quantum field:

$$\delta_\epsilon^{(Q)} \bar{A}_\mu^a = 0 ,$$
$$\delta_\epsilon^{(Q)} a_\mu^a = D_\mu \epsilon^a . \tag{3.55}$$

These are called "quantum gauge transformations". The other is to split the transformation evenly so that the background transforms as a connection and the quantum field as a matter field in the adjoint representation:

$$\delta_\epsilon^{(B)} \bar{A}_\mu^a = \bar{D}_\mu \epsilon^a = \partial_\mu \epsilon^a + f^a{}_{bc} \bar{A}_\mu^b \epsilon^c ,$$
$$\delta_\epsilon^{(B)} a_\mu^a = f^a{}_{bc} a_\mu^b \epsilon^c . \tag{3.56}$$

These are called "background gauge transformations". Note that since the quantum field a_μ^a transforms homogeneously, we are in the general framework of section 3.2.1. The Yang-Mills action is obviously invariant under both quantum and background transformations. The gauge fixing term is meant to break the quantum gauge transformations but it is possible, and in fact extremely advantageous, to choose it in such a way as to to preserve the background gauge invariance. We choose the covariant gauge condition $\bar{D}_\mu a^{\mu a} = 0$, which is implemented in the functional integral by adding to the action the gauge-fixing term

$$S_{GF}(a; \bar{A}) = \frac{1}{2g^2\alpha} \int d^4x (\bar{D}_\mu a^{\mu a})^2 . \tag{3.57}$$

The corresponding ghost operator is obtained by varying the gauge condition under a quantum gauge transformation:

$$\delta_\epsilon^{(Q)} \bar{D}_\mu a^{\mu a} = \Delta_{gh} \epsilon^a ,$$

---

[3]Up to a point the discussion follows [73], section 16.6.

which yields $\Delta_{gh} = \bar{D}_\mu D^\mu$. Then, one adds to the action the ghost term

$$S_{gh} = \int d^4x \, \bar{c}_a \Delta_{gh}^{ab} c_b \ . \tag{3.58}$$

It is convenient to choose the Feynman gauge $\alpha = 1$. Then, a straightforward calculation shows that non-minimal terms of the form $a_\nu \bar{D}_\mu \bar{D}^\nu a^\mu$ cancel out between the Yang-Mills and the gauge-fixing actions. The quadratic part of the action is

$$S_{YM}^{(2)} + S_{GF} = \frac{1}{2g^2} \int d^4x \, a_\mu^a \Delta_{ab}^{\mu\nu} a_\nu^b \ , \tag{3.59}$$

where

$$\Delta_{ab}^{\mu\nu} = -g^{\mu\nu} \bar{D}_{ab}^2 + E_{ab}^{\mu\nu} \ ; \qquad E_{ab}^{\mu\nu} = 2 f_{abc} \bar{F}^{\mu\nu c} \ . \tag{3.60}$$

The one-loop partition fuction is now a Gaussian integral

$$\int (da \, d\bar{c} \, dc) e^{-(S_{YM}^{(2)} + S_{GF} + S_{gh})} . \tag{3.61}$$

The effective action will be a functional of two fields: $\Gamma(a; \bar{A})$, where a denotes here the classical field associated to the quantum field by the same name. This abuse of notation should not cause any confusion. We shall be interested in the special case when $a_\mu^a = 0$. Then, the effective action is

$$\Gamma(0; \bar{A}) = S_{YM}(\bar{A}) + \frac{1}{2} \text{Tr} \log \Delta - \text{Tr} \log \Delta_{gh} \ . \tag{3.62}$$

Note that for $a = 0$, $D_\mu = \bar{D}_\mu$ and the ghost operator is simply $\Delta_{gh} = -\bar{D}^2$. The divergences in this expression can be obtained from the master formula (3.44). We neglect the field-independent quartic divergences. There are no quadratic divergences, since $\text{tr}\mathbf{E} = 0$ and both b_2 coefficients vanish. We are left with the logarithmic divergence

$$-\frac{1}{2} \frac{1}{(4\pi)^2} \int d^4x \sqrt{g} (b_4(\Delta) - 2 b_4(\Delta_{gh})) \log \frac{\Lambda_{UV}^2}{\mu^2}.$$

The heat kernel coefficients can be computed from (3.45). We have $\text{tr}\mathbf{E}^2 = 4 C_2 F_{\mu\nu}^a F^{a\mu\nu}$ where C_2 is defined by $f_{acd} f_{bcd} = C_2 \delta_{ab}$ (e.g. $C_2 = N$ for $G = SU(N)$). Furthermore $\text{tr}\Omega_{\mu\nu}\Omega^{\mu\nu} = -4 C_2 F_{\mu\nu}^a F^{a\mu\nu}$ where the factor 4 counts the number of components of a_μ. Then

$$b_4(\Delta) = \frac{5}{3} C_2 \bar{F}_{\mu\nu}^a \bar{F}^{\mu\nu a} \tag{3.63}$$

$$b_4(\Delta_{gh}) = -\frac{1}{12} C_2 \bar{F}_{\mu\nu}^a \bar{F}^{\mu\nu a} \tag{3.64}$$

and

$$\Gamma(0; \bar{A}) = S_{YM}(\bar{A}) - \frac{1}{(4\pi)^2} \frac{11}{6} C_2 \log \frac{\Lambda_{UV}}{\mu} \int d^4x \, \bar{F}_{\mu\nu}^a \bar{F}^{\mu\nu a} \ . \tag{3.65}$$

This is again of the form of a Yang-Mills action but with a renormalized coupling
defined by

$$\frac{1}{4g_R^2(\mu)} = \frac{1}{4g^2} - \frac{1}{(4\pi)^2}\frac{11}{6}C_2\log\frac{\Lambda_{UV}}{\mu} . \tag{3.66}$$

This can be rewritten

$$g_R(\mu) = \frac{g}{\sqrt{1 - \frac{1}{(4\pi)^2}\frac{22}{3}C_2g^2\log\frac{\Lambda_{UV}}{\mu}}} . \tag{3.67}$$

The beta function is then given by

$$\mu\frac{dg_R}{d\mu} = -\frac{1}{(4\pi)^2}\frac{11}{3}C_2g_R^3 . \tag{3.68}$$

3.2.4   *Other fields in an external metric*

Before discussing the divergences of quantum gravity, we return to the problem of
the divergences due to an external metric, and we consider more general quantum
fields. All that has been said in the preceding sections about a scalar field can be
repeated for higher spin fields with only minor adjustments. The main difference is
in the wave operator and hence in its heat kernel coefficients. Let us list some of
the relevant operators. For spin-$\frac{1}{2}$ fields the square of the Dirac operator is (in any
dimension):

$$\Delta^{(1/2)} = -\nabla_\lambda\nabla^\lambda + \frac{R}{4} . \tag{3.69}$$

The Laplacian acting on one-forms is:

$$\Delta^{(1)\mu}{}_\nu = -\nabla_\lambda\nabla^\lambda\delta^\mu_\nu + R^\mu_\nu . \tag{3.70}$$

We will also need the following operator acting on vectors:

$$\Delta_{(FP)\nu}{}^\mu = -\nabla_\lambda\nabla^\lambda\delta^\mu_\nu - R^\mu_\nu \tag{3.71}$$

and the following operator acting on symmetric rank-two tensors, that we shall
encounter in section 3.5:

$$\Delta_{(h)\rho\sigma}{}^{\mu\nu} = -\nabla_\lambda\nabla^\lambda\delta^{(\mu}_{(\rho}\delta^{\nu)}_{\sigma)} + W_{\rho\sigma}{}^{\mu\nu} , \tag{3.72}$$

where

$$W_{\rho\sigma}{}^{\mu\nu} = R\left(\delta^{(\mu}_{(\rho}\delta^{\nu)}_{\sigma)} - \frac{1}{2}g_{\rho\sigma}g^{\mu\nu}\right) + g^{\mu\nu}R_{\rho\sigma} + R^{\mu\nu}g_{\rho\sigma} - 2\delta^{(\mu}_{(\rho}R^{\nu)}_{\sigma)} - 2R^{(\mu}{}_{(\rho}{}^{\nu)}{}_{\sigma)} . \tag{3.73}$$

The heat kernel coefficients for these operators are given by (3.45). Unlike the
case of a scalar field, considered in section 3.1.4, besides the terms due to the metric
they receive contributions also from the connection and from the endomorphism.

For example, in the case of a vectorfield, Ω coincides with the Riemann ten-
sor, viewed as an endomorphism-valued two-form, and $\operatorname{tr}\Omega_{\mu\nu}\Omega^{\mu\nu} = -R_{\mu\nu\rho\sigma}R^{\mu\nu\rho\sigma}$.
Likewise for covariant symmetric tensors

$$(\Omega_{\mu\nu})^{\rho\sigma}{}_{\alpha\beta} = -\frac{1}{2}\left(\delta^\rho_\alpha R_{\mu\nu}{}^\sigma{}_\beta + \delta^\sigma_\alpha R_{\mu\nu}{}^\rho{}_\beta + \delta^\rho_\beta R_{\mu\nu}{}^\sigma{}_\alpha + \delta^\sigma_\beta R_{\mu\nu}{}^\rho{}_\alpha\right) \tag{3.74}$$

Table 3.1 Heat kernel coefficients of several operators

Operator	Field	b_0	b_2	Coefficients in 180b_4			Coefficients in 180b_4		
				$R^2_{\mu\nu\rho\sigma}$	$R^2_{\mu\nu}$	R^2	C^2	R^2	E
$-\nabla^2$	Scalar	1	$\frac{1}{6}R$	1	-1	$\frac{5}{2}$	$\frac{3}{2}$	$\frac{5}{2}$	$-\frac{1}{2}$
$-\nabla^2 + \frac{R}{6}$	Scalar	1	0	1	-1	0	$\frac{3}{2}$	0	$-\frac{1}{2}$
$\Delta^{(1/2)}$	Dirac sp.	4	$-\frac{1}{3}R$	$-\frac{7}{2}$	-4	$\frac{5}{2}$	-9	0	$\frac{11}{2}$
$-\nabla^2$	Vector	4	$\frac{2}{3}R$	-11	-4	10	-24	5	13
$\Delta^{(1)}$	Vector	4	$-\frac{1}{3}R$	-11	86	-20	21	5	-32
$\Delta_{(FP)}$	Vector	4	$\frac{5}{3}R$	-11	86	40	21	65	-32
$-\nabla^2$	Sym. Tens.	10	$\frac{5}{3}R$	-80	-10	25	-165	-5	85
$\Delta_{(h)}$	Sym. Tens.	10	$-\frac{13}{3}R$	190	-550	295	105	175	85

The last three columns give the coefficients of b_4 in the Weyl basis (as defined in section 2.2.1), the preceding three columns in the Riemann basis, all multiplied by 180 for convenience of presentation.

and

$$\mathrm{tr}\,\Omega_{\mu\nu}\Omega^{\mu\nu} = -(d+2)R_{\mu\nu\rho\sigma}R^{\mu\nu\rho\sigma} . \tag{3.75}$$

For the endomorphism of the operator $\Delta_{(h)}$ we have the following traces:

$$\mathrm{tr}\mathbf{W} = \frac{d(d-1)}{2}R , \tag{3.76}$$

$$\mathrm{tr}\mathbf{W}^2 = 3R_{\mu\nu\rho\sigma}R^{\mu\nu\rho\sigma} + \frac{d^2 - 8d + 4}{d-2}R_{\mu\nu}R^{\mu\nu} + \frac{d^3 - 5d^2 + 8d + 4}{2(d-2)}R^2 . \tag{3.77}$$

Since in all cases under consideration $\mathbf{E}$ is linear in curvature, the new terms can be written as powers of curvatures. In particular b_4 will have the general form (3.45) and can be rewritten as a linear combination of curvature squared terms. For such terms there exist various bases that have been discussed in section 2.2.1. Table (3.1) gives the heat kernel coefficients for the above-listed operators in $d = 4$.[4] For the b_4 coefficient, the result is given in two different bases of invariants.

The first three lines directly give the coefficients of one-loop divergences due to scalar and fermion matter coupled to gravity. In the case of the electromagnetic field one has to deal with the gauge invariance of the theory. The Faddeev-Popov procedure follows the reasoning of section 3.2.3, but since the action is already quadratic in the fields it is not necessary to use the background field method. Alternatively, one can assume that the background field is $\bar{A}_\mu = 0$. The Euclidean functional integral has the form

$$\int (dA d\bar{C} dC) e^{-S_{em}(A,g) - S_{GF}(A,g) - S_{gh}(\bar{C},C,g)} , \tag{3.78}$$

---

[4]For a somewhat similar table for operators acting on arbitrary representations of the Lorentz group, see [74].

where S_{em} is the Maxwell action, S_{GF} a gauge fixing action and S_{gh} a ghost action. Upon integration by parts, the Euclidean action of an electromagnetic field propagating in a metric g is

$$\frac{1}{4} \int d^4x \sqrt{g}\, F_{\mu\nu} F^{\mu\nu} = \frac{1}{2} \int d^4x \sqrt{g}\, A_\mu \left(-\nabla^\rho \nabla_\rho g^{\mu\nu} + \nabla^\mu \nabla^\nu + R^{\mu\nu} \right) A_\nu ,$$

To this one has to add the gauge-fixing term

$$S_{GF} = \frac{1}{2\alpha} \int d^4x \sqrt{g} (\nabla^\mu A_\mu)^2 = -\frac{1}{2\alpha} \int d^4x \sqrt{g} A_\mu \nabla^\mu \nabla^\nu A_\nu .$$

If we choose the gauge fixing parameter $\alpha = 1$ (Feynman gauge) the gauge fixing term cancels the middle term in the Maxwell action, and we remain with the operator $\Delta^{(1)}$ defined in (3.70). For convenience we adopt this gauge choice. In flat spacetime, the ghosts decouple and are usually not considered, but here we are interested in the dependence on the metric, and since the ghosts do couple to the metric we have to take them into account. The ghost operator is obtained from the gauge fixing condition $\nabla^\mu A_\mu$ by replacing the gauge field A_μ by an infinitesimal gauge transformation $\nabla_\mu \epsilon$. This gives the ordinary laplacian on scalars, so

$$S_{gh} = \int d^4x \sqrt{g}\, \bar{C}(-\nabla^2)C . \tag{3.79}$$

The functional integrations in (3.78) are Gaussian, so the contribution of the electromagnetic field to the effective action is

$$\frac{1}{2} \mathrm{Tr} \log \Delta^{(1)} - \mathrm{Tr} \log \Delta^{(0)} . \tag{3.80}$$

The divergent part of this effective action is obtained by using the master formula (3.44) with the coefficients of Table 3.1:

$$-\frac{1}{2} \frac{1}{(4\pi)^2} \int d^4x \sqrt{g} \left[2\Lambda_{UV}^4 - \frac{2}{3} R \Lambda_{UV}^2 + \frac{1}{180} \left(18C^2 - 31E \right) \log \frac{\Lambda_{UV}^2}{\mu^2} \right] . \tag{3.81}$$

A linearized Yang-Mills field consists of n abelian fields, where n is the dimension of the gauge group. So the one-loop divergences due to a Yang-Mills field are given by n times (3.81). (If the background Yang-Mills field is nontrivial there will also be a divergence proportional to the Yang-Mills action, as discussed in the preceding section.)

We close with some remarks on Weyl invariance. In four dimensions a massless scalar with the nonminimal coupling $\frac{1}{12} \phi^2 R$ is Weyl invariant and has kinetic operator $-\nabla^2 + \frac{R}{6}$. Also a minimally coupled massless spinor and a Maxwell field are Weyl invariant. For all these fields the coefficient of the R^2 logarithmic divergence is zero in the Weyl basis. For the first two fields this is seen directly from the second and third rows of Table 3.1, while for the Maxwell field it results from a cancellation between the vector and the ghost. Since in four dimensions C^2 and E are Weyl invariant, this implies that the logarithmically divergent part of the effective action is Weyl invariant.

The finite part is anomalous, however. This can be seen by considering a rescaling of the metric by a constant factor Ω^2. It produces a change of the eigenvalues by a factor Ω^{-2} which, inserted in (3.19), gives

$$
\begin{aligned}
\Gamma(\Omega^2 g) &= \frac{1}{2} \sum_n \log\left(\frac{\Omega^{-2}\lambda_n}{\mu^2}\right) \\
&= \frac{1}{2}\sum_n \log\left(\frac{\lambda_n}{\mu^2}\right) + \frac{1}{2}\log(\Omega^{-2})\mathrm{Tr}\mathbf{1} \\
&= \Gamma(g) - \log\Omega\,\zeta_\Delta(0) ,
\end{aligned}
\tag{3.82}
$$

where in the last step we used formally the definition (3.21). The last term is finite when evaluated by analytic continuation from $\mathrm{Re}(s) > 2$.

The physical meaning of this anomalous non-invariance of the effective action can be understood by considering an infinitesimal rescaling $\Omega = 1 + \omega$. Then, the change in the effective action is

$$
\delta_\omega \Gamma = \int d^4 x \left[\frac{\delta\Gamma}{\delta\phi}d_\phi\omega\phi + \frac{\delta\Gamma}{\delta g^{\mu\nu}}(-2\omega)g^{\mu\nu}\right] = -\omega\int d^4 x\sqrt{g}\,g^{\mu\nu}\langle T_{\mu\nu}\rangle ,
\tag{3.83}
$$

where we have used the equation of motion for ϕ. The trace of the energy-momentum tensor is classically zero for Weyl-invariant actions, so the trace of the VEV of the energy-momentum tensor is a measure of the breaking of Weyl invariance by quantum effects. Comparing with (3.82) we obtain the integrated trace anomaly:

$$
\int d^4 x \sqrt{g}\,\langle T^\mu{}_\mu\rangle = \zeta_\Delta(0) .
\tag{3.84}
$$

The trace anomaly can also be related to the heat kernel:

$$
\zeta_\Delta(0) = \frac{1}{(4\pi)^2}B_4(\Delta) .
\tag{3.85}
$$

One can get some inkling for this by observing that formally we can also write $\mathrm{Tr}\mathbf{1} = \mathrm{Tr}K_\Delta(0)$. This can be evaluated using the small-t asymptotic expansion. All the terms containing B_n with $n > 4$ vanish when $t \to 0$. If by renormalization we get rid of B_0 and B_2, then the remaining finite result is exactly B_4.

One can write the trace anomaly in the form [75]

$$
\langle T^\mu{}_\mu\rangle = \frac{2}{\sqrt{g}}g^{\mu\nu}\frac{\delta\Gamma}{\delta g^{\mu\nu}} = b\,C^2 + b'E + \left(b'' + \frac{2}{3}b\right)\Box R .
\tag{3.86}
$$

The coefficients b and b' for N_S conformal scalars, N_D massless Dirac and N_M Maxwell fields can be read from Table 3.1:

$$
b = \frac{1}{120(4\pi)^2}\,(N_S + 6N_D + 12N_M) ; \qquad b' = -\frac{1}{360(4\pi)^2}\,(N_S + 11N_D + 62N_M) .
\tag{3.87}
$$

These terms cannot be obtained as the variation of a local functional and therefore are unaffected by renormalization ambiguities. The last term of the anomaly can be obtained from the variation of a local counterterm proportional to $\int d^4 x\sqrt{g}R^2$, and therefore the coefficient b'' is arbitrary.

3.3 The gravitational path integral

Our task now is to (formally) define the gravitational path integral. One would like to give some meaning to the expression

$$Z = \int_{\mathcal{M}} (d[g]) e^{iS([g])} , \qquad (3.88)$$

where $[g]$ are *geometries*, *i.e.* equivalence classes of metrics modulo diffeomorphisms, $\mathcal{M}$ is the space of all geometries, $(d[g])$ is some measure on $\mathcal{M}$ and $S(g)$ is a diffeomorphism-invariant action for the metric, which can therefore be viewed as a functional $S([g])$ on $\mathcal{M}$. Later we will specialize S to be the Hilbert action, but the construction is general.

Although the space $\mathcal{M}$ can be given a mathematically precise definition, it is unwieldy. We will therefore use the standard Faddeev-Popov procedure, which is a way of defining the path integral indirectly via the space of metrics. It parallels closely the analogous construction for Yang-Mills theories, with the following difference: in Yang-Mills theory there is no difficulty in talking of a "zero connection" (which is just a representative of a flat connection in a particular gauge) and therefore the use of the background field method is optional. In the case of gravity it is not clear how to make sense of the action for a "zero metric", or more generally for a degenerate metric. In fact it is even debatable whether such configurations should be taken into account or not. As a consequence, the use of the background field method is almost unavoidable.[5] Let us therefore split

$$g_{\mu\nu} = \bar{g}_{\mu\nu} + h_{\mu\nu} , \qquad (3.89)$$

where $\bar{g}$ is the classical background and h the quantum field.

Under an infinitesimal gauge transformation ϵ^{μ} (defined by $x'^{\mu} = x^{\mu} - \epsilon^{\mu}$), the metric transforms by

$$\delta_{\epsilon} g_{\mu\nu} = \mathcal{L}_{\epsilon} g_{\mu\nu} = \nabla_{\mu} \epsilon_{\nu} + \nabla_{\nu} \epsilon_{\mu} ,$$

where ∇ is the Levi-Civita connection of g and the index of ϵ has been lowered with $g_{\mu\nu}$. When the metric is divided in two parts as in (3.89), one can split the infinitesimal variation in different ways between background and fluctuation. One possibility is to assume that the background is invariant and all the change is in the fluctuation:

$$\delta_{\epsilon}^{(Q)} \bar{g}_{\mu\nu} = 0 , \qquad (3.90)$$

$$\delta_{\epsilon}^{(Q)} h_{\mu\nu} = \mathcal{L}_{\epsilon} g_{\mu\nu} . \qquad (3.91)$$

Such transformations are called "quantum gauge transformations". Alternatively, one may evenly split the transformation between background and fluctuation, so that each transforms separately as a tensor:

$$\delta_{\epsilon}^{(B)} \bar{g}_{\mu\nu} = \mathcal{L}_{\epsilon} \bar{g}_{\mu\nu} = \bar{\nabla}_{\mu} \epsilon_{\nu} + \bar{\nabla}_{\nu} \epsilon_{\mu} , \qquad (3.92)$$

$$\delta_{\epsilon}^{(B)} h_{\mu\nu} = \mathcal{L}_{\epsilon} h_{\mu\nu} , \qquad (3.93)$$

---

[5]There is one famous counterexample: three dimensional gravity formulated as a Chern-Simons theory of the Poincaré group [76].

where in the first line $\bar{\nabla}$ is the Levi-Civita connection of the background metric and the index of ϵ has been lowered with the background metric. These are called "background gauge transformations".

The measure on the space of metrics will be denoted (dh) and is assumed to be invariant under both transformations: in particular denoting $h^{1+\epsilon} = h + \delta_\epsilon^{(Q)} h$,

$$(dh^{1+\epsilon}) = (dh) . \tag{3.94}$$

Now we choose a gauge condition $F_\mu(h; \bar{g}) = 0$. It will prove very convenient to assume that F is linear in h. A typical gauge condition of this type is of the form

$$F_\mu = \bar{\nabla}_\rho h^\rho{}_\mu - \frac{1+\beta}{d} \bar{\nabla}_\mu h , \tag{3.95}$$

where indices are raised, lowered and contracted with the background metric. For example, $\beta = \frac{d}{2} - 1$ corresponds to the de Donder condition. These are also called "background gauges" and their virtue will become manifest later.

The "Faddeev-Popov trick" consists of inserting in the functional integral over metrics the formal expression

$$1 = \Psi(h; \bar{g}) \int (df) \delta(F_\mu(h^f; \bar{g})) \tag{3.96}$$

where δ is a functional Dirac delta function and the integral is over the diffeomorphism group. The transform of h under the finite diffeomorphism f is defined as $h^f = g^f - \bar{g}$. The measure on the group is assumed to be invariant, in the sense that $(df' f) = (df)$ for any fixed f'. In particular for an infinitesimal $f' = 1 + \epsilon$,

$$\int (df) \delta(F_\mu(h^{(1+\epsilon)f}; \bar{g})) = \int (d(1+\epsilon)f) \delta(F_\mu(h^{(1+\epsilon)f}; \bar{g})) = \int (df) \delta(F_\mu(h^f; \bar{g})) , \tag{3.97}$$

where we use multiplicative notation for the group composition and in the last step we have just renamed the integration variable. There follows that also the quantity $\Psi(h; \bar{g})$ is invariant:

$$\Psi(h^{1+\epsilon}; \bar{g}) = \Psi(h; \bar{g}) . \tag{3.98}$$

The action is also obviously invariant under quantum and background transformations. We will henceforth write $S(h; \bar{g})$ for $S(g)$ to emphasize its dependence on two arguments. Then

$$S(h^{1+\epsilon}; \bar{g}) = S(h; \bar{g}) . \tag{3.99}$$

With the formal factor of "one" inserted, the integral over metrics reads

$$Z = \int_\mathcal{M} (dh) \Psi(h; \bar{g}) \int (df) \delta(F_\mu(h^f; \bar{g})) e^{iS(h;\bar{g})} . \tag{3.100}$$

We can now use the finite analogues of (3.94,3.98,3.99) to change h to h^f in the measure, in Ψ and in the action, leading to

$$\int_\mathcal{M} (dh^f) \Psi(h^f; \bar{g}) \int (df) \delta(F_\mu(h^f; \bar{g})) e^{iS(h^f;\bar{g})} . \tag{3.101}$$

But now we can change the name of the integration variable from h^f to h and the dependence on f disappears, in such a way that the integral over the diffeomorphism group factors out and can be absorbed in the overall normalization of the functional integral:

$$Z = \int (dh)\Psi(h;\bar{g})\delta(F_\mu(h;\bar{g}))e^{iS(h;\bar{g})} . \tag{3.102}$$

To bring this into a more usable form we have to write Ψ and the delta function as terms in the exponential. To this end we assume that h satisfies the gauge condition: $F_\mu(h;\bar{g}) = 0$. Due to the delta function it is then sufficient to integrate over an infinitesimal neighborhood of the identity. Writing $f = 1 + \eta$ the integral in (3.96) becomes

$$\int (d(1+\eta))\,\delta(F_\mu(h^{1+\eta};\bar{g})) = \int (d\eta)\,\delta\left(F_\mu(h + \delta_\eta^{(Q)}h;\bar{g})\right) = \int (d\eta)\,\delta\left(\frac{\delta F_\mu}{\delta h}\delta_\eta^{(Q)}h\right) . \tag{3.103}$$

We have

$$\frac{\delta F_\mu}{\delta h_{\rho\sigma}}\delta_\eta^{(Q)}h_{\rho\sigma} = 2\,(\Delta_{FP})_\mu{}^\rho\eta_\rho ,$$

where we have defined the Faddeev-Popov operator:

$$(\Delta_{FP})_\mu{}^\rho = \frac{\delta F_\mu}{\delta h_{\rho\sigma}}\nabla_\sigma . \tag{3.104}$$

As in section 3.1.2 we decompose η on a basis of eigenfunctions of Δ_{FP}: $\eta = \sum_n \eta_n\phi_n$, $\Delta_{FP}\phi_n = \lambda_n\phi_n$. Then (3.103) is equal to

$$\Pi_n \int d\eta_n\,\delta(\lambda_n\eta_n) = \Pi_n\lambda_n^{-1} = (\mathrm{Det}\Delta_{FP})^{-1} ,$$

so that $\Psi = \mathrm{Det}\Delta_{FP}$. This can be written as a Gaussian integral over anticommuting fields $\bar{C}^\mu$ and C_μ:

$$\Psi = \int (d\bar{C}dC)e^{iS_{gh}(\bar{C},C;\bar{g})} \tag{3.105}$$

with

$$S_{gh}(\bar{C},C;\bar{g}) = \int dx\sqrt{|\bar{g}|}\bar{C}^\mu(\Delta_{FP})_\mu{}^\nu C_\nu . \tag{3.106}$$

For the delta functional in (3.102), absorbing an irrelevant numerical prefactor in the normalization of the functional integral, we can use the representation

$$\delta(F_\mu(h;\bar{g})) = \lim_{\alpha\to 0} e^{iS_{GF}(h;\bar{g})} \tag{3.107}$$

where the gauge fixing action is given by

$$S_{GF}(h;\bar{g}) = \frac{1}{2\alpha}\int dx\sqrt{|\bar{g}|}\bar{g}^{\mu\nu}F_\mu F_\nu . \tag{3.108}$$

Alternatively, anticipating the gauge independence of the S matrix obtained from the functional integral, one can replace the gauge condition $F_\mu = 0$ by a more general gauge condition $F_\mu = b_\mu$ and perform an average over these gauge conditions with a Gaussian weight for b_μ:

$$\delta(F_\mu(h;\bar{g})) \to \int (db)\delta(F_\mu(h;\bar{g}) - b_\mu)e^{i\frac{1}{2\alpha}\int dx\sqrt{|\bar{g}|}\bar{g}^{\mu\nu}b_\mu b_\nu} = e^{iS_{GF}(h;\bar{g})} \qquad (3.109)$$

which is of the same form as (3.107), but now with a generic finite gauge parameter α. One should pay some attention to the fact that while the S-matrix is not affected by this gauge averaging procedure, the Green's functions, which are not gauge-invariant, will be.

After these manipulations the gravitational partition function can be written in the convenient form

$$e^{iW(j,\bar{\tau},\tau)} = \int (dhd\bar{C}dC)e^{i(S(h;\bar{g})+S_{GF}(h;\bar{g})+S_{gh}(\bar{C},C;\bar{g})+\int dx\sqrt{|\bar{g}|}(j^{\mu\nu}h_{\mu\nu}+\bar{\tau}_\mu\bar{C}^\mu+\tau^\nu C_\nu)},$$

$$(3.110)$$

where we have added source terms that couple linearly to all quantum fields. The corresponding Euclidean functional integral is

$$e^{W(j,\bar{\tau},\tau\bar{g})} = \int (dhd\bar{C}dC)e^{-(S(h;\bar{g})+S_{GF}(h;\bar{g})+S_{gh}(\bar{C},C;\bar{g})+\int dx\sqrt{|\bar{g}|}(j^{\mu\nu}h_{\mu\nu}+\bar{\tau}_\mu\bar{C}^\mu+\tau^\nu C_\nu)}.$$

$$(3.111)$$

The Euclidean gravitational effective action is defined as the Legendre transform of W:

$$\Gamma(h,\bar{C},C;\bar{g}) = -W(j,\bar{\tau},\tau;\bar{g}) + \int dx\sqrt{|\bar{g}|}\left(j^{\mu\nu}h_{\mu\nu} + \bar{\tau}_\mu\bar{C}^\mu + \tau^\nu C_\nu\right) , \qquad (3.112)$$

where, as usual, we have used the names h, $\bar{C}$, C for the classical fields $\langle h \rangle$, $\langle \bar{C} \rangle$, $\langle C \rangle$. In the following we will mostly consider the effective action for vanishing values of these classical fields.

3.4 Perturbations around a general background

Referring to (3.89), we need the expansion of the action around a general background to second order in $h_{\mu\nu}$. This is given by a functional Taylor series of the form

$$S(g) = S(\bar{g}) + S^{(1)}(h;\bar{g}) + S^{(2)}(h;\bar{g}) + \dots \qquad (3.113)$$

where

$$S^{(n)} = \frac{1}{n!}\int dx_1 \dots \int dx_n \frac{\delta S}{\delta g_{\mu_1\nu_1}(x_1)\dots\delta g_{\mu_n\nu_n}(x_n)}\bigg|_{\bar{g}} h_{\mu_1\nu_1}(x_1)\dots h_{\mu_n\nu_n}(x_n) .$$

$$(3.114)$$

The term $S^{(1)}$ is proportional to the equations of motion. We will be interested mainly in the second term.

Let us collect here some intermediate formulae. The inverse metric can be expanded

$$g^{\mu\nu} = \bar{g}^{\mu\nu} - h^{\mu\nu} + h^{\mu\alpha}h_\alpha{}^\nu + \dots . \qquad (3.115)$$

The convention is that indices are raised and lowered with $\bar{g}_{\mu\nu}$. One has to pay attention to the fact that the variation of $g^{\mu\nu}$ is $\delta g^{\mu\nu} = -\bar{g}^{\mu\alpha}\bar{g}^{\nu\beta}\delta g_{\alpha\beta} = -h^{\mu\nu}$. The variations of the volume element are

$$\sqrt{|g|} = \sqrt{|\bar{g}|}\left[1 + \frac{1}{2}h + \left(\frac{1}{8}h^2 - \frac{1}{4}h_{\mu\nu}h^{\mu\nu}\right) + \ldots\right], \tag{3.116}$$

where $h = \bar{g}^{\mu\nu}h_{\mu\nu}$. The first and second variations of the Christoffel symbols are

$$\Gamma^\rho_{\mu\nu} = \bar{\Gamma}^\rho_{\mu\nu} + \Gamma^{\rho(1)}_{\mu\nu} + \Gamma^{\rho(2)}_{\mu\nu}, \tag{3.117}$$

where

$$\Gamma^{\rho(1)}_{\mu\nu} = \frac{1}{2}(\bar{\nabla}_\nu h^\rho{}_\mu + \bar{\nabla}_\mu h^\rho{}_\nu - \bar{\nabla}^\rho h_{\mu\nu}), \tag{3.118}$$

$$\Gamma^{\rho(2)}_{\mu\nu} = -\frac{1}{2}h^{\rho\sigma}(\bar{\nabla}_\nu h_{\mu\sigma} + \bar{\nabla}_\mu h_{\nu\sigma} - \bar{\nabla}_\sigma h_{\mu\nu}). \tag{3.119}$$

From here one gets the first and second variation of the Riemann tensor.

$$R^\mu{}_{\nu\alpha\beta} = \bar{R}^\mu{}_{\nu\alpha\beta} + R^\mu{}_{\nu\alpha\beta}{}^{(1)} + R^\mu{}_{\nu\alpha\beta}{}^{(2)} + \ldots \tag{3.120}$$

where

$$R^\mu{}_{\nu\rho\sigma}{}^{(1)} = \frac{1}{2}(\bar{\nabla}_\rho\bar{\nabla}_\nu h^\mu_\sigma - \bar{\nabla}_\rho\bar{\nabla}^\mu h_{\nu\sigma} - \bar{\nabla}_\sigma\bar{\nabla}_\nu h^\mu_\rho + \bar{\nabla}_\sigma\bar{\nabla}^\mu h_{\nu\rho}) + \frac{1}{2}\bar{R}_{\nu\gamma\rho\sigma}h^{\mu\gamma} + \frac{1}{2}\bar{R}^\mu{}_{\gamma\rho\sigma}h^\gamma_\nu,$$

$$R^\mu{}_{\nu\rho\sigma}{}^{(2)} = \bar{\nabla}_\rho\Gamma^{\mu(2)}_{\nu\sigma} - \bar{\nabla}_\sigma\Gamma^{\mu(2)}_{\nu\rho} + \Gamma^{\mu(1)}_{\lambda\rho}\Gamma^{\lambda(1)}_{\nu\sigma} - \Gamma^{\mu(1)}_{\lambda\sigma}\Gamma^{\lambda(1)}_{\nu\rho}$$

$$= -\frac{1}{2}h^{\mu\gamma}\bar{\nabla}_\rho(\bar{\nabla}_\sigma h_{\nu\gamma} + \bar{\nabla}_\nu h_{\beta\gamma} - \bar{\nabla}_\gamma h_{\nu\sigma}) - \frac{1}{4}\bar{\nabla}_\rho h^{\mu\gamma}(\bar{\nabla}_\sigma h_{\nu\gamma} + \bar{\nabla}_\nu h_{\beta\gamma} - \bar{\nabla}_\gamma h_{\nu\sigma})$$

$$+ \frac{1}{4}\bar{\nabla}_\gamma h^\mu_\rho(\bar{\nabla}_\sigma h^\gamma_\nu + \bar{\nabla}_\nu h^\gamma_\sigma - \bar{\nabla}^\gamma h_{\nu\sigma}) - \frac{1}{4}\bar{\nabla}^\mu h_{\alpha\gamma}(\bar{\nabla}_\sigma h^\gamma_\nu + \bar{\nabla}_\nu h^\gamma_\sigma - \bar{\nabla}^\gamma h_{\nu\sigma})$$

$$-(\rho \leftrightarrow \sigma). \tag{3.121}$$

Contracting once, one obtains the variations of the Ricci tensor

$$R^{(1)}_{\mu\nu} = -\frac{1}{2}(\bar{\nabla}_\mu\bar{\nabla}_\nu h - \bar{\nabla}_\mu\bar{\nabla}_\rho h^\rho{}_\nu - \bar{\nabla}_\nu\bar{\nabla}_\rho h^\rho{}_\mu + \bar{\nabla}^2 h_{\mu\nu})$$

$$-\bar{R}_{\alpha\mu\beta\nu}h^{\alpha\beta} + \frac{1}{2}\bar{R}_{\mu\alpha}h^\alpha{}_\nu + \frac{1}{2}\bar{R}_{\nu\alpha}h^\alpha{}_\mu,$$

$$R^{(2)}_{\mu\nu} = \frac{1}{2}\bar{\nabla}_\mu(h^{\alpha\beta}\bar{\nabla}_\nu h_{\alpha\beta}) - \frac{1}{2}\bar{\nabla}_\alpha\left[h^{\alpha\beta}(\bar{\nabla}_\mu h_{\nu\beta} + \bar{\nabla}_\nu h_{\mu\beta} - \bar{\nabla}_\beta h_{\mu\nu})\right]$$

$$-\frac{1}{4}(\bar{\nabla}_\mu h^\beta_\alpha + \bar{\nabla}_\alpha h^\beta_\mu - \bar{\nabla}^\beta h_{\alpha\mu})(\bar{\nabla}_\beta h^\alpha_\nu + \bar{\nabla}_\nu h^\alpha_\beta - \bar{\nabla}^\alpha h_{\beta\nu})$$

$$+\frac{1}{4}\bar{\nabla}_\alpha h(\bar{\nabla}_\mu h^\alpha_\nu + \bar{\nabla}_\nu h^\alpha_\mu - \bar{\nabla}^\alpha h_{\mu\nu}). \tag{3.122}$$

The variations of the Ricci scalar are:

$$R^{(1)} = -h^{\mu\nu}\bar{R}_{\mu\nu} + \bar{g}^{\mu\nu}R^{(1)}_{\mu\nu} = \bar{\nabla}_\mu\bar{\nabla}_\nu h^{\mu\nu} - \bar{\nabla}^2 h - \bar{R}_{\mu\nu}h^{\mu\nu},$$

$$R^{(2)} = h^{\mu\rho}h_\rho{}^\nu\bar{R}_{\mu\nu} - h^{\mu\nu}R^{(1)}_{\mu\nu} + \bar{g}^{\mu\nu}R^{(2)}_{\mu\nu}$$

$$= \frac{3}{4}\bar{\nabla}_\alpha h_{\mu\nu}\bar{\nabla}^\alpha h^{\mu\nu} + h_{\mu\nu}\bar{\nabla}^2 h^{\mu\nu} - \bar{\nabla}_\rho h^\rho{}_\mu\bar{\nabla}_\sigma h^{\sigma\mu} + \bar{\nabla}_\rho h^\rho{}_\mu\bar{\nabla}^\mu h - 2h_{\mu\nu}\bar{\nabla}^\mu\bar{\nabla}_\rho h^{\rho\nu}$$

$$+ h_{\mu\nu}\bar{\nabla}^\mu\bar{\nabla}^\nu h - \frac{1}{2}\bar{\nabla}_\mu h_{\nu\alpha}\bar{\nabla}^\alpha h^{\mu\nu} - \frac{1}{4}\bar{\nabla}_\mu h\bar{\nabla}^\mu h + \bar{R}_{\alpha\beta\gamma\delta}h^{\alpha\gamma}h^{\beta\delta}. \tag{3.123}$$

Neglecting total derivatives,

$$R^{(2)} \approx \frac{1}{4}(h_{\mu\nu}\Box h^{\mu\nu} + h\Box h + 2h_\mu^2 + 2\bar{R}_{\alpha\beta}h^{\alpha\gamma}h_\gamma^\beta + 2\bar{R}_{\alpha\beta\gamma\delta}h^{\alpha\gamma}h^{\beta\delta}). \quad (3.124)$$

We now have all the variations that are needed to expand a gravitational action to second order in h. The action to be expanded is the Euclidean Hilbert action:

$$S(g) = \frac{1}{2\kappa^2} \int d^d x \sqrt{g}(2\Lambda - R) . \quad (3.125)$$

(The analogous calculation for Lorentzian metrics only differs by an overall sign.) Collecting all terms one gets:

$$S^{(2)}(h;\bar{g}) = \frac{1}{4\kappa^2} \int d^d x \sqrt{\bar{g}} \left[-\frac{1}{2}h_{\mu\nu}\bar{\nabla}^2 h^{\mu\nu} + h_{\mu\nu}\bar{\nabla}^\mu\bar{\nabla}^\rho h_\rho^{\ \nu} - h\bar{\nabla}^\mu\bar{\nabla}^\nu h_{\mu\nu} + \frac{1}{2}h\bar{\nabla}^2 h \right.$$

$$\left. + h\bar{R}^{\mu\nu}h_{\mu\nu} - h_{\mu\nu}\bar{R}^{\nu\sigma}h^\mu_{\ \sigma} - h_{\mu\nu}\bar{R}^{\mu\rho\nu\sigma}h_{\rho\sigma} + \frac{2\Lambda - \bar{R}}{2}\left(\frac{1}{2}h^2 - h_{\mu\nu}h^{\mu\nu}\right) \right] . \quad (3.126)$$

Note that if in the second term we commute the covariant derivatives, the term with the Riemann tensor will be removed. It is nevertheless convenient to leave the covariant derivatives in this form, since the combination $\bar{\nabla}^\rho h_\rho^{\ \nu}$ is a vector.

Note that these formulas hold "off-shell": we have not used, at this stage, that $\bar{g}_{\mu\nu}$ is a stationary point of S. The equations of motion for pure gravity with cosmological constant are

$$\bar{R}_{\mu\nu} = \frac{2}{d-2}\Lambda g_{\mu\nu} . \quad (3.127)$$

Spaces satisfying these equation are called Einstein spaces. They have the special property that the Ricci tensor is covariantly constant. In the special case $\Lambda = 0$, the space is said to be Ricci flat. The trace of the equation of motion is

$$\bar{R} = \frac{2d}{d-2}\Lambda, \quad (3.128)$$

so we can also write

$$\bar{R}_{\mu\nu} = \frac{1}{d}\bar{R}g_{\mu\nu} . \quad (3.129)$$

This, however, is a weaker equation than (3.127), because it leaves the Ricci scalar undetermined.

If we use the full equation of motion (3.127), the second variation (3.126) simplifies to

$$S^{(2)}(h;\bar{g}) = \frac{1}{4\kappa^2} \int d^d x \sqrt{\bar{g}} \left[-\frac{1}{2}h_{\mu\nu}\bar{\nabla}^2 h^{\mu\nu} + h_{\mu\nu}\bar{\nabla}^\mu\bar{\nabla}^\rho h_\rho^{\ \nu} - h\bar{\nabla}^\mu\bar{\nabla}^\nu h^{\mu\nu} + \frac{1}{2}h\bar{\nabla}^2 h \right.$$

$$\left. - h_{\mu\nu}\bar{R}^{\mu\rho\nu\sigma}h_{\rho\sigma} + \frac{\Lambda}{d-2}h^2 \right] . \quad (3.130)$$

To proceed further we choose the gauge fixing term

$$S_{GF}(h;\bar{g}) = \frac{1}{4\alpha\kappa^2} \int d^dx\sqrt{\bar{g}}\, F_\mu \bar{g}^{\mu\nu} F_\nu \,, \tag{3.131}$$

where

$$F_\nu = \bar{\nabla}^\mu h_{\mu\nu} - \frac{1}{2}\bar{\nabla}_\nu h \,. \tag{3.132}$$

The gauge condition $F_\nu = 0$ is the de Donder (harmonic) gauge condition that we used already in section 2.1.3. The factor in front of the gauge fixing action is $1/2\kappa^2$, the same that is in front of the Hilbert action, times a factor $1/2\alpha$, where α is a gauge parameter. Expanding and integrating by parts, the gauge fixing term is

$$S_{GF}(h;\bar{g}) = \frac{1}{4\alpha\kappa^2} \int d^dx\sqrt{\bar{g}} \left(-h_{\mu\nu}\bar{\nabla}^\mu\bar{\nabla}^\rho h_\rho{}^\nu + h\bar{\nabla}^\mu\bar{\nabla}^\nu h_{\mu\nu} - \frac{1}{4}h\bar{\nabla}^2 h \right) \,. \tag{3.133}$$

As in Yang-Mills theory (section 3.2.3), in the "Feynman gauge" $\alpha = 1$ the non-minimal terms in the Hessian and in the gauge fixing term cancel out. This leads to a significant simplification of the kinetic operator. In what follows we shall make this choice. The full quadratic part of the action, including the gauge fixing term, can then be written in the form

$$S^{(2)} + S_{GF} = \frac{1}{4\kappa^2}\mathcal{H}(h,h) = \frac{1}{4\kappa^2} \int d^dx\sqrt{\bar{g}}\, h_{\mu\nu} H^{\mu\nu\rho\sigma} h_{\rho\sigma} \,, \tag{3.134}$$

where

$$H^{\mu\nu\rho\sigma} = K^{\mu\nu\rho\sigma}(-\nabla^2 - 2\Lambda) + U^{\mu\nu\rho\sigma} \,, \tag{3.135}$$

with

$$K^{\mu\nu\alpha\beta} = \frac{1}{4}\left(\bar{g}^{\mu\alpha}\bar{g}^{\nu\beta} + \bar{g}^{\mu\beta}\bar{g}^{\nu\alpha} - \bar{g}^{\mu\nu}\bar{g}^{\alpha\beta} \right) \tag{3.136}$$

and

$$U^{\mu\nu}_{\rho\sigma} = R\,K^{\mu\nu}_{\rho\sigma} + \frac{1}{2}\left(g^{\mu\nu}R_{\rho\sigma} + R^{\mu\nu}g_{\rho\sigma} \right) - \delta^{(\mu}_{(\rho}R^{\nu)}_{\sigma)} - R^{(\mu}{}_{(\rho}{}^{\nu)}{}_{\sigma)} \,.$$

Now comes a point where gravity presents a new aspect that we did not need to discuss in the preceding examples. The general formula for the one-loop effective action is given in terms of the determinant of a differential operator. In the present context, a differential operator is a linear map from the space of symmetric tensors to itself. However, the Hessian itself is not a differential operator but rather a bilinear form, mapping two copies of the space of symmetric tensors to real numbers. In the standard language of differential geometry, applied to the functional space of symmetric tensors, the Hessian is a covariant symmetric tensor, whereas a differential operator is a tensor with one covariant and one contravariant index. The reason why we need a differential operator is that the determinant of a linear map is a basis-independent notion whereas the determinant of a covariant tensor is not. In order to transform a Hessian into a differential operator we need a metric in function space.

The reason why this can be confusing is that the position of indices in the sense of four-dimensional tensors may (and in the case of gravity actually is) opposite to the one in the functional sense. Thus the components of the metric, $g_{\mu\nu}$, are coordinates on the function space of metrics and therefore are regarded conventionally as carrying a *contravariant* index. The infinitesimal variation $h_{\mu\nu}$ is a contravariant vector tangent to the space of metrics at $\bar{g}_{\mu\nu}$, the Hessian (3.135) is a covariant two-tensor, and so on. A covariant metric would be an object that maps two symmetric tensors into a real number, hence an object of the form $\mathcal{G}^{\mu\nu\rho\sigma}$. In principle it could involve also derivatives or even a non-local kernel, but it is convenient to choose the metric to be ultralocal. If we denote the inner product

$$\mathcal{G}(v, w) = \int d^d x \sqrt{\bar{g}(x)} \int d^d y \sqrt{\bar{g}(y)} \, \mathcal{G}^{\mu\nu\rho\sigma}(x, y) v_{\mu\nu}(x) w_{\rho\sigma}(y)$$

then a metric is ultralocal if it is proportional to the delta function $\delta(x - y)$. Up to an overall normalization, there is only a one-parameter family of ultralocal metrics, namely the DeWitt metric[6]

$$\mathcal{G}^{\mu\nu\alpha\beta} = \frac{1}{2} \left(\bar{g}^{\mu\alpha} \bar{g}^{\nu\beta} + \bar{g}^{\mu\beta} \bar{g}^{\nu\alpha} + a \bar{g}^{\mu\nu} \bar{g}^{\alpha\beta} \right) . \tag{3.137}$$

Note that $\det \mathcal{G} = (1 + da/2)(\det g)^{-(d+1)}$, so that it becomes degenerate for $a = -2/d$. This value should thus be avoided.

Why did we not encounter this issue in the examples discussed earlier in this chapter? In fact, the issue is present in any field theory, and also in the preceding examples a choice of metric has been implicitly made, but it was particularly simple. In the case of a real scalar field an ultralocal metric is just a number that can be normalized to one, and in the case of a vectorfield it is just $g_{\mu\nu}$, again up to an irrelevant overall normalization. Thus gravity is the first nontrivial example where an ultralocal metric could be nontrivial.

With an ultralocal metric, we transform the Hessian into an operator $\Delta^\Lambda_{(h)}$ by writing (3.134) in the form

$$S^{(2)} + S_{GF} = \frac{1}{4\kappa^2} \mathcal{G}(h, \Delta^\Lambda_{(h)} h) . \tag{3.138}$$

In components

$$\Delta^\Lambda_{(h)\mu\nu}{}^{\alpha\beta} = \mathcal{G}^{-1}_{\mu\nu\rho\sigma} H^{\rho\sigma\alpha\beta} , \tag{3.139}$$

where $\mathcal{G}^{-1}_{\mu\nu\rho\sigma}$ is the *contravariant*, or inverse metric defined by

$$\mathcal{G}^{-1}_{\mu\nu\rho\sigma} \mathcal{G}^{\rho\sigma\alpha\beta} = 1^{\alpha\beta}_{\mu\nu} \equiv \frac{1}{2} \left(\delta^\alpha_\mu \delta^\beta_\nu + \delta^\alpha_\nu \delta^\beta_\mu \right) . \tag{3.140}$$

The Hessian (3.135) has a second order part whose tensor structure is given by $K^{\mu\nu\rho\sigma}$, so it is natural to choose $\mathcal{G}^{\mu\nu\rho\sigma} = K^{\mu\nu\rho\sigma}$.[7] This is the choice we shall

---

[6] Originally the DeWitt metric has been defined in the Hamiltonian formalism for the three-metrics in the ADM decomposition, but by a slight abuse we may also extend the term to the metric on the space of four-metrics.

[7] This metric is not normalized as in (3.137), but this does not cause any issue in the following.

tacitly make in the following. It is important, however, to be aware that a choice has been made and that certain results could depend on it. With this choice, the definition of the kinetic operator Δ_h^Λ boils down to factoring an overall K in the Hessian:

$$H^{\mu\nu\rho\sigma} = K^{\mu\nu\alpha\beta} \Delta_{(h)\alpha\beta}^{\Lambda}{}^{\rho\sigma} \,, \tag{3.141}$$

or equivalently

$$\Delta_{(h)\mu\nu}^{\Lambda}{}^{\rho\sigma} = K_{\mu\nu\alpha\beta}^{-1} H^{\alpha\beta\rho\sigma} \,, \tag{3.142}$$

where

$$K_{\mu\nu\alpha\beta}^{-1} = \bar{g}_{\mu\alpha}\bar{g}_{\nu\beta} + \bar{g}_{\mu\beta}\bar{g}_{\nu\alpha} - \frac{2}{d-2}\bar{g}_{\mu\nu}\bar{g}_{\alpha\beta} \,. \tag{3.143}$$

(Note the difference between $K_{\mu\nu\alpha\beta}^{-1}$ and $K_{\mu\nu\alpha\beta}$, which is the same as (3.136), but with all indices lowered.) The operator $\Delta_{(h)}^\Lambda$ is

$$\Delta_{(h)\rho\sigma}^{\Lambda}{}^{\mu\nu} = -\bar{\nabla}^2 1_{\rho\sigma}^{\mu\nu} + W_{\rho\sigma}{}^{\mu\nu} - 2\Lambda 1_{\rho\sigma}^{\mu\nu} \,, \tag{3.144}$$

where

$$W_{\rho\sigma}{}^{\mu\nu} = 2U_{\rho\sigma}^{\mu\nu} - \frac{d-4}{d-2}g_{\rho\sigma}\left(R^{\mu\nu} - \frac{1}{2}Rg^{\mu\nu}\right) \,. \tag{3.145}$$

Thus, in four dimensions,

$$W_{\rho\sigma}{}^{\mu\nu} = 2\bar{R}K_{\rho\sigma}^{\mu\nu} + \bar{g}^{\mu\nu}\bar{R}_{\rho\sigma} + \bar{R}^{\mu\nu}\bar{g}_{\rho\sigma} - 2\delta_{(\rho}^{(\mu}\bar{R}_{\sigma)}^{\nu)} - 2\bar{R}^{(\mu}{}_{(\rho}{}^{\nu)}{}_{\sigma)} \,. \tag{3.146}$$

For $\Lambda = 0$ this coincides with the operator $\Delta_{(h)}$ given in (3.72). Thus

$$\Delta_{(h)}^{\Lambda} = \Delta_{(h)} - 2\Lambda \,. \tag{3.147}$$

From (3.104) one finds that the ghost operator is

$$\delta_\mu^\rho \bar{\nabla}^\sigma \nabla_\sigma + \bar{\nabla}^\rho \nabla_\mu - \bar{\nabla}_\mu \nabla^\rho \,. \tag{3.148}$$

Anticipating that we will compute the effective action for vanishing expectation value of $h_{\mu\nu}$, we can identify the connections ∇ and $\bar{\nabla}$. Then the ghost action is

$$S_{\text{ghost}} = -\int d^d x \sqrt{\bar{g}}\, \bar{C}_\mu \left(-\bar{\nabla}^2 \delta^\mu_\nu - \bar{R}^\mu{}_\nu\right) C^\nu \,. \tag{3.149}$$

We observe the appearance of the operator $\Delta_{(FP)}$ defined in (3.71).

3.5 One-loop divergences in quantum GR

We have shown in section 2.3 that GR is power-counting non-renormalizable: new divergences are expected to arise at each order of perturbation theory. However, those arguments do not guarantee that the divergences are present: they leave open the possibility of unexpected cancellations. Armed with the tools developed in the preceding sections, we shall now compute the divergent part of the effective action at one loop and in particular establish whether there are divergences of a form that is not already present in the classical action.

The one-loop effective action can be evaluated expanding the action to second order around a classical background field and evaluating the appropriate Gaussian integrals, whose divergences can be isolated using the methods of sections 3.1–2. In this way, one-loop quantum gravity is equivalent to studying the quantum effects of a tensor field $h_{\mu\nu}$ and its ghosts $\bar{C}_\mu$, C^μ, all of which can be treated like matter fields propagating in a background metric $\bar{g}_{\mu\nu}$. With the groundwork laid down in the preceding sections, the calculation of the one-loop divergences in GR is now quite straightforward.

To second order in h, the relevant terms in the action are

$$S^{(2)}(h;\bar{g}) + S_{GF}(h;\bar{g}) = \frac{1}{4\kappa^2} \int d^4x \sqrt{\bar{g}}\, h_{\mu\nu} K^{\mu\nu\rho\sigma} \Delta^\Lambda_{(h)\rho\sigma}{}^{\alpha\beta} h_{\alpha\beta}$$

with $\Delta^\Lambda_{(h)}$ given by (3.144), and the ghost action $S_{gh}(\bar{C}, C; \bar{g})$ given by (3.149). Thus the effective action for vanishing expectation values of $h_{\mu\nu}$ and ghosts is

$$\Gamma(0,0,0;\bar{g}) = S(\bar{g}) + \frac{1}{2}\mathrm{Tr}\log\left(\frac{\Delta^\Lambda_{(h)}}{\mu^2}\right) - \mathrm{Tr}\log\left(\frac{\Delta_{FP}}{\mu^2}\right). \tag{3.150}$$

The divergences of this expression are given by the master formula (3.44). For $\Lambda = 0$ the heat kernel coefficients to be used in that formula are listed directly in Table 3.1. When $\Lambda \neq 0$, for the heat kernel coefficients of the operator $\Delta^\Lambda_{(h)}$, we must do a little extra work. Using (3.147) and the general argument of Section 3.7.1, they are given by

$$b_0(\Delta^\Lambda_{(h)}) = b_0(\Delta_{(h)}) = 10\,, \tag{3.151}$$

$$b_2(\Delta^\Lambda_{(h)}) = b_2(\Delta_{(h)}) - (-2\Lambda)b_0(\Delta_{(h)}) = -\frac{13}{3}\bar{R} + 20\Lambda\,, \tag{3.152}$$

$$b_4(\Delta^\Lambda_{(h)}) = b_4(\Delta_{(h)}) - (-2\Lambda)b_2(\Delta_{(h)}) + \frac{1}{2}(-2\Lambda)^2 b_0(\Delta_{(h)})$$

$$= \frac{7}{12}\bar{C}^2 + \frac{35}{36}\bar{R}^2 + \frac{17}{36}\bar{E} - \frac{26}{3}\bar{R}\Lambda + 20\Lambda^2. \tag{3.153}$$

Putting together the graviton and ghost contributions, the quartic divergences are given by

$$-\frac{1}{2}\frac{1}{(4\pi)^2}\Lambda_{UV}^4 \int d^4x \sqrt{\bar{g}}\,, \tag{3.154}$$

where we denote Λ_{UV} the UV cutoff, to distinguish it from the cosmological constant. The quadratic divergences are given by

$$-\frac{1}{(4\pi)^2}\Lambda_{UV}^2 \int d^4x \sqrt{\bar{g}}\left(-\frac{23}{6}\bar{R} + 10\Lambda\right). \tag{3.155}$$

Most important are the logarithmic divergences. In the "Weyl basis" they are

$$-\frac{1}{(4\pi)^2}\log\left(\frac{\Lambda_{UV}^2}{\mu^2}\right)\left(\frac{1}{2}B_4(\Delta_{(h)}) - B_4(\Delta_{(FP)})\right) \tag{3.156}$$

$$= -\frac{1}{(4\pi)^2}\log\left(\frac{\Lambda_{UV}^2}{\mu^2}\right)\int d^4x \sqrt{\bar{g}}\left(\frac{7}{40}\bar{C}^2 + \frac{1}{8}\bar{R}^2 + \frac{149}{360}\bar{E} - \frac{13}{3}\bar{R}\Lambda + 10\Lambda^2\right).$$

This result was found originally by 't Hooft and Veltman [13] for the case when the bare cosmological constant is zero, and by Christensen and Duff [77] in the presence of a cosmological constant. In both cases dimensional regularization was used, where $\log(\Lambda_{UV}/\mu)$ corresponds to a simple pole $1/\epsilon$, with $\epsilon = 4 - d$. This has the additional advantage that one does not have to worry about the power divergences, because they are simply set to zero.

In our case, the power divergences can be absorbed into renormalizations of the cosmological constant and Newton constant. The simplest procedure is to define renormalized Newton's constant G_R and cosmological constant Λ_R by:

$$\frac{1}{G_R} = \frac{1}{G} - \frac{1}{\pi}\left(\frac{23}{6}\Lambda_{UV}^2 + \frac{13}{3}\Lambda \log\left(\frac{\Lambda_{UV}^2}{\mu^2}\right)\right) , \qquad (3.157)$$

$$\frac{\Lambda_R}{G_R} = \frac{\Lambda}{G} - \frac{1}{4\pi}\left(\Lambda_{UV}^4 + 10\Lambda\Lambda_{UV}^2 + 10\Lambda^2 \log\left(\frac{\Lambda_{UV}^2}{\mu^2}\right)\right) . \qquad (3.158)$$

The bare couplings G and Λ have to be adjusted so that the renormalized G_R and Λ_R correspond to the observed values. In particular, to match observations, Λ has to be chosen so that Λ_R is zero, or in any case much smaller than the renormalized Planck mass $(8\pi G_R)^{-1/2}$. This is the usual fine-tuning problem of the cosmological constant.

The remaining logarithmic divergences consist of terms of a form that is not already present in the bare action. At first sight their occurrence seems to imply that the theory is non-renormalizable. This, however, is not necessarily the case. Physical results are obtained by going on-shell, which means expanding the metric about a stationary point of the action, which means that the background metric has to satisfy the equations of motion. In order to understand this point it is convenient to present the logarithmic divergences in the "Ricci basis" defined by (2.79):

$$-\frac{1}{(4\pi)^2}\log\left(\frac{\Lambda_{UV}^2}{\mu^2}\right)\int d^4x \sqrt{\bar{g}}\left(\frac{7}{20}\bar{R}_{\mu\nu}\bar{R}^{\mu\nu} + \frac{1}{120}\bar{R}^2 + \frac{53}{90}E\right) . \qquad (3.159)$$

Omitting the last term, which is locally a total derivative, this is the form originally derived in [13]. Let us now see what conclusions one should derive from this result. Things depend on whether matter fields are present or not, and whether the renormalized cosmological constant is zero or not.

Let us start from pure gravity with zero cosmological constant. The equations of motion say that the Ricci tensor is zero, so the potentially divergent terms are actually zero. This implies that the theory is one-loop renormalizable. It is worth pointing out that although the coefficient of the logarithmically divergent terms was not known until 't Hooft and Veltman calculated it, this conclusion was already known beforehand. In the absence of a cosmological constant the field equations for pure gravity imply that the Ricci tensor vanishes on-shell. Therefore, working in the Riemann basis (2.75), the only term quadratic in the curvature that does not vanish on-shell is $R_{\mu\nu\rho\sigma}R^{\mu\nu\rho\sigma}$. However, in four dimensions the identity (2.107) implies that this term can be rewritten locally as a total derivative plus terms that are again quadratic in the Ricci tensor and Ricci scalar. Thus all potential divergences at one loop are of a type that does not affect local physical quantities.

Since the condition of being on-shell is so important, it is worthwhile to look at it also from another point of view. In quantum field theory in flat space the S-matrix is not affected by field redefinitions. So if a term in the action can be eliminated by a field redefinition, it does not have any effect on physically measurable quantities such as cross sections. How can we tell whether a term can be eliminated by a field redefinition? Consider the effect on the Lagrangian of an infinitesimal field redefinition $\phi' = \phi + \delta\phi$, with $\delta\phi = f(\phi, \partial_\mu \phi, \dots)$, where the dots stand for terms containing second and higher derivatives of the field. Varying the action and integrating by parts as usual, we obtain:

$$\delta\mathcal{L} = \left(\frac{\partial \mathcal{L}}{\partial \phi} - \partial_\mu \frac{\partial \mathcal{L}}{\partial \partial_\mu \phi} + \dots \right) f + \partial_\mu \left(\frac{\partial \mathcal{L}}{\partial \partial_\mu \phi} f + \dots \right) .$$

The first term is proportional to the equation of motion, so we see that any term in the action that either vanishes or reduces to a total derivative on-shell can be eliminated by performing a field redefinition, up to terms of higher order. Thus, at any finite order in perturbation theory, divergences that vanish on-shell are harmless.

In the case of pure gravity, the logarithmic divergences vanish on-shell, so there must be a field redefinition that removes those terms. Indeed one can check that in the case of (3.159) the appropriate redefinition is[8]

$$\delta g_{\mu\nu} = \frac{1}{(4\pi)^2} \log \left(\frac{\Lambda_{UV}^2}{\mu^2} \right) \frac{1}{20} \left(7 \bar{R}_{\mu\nu} - \frac{11}{3} \bar{g}_{\mu\nu} \bar{R} \right) . \tag{3.160}$$

The situation is different when matter is present. In this case the equation of motion

$$\bar{R}_{\mu\nu} = 8\pi G \left(T_{\mu\nu} - \frac{1}{2} \bar{g}_{\mu\nu} T \right) , \tag{3.161}$$

implies that the coefficient of the logarithm is proportional to the square of the energy-momentum tensor, and such terms are not present in the bare Lagrangian. Of course quantum matter will itself contribute to the coefficients of the logarithmically divergent term. The contributions of matter are additive and can be read again from table (3.1), as we have seen in section 3.2.4. In the literature, similar calculations have been done for scalar fields [13], for abelian and non-abelian gauge fields [14,15], for Dirac spinors [16] and for Majorana spinors [17]. We do not need to give details here. Generically, all three coefficients will be nonzero and so there will be genuinely divergent terms that are not of the form originally present in the action. Therefore gravity coupled to matter is generically non-renormalizable at one loop. The possible exceptions correspond to special matter choices that would make the coefficient of the logarithm equal to zero. We shall briefly discuss this potential loophole in section 4.4.

---

[8]It is actually easiest to see that $-\delta g_{\mu\nu}$, applied to the Hilbert action, *generates* the terms (3.159), so that $\delta g_{\mu\nu}$ applied to Γ *removes* them, modulo terms of higher order.

If the cosmological constant is not zero, the equation of motion (3.127) implies that $\bar{R}_{\mu\nu}\bar{R}^{\mu\nu} = 4\Lambda^2$ and $\bar{R}^2 = 16\Lambda^2$. Therefore the first two terms in (3.159) are not zero, but contribute to the logarithmic renormalization of the cosmological constant. We can rewrite the on-shell logarithmic divergence in the Riemann basis (2.75):

$$-\frac{1}{(4\pi)^2}\log\left(\frac{\Lambda_{UV}^2}{\mu^2}\right)\int d^4x\,\sqrt{\bar{g}}\left(\frac{53}{90}\bar{R}_{\mu\nu\rho\sigma}\bar{R}^{\mu\nu\rho\sigma} - \frac{29}{5}\Lambda^2\right). \tag{3.162}$$

This is the form of the divergence derived in [77]. The rest of the argument is unchanged, so pure gravity with cosmological constant is also one-loop renormalizable. On the other hand, in the presence of matter, the equation of motion

$$\bar{R}_{\mu\nu} = 8\pi G\left(T_{\mu\nu} - \frac{1}{2}\bar{g}_{\mu\nu}T\right) + \Lambda\bar{g}_{\mu\nu}, \tag{3.163}$$

when used in the logarithmic divergence, will again generate, besides terms proportional to the cosmological constant, also terms quadratic in the energy-momentum tensor which render the theory non-renormalizable. Thus the conclusions are the same independently of the value of the cosmological constant.

We observe that in dimensional regularization the power divergences are absent, so that in (3.157), (3.158) only the logarithmic terms are present. The preceding arguments regarding renormalizability go through in the same way.

In conclusion let us stress once more that off-shell quantities are in general renormalization scheme- as well as gauge-dependent and that such dependences cannot enter in physical observables. The off-shell gauge-dependence of the logarithmic divergence has been exploited by Kallosh, Tarasov and Tyutin [78], who found a gauge where it is zero. This is another way of seeing that the logarithmic divergence of pure gravity is unphysical. We shall reproduce part of their calculation in section 7.3.3.

3.6   Two loop divergences in quantum GR

Since pure gravity is one-loop renormalizable, the question of the existence of non-renormalizable divergences is postponed to the next order of the expansion.

The two loop effective action can be represented diagrammatically by the diagrams in Fig. 3.1. By power counting, these diagrams are all proportional to

Fig. 3.1   Two loop contributions to the effective action.

G and may exhibit sixtic divergences renormalizing the cosmological term, quartic

divergences multiplying the Hilbert action, quadratic divergences multiplying terms with four derivatives, and logarithmic divergence multiplying terms with six derivatives. Taking into account the symmetries of the Riemann and Ricci tensors, the list of independent invariants of dimension six is

$$\bar{\nabla}_\mu \bar{R} \bar{\nabla}^\mu \bar{R}; \ \bar{\nabla}_\rho \bar{R}_{\mu\nu} \bar{\nabla}^\rho \bar{R}^{\mu\nu}; \ \bar{R}^3; \ \bar{R}\bar{R}_{\mu\nu}\bar{R}^{\mu\nu}; \ \bar{R}\bar{R}_{\mu\nu\rho\sigma}\bar{R}^{\mu\nu\rho\sigma}; \ \bar{R}^\mu{}_\nu \bar{R}^\nu{}_\rho \bar{R}^\rho{}_\mu;$$
$$\bar{R}_{\mu\nu}\bar{R}_{\rho\sigma}\bar{R}^{\mu\rho\nu\sigma}; \ \bar{R}^\mu{}_\nu \bar{R}^{\nu\alpha\beta\gamma}\bar{R}_{\mu\alpha\beta\gamma}; \ \bar{R}^{\mu\nu}{}_{\rho\sigma}\bar{R}^{\rho\sigma}{}_{\alpha\beta}\bar{R}^{\alpha\beta}{}_{\mu\nu}; \ \bar{R}_{\mu\nu\rho\sigma}\bar{R}^\mu{}_\alpha{}^\rho{}_\beta\bar{R}^{\nu\alpha\sigma\beta} \ .$$

All of them except the last two contain the Ricci tensor and therefore vanish on-shell. The last two invariants are related by a Schouten-like identity. In four dimensions the total antisymmetrization over five indices must be zero:

$$0 = \bar{R}^{[\mu\nu}{}_{\rho\sigma}\bar{R}^{\rho\sigma}{}_{\alpha\beta}\bar{R}^{\alpha]\beta}{}_{\mu\nu} \ .$$

The permutations for which either α is in the third or fourth position, or μ or ν are in the fifth position, or ρ or σ are in the first two positions, all give terms containing the Ricci tensor and therefore can be discarded on-shell. The remaining permutations give the invariant $\bar{R}^{\mu\nu}{}_{\rho\sigma}\bar{R}^{\rho\sigma}{}_{\alpha\beta}\bar{R}^{\alpha\beta}{}_{\mu\nu}$ with total weight 4 and the invariant $\bar{R}_{\mu\nu\rho\sigma}\bar{R}^\mu{}_\alpha{}^\rho{}_\beta\bar{R}^{\nu\alpha\sigma\beta}$ with total weight -8. Altogether one finds that, modulo Ricci terms

$$\bar{R}^{\mu\nu}{}_{\rho\sigma}\bar{R}^{\rho\sigma}{}_{\alpha\beta}\bar{R}^{\alpha\beta}{}_{\mu\nu} = 2\bar{R}_{\mu\nu\rho\sigma}\bar{R}^\mu{}_\alpha{}^\rho{}_\beta\bar{R}^{\nu\alpha\sigma\beta} \ .$$

There is therefore a single invariant that one has to worry about at two loops, and the question is whether the coefficient of this logarithmic divergence is zero or not.

In dimensional regularization, the invariant could appear either as a single pole $(1/\epsilon)$ or a double pole $(1/\epsilon^2)$. Several authors have shown on general grounds that if a theory is finite at n loops, then at $n + 1$ loops it can have at most a single pole [79–81].[9] On the basis of one-loop finiteness of pure gravity there should be at most a single pole at two loops. The calculation of this coefficient is rather cumbersome and can only be done with a computer. It was done for the first time by Goroff and Sagnotti [20] who verified explicitly the cancellation of the double pole and found the single pole

$$\frac{1}{\epsilon}\frac{1}{(4\pi)^4}\int d^4x\sqrt{|g|}\frac{209}{2880}\bar{R}^{\mu\nu}{}_{\rho\sigma}\bar{R}^{\rho\sigma}{}_{\alpha\beta}\bar{R}^{\alpha\beta}{}_{\mu\nu} \ . \tag{3.164}$$

This result was confirmed by van de Ven, using different techniques [21]. This result buried all hopes that GR would have good renormalizability properties due to some unknown feature. Very recently, it has been rederived using modern amplitude methods [82]. (This is related to the discussion of supergravity in section 4.4.)

---

[9] Here by finiteness one means absence of logarithmic divergences on-shell. We recall that in dimensional regularization there are no power divergences.

3.7 Appendix: Calculations of heat kernel coefficients

The general structure of the heat kernel coefficients in (3.45) comes from the assumption of invariance and dimensional analysis. We collect here some calculations that determine the numerical coefficients. Instead of performing a single calculation for the most general operator Δ, it is easier to consider special cases where one can evaluate separately the effect of the metric, of the connection and of the endomorphism $\mathbf{E}$. We begin with the latter, which is easiest.

3.7.1 *Terms due to a potential*

We assume that the metric and connection are flat, but the Laplacian contains an endomorphism acting on the fields $\Delta = -\partial_\mu \partial^\mu + \mathbf{E}$. The simplest example is when $\mathbf{E}$ is constant, *i.e.* a mass matrix. This is actually sufficient for our purposes. Then $\mathbf{E}$ commutes with $\Delta_0 = -\partial_\mu \partial^\mu$ and the solution of the heat equation is

$$K = \operatorname{Tr} e^{-t\Delta} = \operatorname{Tr} e^{-t\mathbf{E}} e^{-t\Delta_0} = \operatorname{Tr} e^{-t\mathbf{E}} K_0 \tag{3.165}$$

where K_0 is the trace of the heat kernel of Δ_0. In this case the heat kernel coefficients are just the coefficients of the expansion of the first exponential:

$$b_0 = \operatorname{tr}\mathbf{1} ; \tag{3.166}$$

$$b_2 = -\operatorname{tr}\mathbf{E} ; \tag{3.167}$$

$$b_4 = \frac{1}{2}\operatorname{tr}\mathbf{E}^2 . \tag{3.168}$$

If $\mathbf{E}$ is not constant, dimensional considerations permit a term of the form $\partial^2 \mathbf{E}$ in b_4. The determination of its coefficient requires more work. This term, however, is a total derivative and is not important for our purposes.

3.7.2 *Terms due to a connection*

Again we assume that the metric is flat, but this time the quantum field carries a representation of a group G and is minimally coupled to an external gauge field A. We are interested in the heat kernel coefficients of the operator $\Delta = -D_\mu D^\mu$. For the trace of the heat kernel (3.31) we need the diagonal matrix element

$$K(x, x; t) = \int \frac{d^d p}{(2\pi)^d} e^{-ip \cdot x} e^{-t\Delta} e^{ip \cdot x} . \tag{3.169}$$

We have $\Delta e^{ip \cdot x} = -e^{ip \cdot x}(ip_\mu + D_\mu)(ip^\mu + D^\mu) = e^{ip \cdot x}(p^2 - 2i\hat{p} - D^2)$, where $\hat{p} = p^\mu D_\mu$, so (3.169) can be rewritten

$$K(x, x; t) = \int \frac{d^d p}{(2\pi)^d} e^{-tp^2} e^{t(D^2 + 2i\hat{p})} . \tag{3.170}$$

The first term under the integral is the heat kernel of $-\partial^2$ in flat space. The heat kernel coefficients come from the expansion of the second exponential. Terms with odd powers of $\hat{p}$ give zero upon integration over p. The even terms give

$$1 + tD^2 + \frac{t^2}{2}\left((D^2)^2 - 4\hat{p}^2\right) + \frac{t^3}{3!}\left((D^2)^3 - 4D^2\hat{p}^2 - 4\hat{p}D^2\hat{p} - 4\hat{p}^2 D^2\right) + \frac{t^4}{4!}\left((D^2)^4\right.$$

$$\left. - 4(D^2)^2\hat{p}^2 - 4D^2\hat{p}D^2\hat{p} - 4D^2\hat{p}^2 D^2 - 4\hat{p}(D^2)^2\hat{p} - 4\hat{p}D^2\hat{p}D^2 - 4\hat{p}^2(D^2)^2 + 16\hat{p}^4\right) + \ldots$$

In the first line we encounter the following momentum integrals:

$$\int \frac{d^d p}{(2\pi)^d} e^{-tp^2} = \frac{1}{(4\pi t)^{d/2}} \tag{3.171}$$

$$\int \frac{d^d p}{(2\pi)^d} e^{-tp^2} p_\mu p_\nu = \frac{g_{\mu\nu}}{d} \int \frac{d^d q}{(2\pi)^d} e^{-tp^2} p^2$$

$$= \frac{g_{\mu\nu}}{d}\left(-\frac{\partial}{\partial t}\right)\int \frac{d^d p}{(2\pi)^d} e^{-tp^2} = \frac{1}{2}g_{\mu\nu}\frac{t^{-1}}{(4\pi t)^{d/2}} \tag{3.172}$$

The terms with prefactor t^3 and two $\hat{p}$'s are seen to actually be of order t^2 and therefore contribute to b_4. The terms with prefactor t^4 and two $\hat{p}$'s are of order t^3 and can be dropped. However, the last term, which contains four $\hat{p}$'s, is again important. Integrating, and using the rules of symmetric integration

$$\int \frac{d^d p}{(2\pi)^d} e^{-tp^2} p_\mu p_\nu p_\rho p_\sigma = \frac{g_{\mu\nu}g_{\rho\sigma} + g_{\mu\rho}g_{\nu\sigma} + g_{\mu\sigma}g_{\nu\rho}}{d(d+2)} \int \frac{d^d p}{(2\pi)^d} e^{-tp^2} p^4$$

$$= \frac{g_{\mu\nu}g_{\rho\sigma} + g_{\mu\rho}g_{\nu\sigma} + g_{\mu\sigma}g_{\nu\rho}}{4}\frac{t^{-2}}{(4\pi t)^{d/2}} \tag{3.173}$$

which lowers the overall degree to t^2. Collecting, we find

$$K(x,x;t) = \frac{1}{(4\pi t)^{d/2}}\left\{1 + tD^2 + \frac{1}{2}t^2\left((D^2)^2 - 4\frac{1}{2t}D^2\right)\right.$$

$$+ \frac{1}{3!}t^3\left(\ldots - \frac{2}{t}\left(2(D^2)^2 - D_\mu D^2 D^\mu\right)\right)$$

$$\left. + \frac{1}{4!}t^4\left(\ldots - 16\frac{4}{t^2}\left((D^2)^2 + D_\mu D_\nu D^\mu D^\nu + D_\mu D^2 D^\mu\right)\right) + \ldots\right\}, \tag{3.174}$$

where the dots stand for terms that do not contribute to order t^2. Collecting the terms of the same order in t and using the relations

$$D_\mu D^2 D^\mu = (D^2)^2 + (D_\mu \Omega^{\mu\nu})D_\nu - \Omega_{\mu\nu}\Omega^{\mu\nu}$$

$$D_\mu D_\nu D^\mu D^\nu = (D^2)^2 - (D_\mu \Omega^{\mu\nu})D_\nu - \frac{1}{2}\Omega_{\mu\nu}\Omega^{\mu\nu}$$

one finds that all terms involving derivative operators cancel out, as they must. Finally, tracing over the (hidden) representation indices, one remains with $b_2 = 0$ and

$$b_4 = \frac{1}{12}\text{tr}\,\Omega_{\mu\nu}\Omega^{\mu\nu}\,. \tag{3.175}$$

3.7.3 *Terms due to an external metric*

Finally we consider here the case of the Laplacian $-\nabla^2$ acting on a scalar field minimally coupled to a metric g. Our aim here is to derive (3.36), or equivalently the terms proportional to tr1 in (3.45). One can do that by methods similar to those of the preceding section, but here we shall follow a shortcut. As discussed in section 3.1.4, the heat kernel expansion coefficient b_n (3.35) must be a scalar formed with $n/2$ curvatures or other combinations of covariant derivatives of curvatures. Thus b_0 must be just a constant, b_2 must be a multiple of R and b_4 must have the general form

$$aR_{\mu\nu\rho\sigma}R^{\mu\nu\rho\sigma} + bR_{\mu\nu}R^{\mu\nu} + cR^2 \ ,$$

up to total derivatives. In order to determine the coefficients a, b, c, one could compute b_4 on a manifold for which all the three curvature invariants are nonzero and linearly independent. However, such manifolds are typically rather complicated. One manifold where one can easily compute the heat kernel coefficients is the sphere, since the spectrum of the Laplacian on the sphere is known and then the asymptotic expansion of the trace of the heat kernel can be computed by means of the Euler-Maclaurin formula. This calculation will be done in section 5.6.1 and we will use those results here. The sphere is conformally flat, so the Riemann and Ricci tensors are fully determined by the Ricci scalar. At first sight, it is not possible to disentangle the three invariants, because they are all proportional to R^2. However, one can use the fact that the relation between the three invariants depends on the dimension.[10] On the sphere we have

$$R_{\mu\nu\rho\sigma} = \frac{R}{d(d-1)}(g_{\mu\rho}g_{\nu\sigma} - g_{\nu\rho}g_{\mu\sigma}) \ ; \qquad R_{\mu\nu} = \frac{R}{d}g_{\mu\nu} \ , \qquad (3.176)$$

so that the curvature invariants are related by

$$R_{\mu\nu\rho\sigma}R^{\mu\nu\rho\sigma} = \frac{2R^2}{d(d-1)} \ ; \qquad R_{\mu\nu}R^{\mu\nu} = \frac{R^2}{d} \ , \qquad (3.177)$$

The first three terms in the expansion of the heat kernel of the sphere are calculated in section 5.6.1. In two dimensions we find

$$\frac{1}{4\pi t}\left(1 + \frac{1}{6}Rt + \frac{1}{60}R^2 t^2 + \dots\right) \ ,$$

in three dimensions

$$\frac{1}{(4\pi t)^{3/2}}\left(1 + \frac{1}{6}Rt + \frac{1}{72}R^2 t^2 + \dots\right) \ ,$$

and in four dimensions

$$\frac{1}{(4\pi t)^2}\left(1 + \frac{1}{6}Rt + \frac{29}{2160}R^2 t^2 + \dots\right) \ .$$

---

[10]I owe this trick to A. Codello and C. Pagani.

This implies that $b_0 = 1$ and $b_2 = R/6$, while for the coefficients a, b, c one gets the following three equations:

$$a + \frac{b}{2} + c = \frac{1}{60}$$

$$\frac{a}{3} + \frac{b}{3} + c = \frac{1}{72}$$

$$\frac{a}{6} + \frac{b}{4} + c = \frac{29}{2160}$$

Assuming that a, b, c do not depend on d, the solution is

$$a = \frac{1}{180} \; ; \qquad b = -\frac{1}{180} \; ; \qquad c = \frac{1}{72} \; .$$

This determines the coefficients in (3.36), and hence the coefficients of the terms proportional to tr**1** in (3.45).

Chapter 4

Other perturbative approaches

4.1 Options

The results of the preceding chapter provide us with a proof that a perturbative quantization of Einstein's General Relativity will not yield a renormalizable theory. In this section we make a list of options that remain open.

The proof of non-renormalizability has been given for the standard "second order" formulation of GR as a theory of a metric, but there are several versions of the theory that use other variables, for example the "first order" formulations with independent metric and connection. The one-loop divergences of these off-shell extensions of GR have been discussed in [83,84] and have been found to agree with those of the metric formulation. This is expected: Since the coupling is the same, one expects the power counting arguments to work in the same way in all these reformulations. (The verdict is still open for the more radical reformulation of gravity as a theory of a connection, discussed by Krasnov in [85].) From now on we assume that all these formulations are equivalent also in their quantum properties.

The remaining options can be classified according to several binary choices: is the metric (or the vierbein, or some other degree of freedom that is used to describe General Relativity) a fundamental degree of freedom or the manifestation of a collective behavior? Is quantum gravity described by a QFT or do we need to use some broader framework? Does the theory hold (at least formally) up to arbitrarily high energies or does it break down at some finite UV scale?

Let us consider first the case when the degrees of freedom of classical GR can also be used as degrees of freedom of a quantum theory. Then we can try to classify the possible subcases as follows:

- Give up UV completeness: the resulting QFT is called an

    (1) Effective Field Theory (EFT)

- Require UV completeness: this leaves several sub-options

    (2) change the action, e.g. include higher curvature terms
    (3) couple gravity to a special choice of matter
    (4) non-perturbative renormalizability *a.k.a.* asymptotic safety

Since, as we shall see, QFT works well provided one gives up the requirement of UV completeness, there does not seem to be much point in giving up QFT *and* UV completeness, so this option is left out. If we seek a UV complete theory of gravity and we are ready to go beyond QFT, by far the most developed approach is superstring theory. Note that superstring theory can actually be seen as a QFT with infinitely many fields, one of which is the standard spin-2 graviton. It can therefore be seen as an extreme example of the general philosophy of point (3) above.

Finally there is the possibility that at a fundamental level gravity is not described by a metric, perhaps not even by a QFT. In a broad sense, GR would then be an "emergent" theory. There are many possible variations on this theme, ranging from the rather conservative to the extremely speculative.

Note the following alternative useful classification: options (1,2,3) and superstrings remain largely within the domain of perturbation theory, while (4) requires non-perturbative methods. Also, options (1,2,4) work for gravity alone, while (3) and superstrings require the presence of matter degrees of freedom.

It is also customary to divide approaches to quantum gravity into canonical and covariant ones. This could be merely a methodological issue, were it not for the fact that the issues encountered by these two approaches are so widely different. In particular, the conceptual issues raised by the "problem of time" in GR and its quantum ramifications make it conceivable that the usual correspondence between Hamiltonian and Lagrangian descriptions could break down.

Item (4) in the preceding list will be discussed at length in later chapters. In particular, the general definition of asymptotic safety will be given in section 7.1. In the rest of this chapter we will discuss briefly the other alternatives.

4.2 Emergent gravity

If we adopt the strict definition of calling "fundamental interactions" the interactions between the elementary particles, then it is not obvious that gravity is a fundamental interaction. We know that gravity is a force between macroscopic bodies and we also know that an elementary particle falls in the gravitational field produced by a macroscopic body, but nobody has ever been able to observe a gravitational interaction between two elementary particles. It is therefore possible in principle that gravity is not a fundamental interaction but rather an "emergent" phenomenon.

This is a very broad notion that could have many different meanings, depending on which aspects of the theory of gravity are assumed to be "fundamental" and which ones "emergent". The most conservative possibility is to assume that spacetime and a metric are given *a priori* and only the dynamics of gravity is emergent. The oldest incarnation of this notion is Sakharov's "induced gravity" [86–88]: the gravitational action is not present at the fundamental level but is generated by the quantum fluctuations of matter fields, essentially by the mechanism discussed in section 3.2.4. A modern and mathematically more sophisticated incarnation of this idea is the spectral action [89]. Also in this category are models where Einstein's equations emerge from thermodynamics [90, 91].

A somewhat more ambitious possibility is to assume the existence of some basic geometric structure (a topological space, a manifold, a lattice...) where some degrees of freedom are defined, and to have the metric emerge as a kind of condensate. Old attempts in this direction went under the name of "pregeometry" [92–95]. The analogy between gravity and chiral models, that we shall discuss in section 4.5.3, suggests that before even attempting to formulate equations for the metric, the basic question should be "why is the metric nondegenerate and why does it have Lorentzian signature?". Attempts to answer this question in a self-consistent manner using standard tools of QFT have been made in [96–100] and have also been related to the unification of gravity with the other interactions [101]. Possible dynamical origins of the Lorentzian signature have been discussed in [102–104]. Modern versions go under the name of "analog models" and are strongly motivated by condensed matter phenomena such as fluid flows [105], superfluids [106] etc.

An even more radical possibility is to assume that the whole spacetime structure is emergent. In these models one would also assume the existence of a large number of basic degrees of freedom, which are however not organized according to "location" as in an ordinary QFT. The "location" would appear as a result of the dynamics. Straightforward implementations of this idea are [107, 108]; more subtle ones are Group Field Theory and certain matrix models [42, 109]. Such models are obviously further removed from straightforward phenomenology, so that the appearance of a macroscopic spacetime should already be counted as a big success.

Difficulties encountered by models of emergent gravity have been discussed by Carlip [110]. Still, the general idea of emergence is a very attractive one and may very well contain some truth. In any case, it should not be counted as an argument for *not* quantizing the metric. From the scale of emergence downwards, gravity can be described by the metric and, as we shall discuss later in this chapter, it has to be treated by methods of EFT. If the scale of emergence is the Planck scale, this leaves a very large range of energies for the validity of a QFT of the metric.

4.3 Higher derivative gravities

One of the earliest attempts to modify the dynamics of gravity was to consider actions quadratic in curvature, instead of linear. We will refer to these theories, whose actions can be written alternatively in one of the forms (2.75) or (2.79) or (2.83), as "quadratic gravity". Part of the motivation for this was to make the theory more similar to other gauge theories [111–114]. With an action quadratic in the Riemann tensor, the action is superficially very similar to the Yang-Mills action, with a dimensionless coupling and classical scale-invariance. This analogy with

Yang-Mills theory is a bit superficial, however, because in the Yang-Mills case the action contains only two derivatives, in the case of quadratic gravity four derivatives. In order to improve it, one could treat the connection as an independent variable, so that terms quadratic in curvature contain again only two derivatives. This leads to "gauge theories of gravity", a vast subject whose quantum properties are technically very hard to work out because of the large number of field components. Instead of going that way, one can try to exploit the fact that the propagator of a four-derivative theory leads to enhanced suppression of loops in Feynman diagrams relative to ordinary two-derivative propagators, so that one may expect improved quantum properties. It was shown by Stelle [22,23], even before the proof of two-loops non-renormalizability of GR, that quadratic gravity is perturbatively renormalizable. The power counting argument for this has already been given in section 2.3. Later, it was also shown that the dimensionless couplings of this theory are asymptotically free [115–117]. We will calculate these beta functions in section 7.4.

These properties make quadratic gravity very attractive. The price one pays is that the asymptotic states of the linearized theory contain, in addition to the massless spin-2 graviton, also a massive spin-2 ghost. This has also already been discussed in section 2.2.2. For this reason, quadratic gravity has never been accepted as a viable solution to the problem of quantum gravity. Over time, several arguments have been given to get rid of the ghosts. Here are some:

- The mass of the ghost is not a fixed parameter but is rather subject to strong (quadratic) running above the Planck threshold. Then, the equation for the pole mass $m_{phys}^2 = m^2(k = m_{phys})$ (where $m(k)$ is the running mass) may not have a solution [99,115,118].
- The ghost may be an artifact of expanding around the wrong vacuum. The true vacuum of quadratic gravity (in the presence also of a Hilbert term) is not flat space but rather a kind of wave with wavelength of the order of the Planck length [119].
- The quadratic term is one of an infinite series and the sum of the series is a function that has no massive ghost pole. The ghost pole is an artifact of Taylor expanding this function to second order. For some concrete work along these lines see *e.g.* [120,121].

Other arguments have been given by Tomboulis [122]. So far, none of these arguments has convinced the community at large, so the issue of the ghosts remains open for the time being.

Much more recently, it has been observed that one can have a theory that is both perturbatively renormalizable *and* free of ghosts, provided it contains higher spatial derivatives but only two time derivatives. In order to save dimensional analysis, the different treatment of the space and time coordinate requires different dimensions (anisotropic scaling). This is known as Hořava-Lifshitz (HL) gravity [123]. The different treatment of space and time breaks local Lorentz invariance, at least at the microscopic level.

The natural framework to write actions for such theories is the ADM decomposition. The actions will not be invariant under diffeomorphisms but only under foliation-preserving diffeomorphisms. The smaller invariance group of this theory allows a much larger number of invariants in the action. In order to restrict the number of admissible invariants, several additional conditions have been tried, but they often lead to pathologies [124]. The result of this is that the analysis of the quantum properties of these theories has been slow. For some partial results see [125–128], and numerical simulations using Causal Dynamical Triangulations (CDT) [129–132].

One of the major outstanding problems of this approach is recovering Lorentz symmetry at low energies. One necessary condition for this is that all fields propagate with a single "speed of light" at low energy. The analysis of scalar toy models where the issue can be treated does not seem to bode well [133]. It is nevertheless too early to draw firm conclusions, and HL gravity remains an interesting candidate for a perturbatively renormalizable and unitary theory of gravity. It is also worth observing that (in the language that we shall use in section 7.1) HL gravity and asymptotically safe gravity can be made to coexist in a suitable "theory space". The numerical simulations of CDT contain several phases, critical lines and critical points that could be used to describe one or the other, or both.

4.4   Special matter choice: Supergravity

It has been shown in the original paper by 't Hooft and Veltman [13] that GR coupled to a scalar field is non-renormalizable at one-loop. This result was later extended to GR coupled to spinor fields [16], gauge fields [14, 15, 18], and antisymmetric tensors [19]. In all cases, the squares of the Ricci tensor and Ricci scalar become, on shell, squares of the energy-momentum tensor, and such terms are not present in the original action. It is in principle possible that a special choice of matter fields will yield a renormalizable theory. For example at one loop one may try to tune the number of matter fields so as to cancel the logarithmic divergence on-shell. Using (3.159) and the coefficients in Table 3.1, we find that the one-loop logarithmic divergence for gravity minimally coupled to N_S scalars, N_D Dirac and N_M Maxwell fields is

$$-\frac{1}{(4\pi)^2}\log\left(\frac{\Lambda_{UV}^2}{\mu^2}\right)\int d^4x\,\sqrt{\bar{g}}\left(\frac{42+N_S+6N_D+12N_M}{120}\bar{R}_{\mu\nu}\bar{R}^{\mu\nu}\right.$$

$$\left.+\frac{2+N_S-4N_D-8N_M}{240}\bar{R}^2+\frac{424+2N_S+7N_D-26N_M}{720}E\right). \qquad (4.1)$$

It is easy to see that there are no values of N_S, N_D, N_M such that the coefficients of $\bar{R}_{\mu\nu}\bar{R}^{\mu\nu}$ and $\bar{R}^2$ are zero.

This is a bit disappointing, but one has to bear in mind that the value of such a result would anyhow be limited, because any cancellation is unlikely to hold also at higher loops. The most promising route seems to be a symmetry principle. By far the most successful implementation of this idea is supergravity (SUGRA). In SUGRA, the graviton is accompanied by a spin-$\frac{3}{2}$ particle, called gravitino, and the action is invariant under local supersymmetry. It was shown early on that even the simplest SUGRA is indeed free of the one-loop [134] and two loop [135] divergences that appear in pure gravity. Assuming that SUSY can be preserved in the quantum theory, this can be traced to the fact that there is no supersymmetric counterterm containing as its bosonic part the combinations of curvatures that appear in the one- and two-loop divergences of pure gravity. A suitable term with the structure R^4 (actually, the square of the so-called Bel-Robinson tensor) can be constructed and by power counting it could appear at three loops [136]. Thus, supersymmetry delays the appearance of divergences to at least three loops.

In the 1980's much more work was done on the existence of SUSY counterterms in various SUGRAs in diverse dimensions. The best behaved theory is expected to be the most symmetric one, namely $N = 8$ SUGRA in $d = 4$, which is related to $N = 1$ SUGRA in $d = 11$. While delayed, it was believed that as soon as a SUSY counterterm existed, divergences would appear, and therefore that the theory would be non-renormalizable. Until the 1980's it was thought that all SUGRAS in $d = 4$ would be divergent at three loops.

In the 1990's Bern, Dixon and others started developing new unitarity-based methods for the calculation of amplitudes that largely bypass the standard calculations of Feynman diagrams. These methods were first used in gauge theories, but it was realized that gravitational amplitudes could be constructed by "doubling" gauge theory amplitudes in a suitable sense. With these new and powerful methods it has become possible to perform previously unthinkable calculations in perturbative SUGRA. In 2007 Bern et al. showed by explicit calculation the three-loop finiteness of $N = 8$ SUGRA in $d = 4$ [137]. This result was later extended to four loops in [138]. The appearance of unexpected cancellations led the authors to suggest that the theory may even by finite.

It has subsequently been understood that additional cancellations are due to the $E_{7(7)}$ duality symmetry of the theory, and it has been shown that enforcing this symmetry, no divergences can occur below seven loops [139]. A counterterm respecting all symmetries including $E_{7(7)}$ is proportional to $\nabla^8 R^4$ and could be expected to appear at seven loops.

At present it seems reasonable to expect that this divergence will be present. There are however at least two examples in $d = 4$ ($N = 4$ SUGRA at three loops [137] and $N = 5$ SUGRA at four loops [140]) where symmetry arguments do not prevent a divergence, and yet a divergence is seen not to be present by explicit calculation. It is thus possible that some more subtle mechanism is at work. Recent hints, in a non-SUSY context, have appeared in [82].

4.5   The Effective Field Theory approach

Our understanding of renormalizability has undergone profound changes over the years [141]. In the beginning it was thought that only renormalizable quantum field theories such as QED or ϕ_4^4 could make sense and be useful. In a non-renormalizable theory, every local term compatible with the symmetries of the theory will appear at some order with a divergent coefficient requiring renormalization, so all local terms come with an uncalculable coefficient which has to be determined from experiment. This seems to make the theory completely useless, if by useful we mean a theory that can be used to make predictions. Consequently, nonrenormalizable theories were regarded as being essentially unworthy of consideration. Long before direct experimental tests were available, the proof of (perturbative) renormalizability of the Weinberg-Salam Model by 't Hooft was crucial to make it a popular candidate for a theory of weak interactions.

With time, however, a different view emerged. On one hand there were persistent questions on the viability of the two paradigmatic examples, QED and ϕ_4^4. These theories have positive beta functions and their couplings diverge at some finite energy scale. This is known as a "Landau pole". Of course the perturbative beta function ceases to be reliable when the coupling becomes of order one, so one may think that perhaps the "true" beta function has no pole, but there is much evidence to the contrary from lattice simulations. So it appears that renormalizability, by itself, is not sufficient to make a theory UV complete. The discovery of asymptotic freedom provided a subclass of theories that are free of such issues, and QCD is now the best example of a physical theory that is UV complete.

On the other hand work by Gasser, Leutwyler and others on the chiral model, which is used as a theory of mesons at low energy, showed that one could extract quantum predictions even from nonrenormalizable theories. It thus appears that renormalizability is neither sufficient for a theory to be complete, nor necessary for a theory to be useful. We shall now discuss the effective field theory point of view and show how it can be used to perform reliable computations in quantum gravity at low energies.

4.5.1   *The general idea*

In particle physics one encounters many particles of widely different masses. For simplicity consider just one light particle with mass m, interacting with a heavy particle with mass M. When one performs experiments at energies $E < M$ the heavy particles cannot be produced as final states. Still, they make their presence felt through virtual effects in the scattering of the light particles. If we describe the dynamics of the light field by a Lagrangian, we must include in it these effects. This is done by "integrating out" the heavy particles and in this way one obtains an effective Lagrangian for the light particles. This effective Lagrangian will not be renormalizable, because quantum effects will generate all possible terms which are compatible with the symmetries of the system. This non-renormalizability is not problematic, however, since the effective Lagrangian should only be used at energies lower than M. The mass of the heavy states can be taken as a UV cutoff for the effective theory of the light states.

The prototypical example is given by Fermi's theory of weak interactions, (1.1). The fermionic currents have mass dimension three, so G_F must have dimension minus two. This is a non-renormalizable interaction. The value of the Fermi constant, which can be measured from muon decay, is $G_F = 1.16 \times 10^{-5}$ GeV^{-2}. In the Weinberg-Salam model there is no four-fermion interaction but there is a renormalizable interaction between the fermions and a gauge field. Four fermions can interact through the exchange of a massive W boson. In the limit when the momentum transfer is much smaller than the gauge boson's mass m_W, its propagator is $1/m_W^2$ and this process reproduces the Fermi interaction, provided we identify $G_F = \sqrt{2}g^2/8m_W^2$, where g is the gauge coupling. As long as the fermion momenta are much smaller than 100 GeV, the Fermi theory is a good description.

It is generically true that the coefficients of non-renormalizable terms in an effective Lagrangian are inverse powers of the heavy mass, modulo dimensionless numerical factors, typically of order one. The contribution of these terms to a low-energy scattering process of the light particles is suppressed, relative to the contribution of renormalizable interactions, by powers of E/M. They are therefore negligible at sufficiently low energy. When one goes to energies higher than M one has to consider the full theory involving both fields. This may be a renormalizable theory, or perhaps it will be another effective theory where still heavier states have been integrated out.

Particle physics can be described by a sequence of effective Lagrangians describing heavier and heavier states. As the energy available in accelerators increases, today's fundamental theory turns into tomorrow's effective theory. When we consider the current energy frontier, it is tempting to think that this will go on: the LHC or perhaps some future accelerator will discover new weakly coupled states that we will describe by some effective theory. From this point of view the demand of UV completeness may seem to be excessive. After all we will never test any theory up to infinite energy, neither directly, since accelerators have finite energy, nor indirectly through the induced nonrenormalizable terms, since measurements have finite precision. These dual limitations are the basis of a general recipe to make good use of non-renormalizable theories. It is called the "Effective Field Theory" (EFT) approach.

Suppose for example that we have to compute some cross section for a process involving only the light particles that will be measured in a new accelerator. As in the example of the four fermion interactions mentioned above, the QFT describing the light particles contains a hint of the "scale of new physics" via some large mass scale M that appears in its non-renormalizable interactions. The information we need about the experiment is: the energy of the beam, E, and the precision of the apparatus.

Since $E \ll M$, we can try to use the small ratio E/M as an expansion parameter. For example, if $E = M/10$ and the cross-section is going to be measured with a 1% precision, we will need to compute the cross section in the EFT at order $(E/M)^2$. Power-counting arguments show that at any finite order in E/M there will be only a finite number of terms contributing to the process [142]. Rather than giving here the argument in general, we shall see some specific examples later. If the underlying fundamental theory is known, one may try to calculate the couplings of the EFT from first principles. It is more frequently the case that either the fundamental theory is unknown, or if it is known, this calculation proves too hard. In these cases, the coefficients of these terms can be measured by a finite number of experiments and these values can then be used in the formula for the cross-section. The theoretical prediction for the cross section can be compared to the result of the experiment. The cross section is only measured at finitely many data points, but it is clear that in principle there can be many more data points than undetermined coefficients. In this way even a non-renormalizable EFT can be predictive.

4.5.2 *Example: Chiral perturbation theory*

The classic example of an EFT is chiral perturbation theory, which was invented to describe the dynamics of pions. In QCD with two massless quarks there is an $SU(2)_L \times SU(2)_R$ global "chiral" symmetry generated by suitable linear combinations of the conserved vector and axial currents. The vector current is generated by the sum, and the axial current by the difference of the left and right currents. To the extent that the up and down quark masses can be neglected, this is a good approximation in the real world. This is a symmetry of the Lagrangian, but not of the QCD ground state. If the ground state was chirally symmetric, for every mesonic multiplet we would see another multiplet with the same mass but opposite parity. Since such multiplets do not exist, not even with approximately similar masses, it means that chiral symmetry must be spontaneously broken. Only the vector subgroup $SU(2)_V$, which corresponds to isospin, is a symmetry of the QCD vacuum. Then, by Goldstone's theorem, there must exist three massless scalar fields. The pion fields are quite light and can be identified as the Goldstone bosons of spontaneously broken chiral symmetry. These Goldstone bosons are nonlinear fields having values in the coset space $SU(2)_L \times SU(2)_R/SU(2)_V$, which is diffeo-morphic to a copy of $SU(2)$. Thus we can describe the elements of the coset by a group-valued field

$$U(x) = \exp\left(i\frac{\pi^a(x)\sigma_a}{2F_\pi}\right) \tag{4.2}$$

where π^a are the canonically normalized pion fields, σ_a are the Pauli matrices and $F_\pi \approx 92\text{MeV}$ is the pion decay constant.

The most general chirally invariant action can be written as traces of the Lie algebra-valued field $U^{-1}\partial_\mu U$. The first terms in an expansion in the number of derivatives are

$$S = \int dx \left[L_2 + L_4 + O(\partial^6) \right] \tag{4.3}$$

$$L_2 = -\frac{F_\pi^2}{4} \text{tr}(U^{-1}\partial U)^2 \tag{4.4}$$

$$L_4 = \ell_1 \text{tr}(((U^{-1}\partial U)^2)^2) + \ell_2 (\text{tr}(U^{-1}\partial U)^2)^2 \ . \tag{4.5}$$

Let us concentrate for a moment on the first term, containing two derivatives. It is manifestly non-polynomial. If we expand the exponential, it gives rise to a canonical pion kinetic term plus infinitely many interaction terms, all involving two derivatives and increasing powers of $g = 1/F_\pi$:

$$L_2 = \frac{1}{2} \left[(\partial_\mu \pi^\alpha)^2 - \frac{1}{12} g^2 \left[\pi^\alpha \pi_\alpha (\partial_\mu \pi^\beta)^2 - (\pi^\alpha \partial_\mu \pi_\alpha)^2 \right] \right.$$

$$\left. + \frac{1}{360} g^4 \left[(\pi^\alpha \pi_\alpha)^2 (\partial_\mu \pi^\beta)^2 - \pi^\alpha \pi_\alpha (\pi^\beta \partial_\mu \pi_\beta)^2 \right] + O(\pi^8) \right] . \tag{4.6}$$

We see that the coupling in this action is g. It has dimension of length, so by power counting this theory must be non-renormalizable. Even if we didn't know QCD, but only this pion theory, we could make a good guess of the "scale of new physics": by dimensional analysis it must be related to the pion decay constant. A more accurate diagnostic for the breakdown of the effective theory is the violation of unitarity by the tree level scattering cross section. This gives the scale $M = 16\pi F_\pi$ which is of the order of the GeV.

Let us make some rough estimates for the contribution of various terms in (4.3) to a $2\pi \to 2\pi$ scattering process. The crucial point to observe is that all interaction terms contain derivatives. Assuming that all the momenta of the external particles are of order p, L_2 will give at tree level a contribution of order $g^2 p^2 \approx (p/M)^2$, while L_4 give a contribution of order $\ell g^4 p^4 \approx \ell(p/M)^4$, which is evidently subleading.

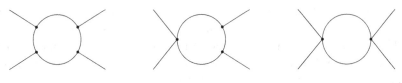

Fig. 4.1

Now we may try to estimate the effect of the diagrams in Fig. 4.1, constructed with two, three or four vertices taken from L_2. The integrands are all of the form $g^4 \int d^4 q F(q,p)$ where $F(q,p)$ is a fraction involving combinations of q or p to fourth power in the numerator (coming from the vertices) and combinations of q or p to fourth power in the denominator (coming from the propagators). It is therefore quartically divergent. When the integral is regulated, for example by means of dimensional regularization, it leaves behind something that for dimensional reasons can only involve p^4. Thus the diagram gives a contribution to the process of order $g^4 p^4 = (p/M)^4$.

A systematic analysis [142] shows that at order n in $(p/M)^2$ one must take into account diagrams with $n - 1$ loops constructed from L_2, $n - 2$ loops constructed from L_4, down to tree diagrams from L_n. In practice for low-energy meson physics one needs F_π, ℓ_1, ℓ_2 and a bunch of other parameters that are related to the quark masses. Calculations at one loop in F_π and at tree level in ℓ_1, ℓ_2 successfully describe a rich phenomenology [143].

Another important application of the same formalism is electroweak physics. Before the discovery of the Higgs particle (July 4, 2012), the Higgs sector of the SM could be described by a Lagrangian of the form (4.3), with suitable couplings to the gauge fields and fermions. The reason is that the Higgs doublet can be parametrized by four real fields, and suppressing the hitherto unobserved radial mode leaves one with three scalars parametrizing a three-sphere. The three-sphere is both topologically and geometrically equivalent to $SU(2)$. These three nonlinear degrees of freedom are the electroweak Goldstone bosons, which, via the Higgs mechanism, manifest themselves through the longitudinal components of the W and Z bosons. The existence of these degrees of freedom had been known since the discovery of the W and Z in 1983. The main difference between the electroweak chiral model and the QCD one is the value of the coupling g^{-1}, which in the electroweak case is equal to the Higgs VEV, 246 GeV. This "scale of new physics" was thus known since the earliest days of weak interaction theory.

4.5.3 *Gravity*

There are deep similarities between the chiral models and gravity, that had been noticed from early days. Starting from the kinematics, a metric of signature (p, q) in an n-dimensional vector space (we consider ony non-degenerate metrics, so $n = p+q$) can be viewed as an element of the coset $GL(n)/O(p, q)$. This makes the space of metrics of any given signature very non-linear, so that the term "metric tensor" is somewhat misleading. A Riemannian (*i.e.* positive definite) metric on a manifold M is a section of a bundle with fiber $GL(n)/O(n)$ associated to the bundle of frames. It is therefore a "gauged nonlinear sigma model", pretty much like the electroweak chiral model.

A general diffeomorphism-invariant action for a metric g, expanded in derivatives, reads

$$S = \int dx \sqrt{g} \left[L_0 + L_2 + L_4 + O(\partial^6) \right] \tag{4.7}$$

$$L_0 = m_P^2 \Lambda \tag{4.8}$$

$$L_2 = -\frac{1}{2}m_P^2 R \tag{4.9}$$

$$L_4 = \alpha R^2 + \beta R_{\mu\nu}R^{\mu\nu} + \gamma R_{\mu\nu\rho\sigma}R^{\mu\nu\rho\sigma} \tag{4.10}$$

If we recall that the Christoffel symbols have the structure $\Gamma \sim g^{-1}\partial g$ and that the curvature tensors contains terms of the form $\Gamma^2 \sim (g^{-1}\partial g)^2$, this is very similar to the chiral action (4.3), with m_P in place of F_π. In particular, both actions are nonpolynomial and describe massless particles that have derivative interactions.

The power counting is also similar, with E/m_P playing the role of expansion parameter [144–146]. Thus tree level diagrams with vertices from L_2 give contributions of order $(E/m_P)^2$ while those with vertices from L_4 are of order $(E/m_P)^4$; one-loop diagrams with vertices from L_2 give contributions of order $(E/m_P)^4$, those with vertices from L_4 give contributions of order $(E/m_P)^6$, and so on. In this way, using this expansion in E/M, it is possible to reliably compute quantum effects for energies up to the Planck scale. Actually, even at the highest energies presently available in accelerators, the ratio E/M is of order 10^{-16}, so the expansion in E/M is a very good one, in fact better than in any other QFT.

One can also get some feeling for the effect of L_4 by solving the linearized equation for the static potential. Following [147] one finds that in the Newtonian limit the gravitational potential generated by a point mass M has corrections that decay exponentially:

$$\Phi(r) = GM\left[-\frac{1}{r} + \frac{4}{3}\frac{e^{-rm_2}}{r} - \frac{1}{3}\frac{e^{-rm_0}}{r}\right], \tag{4.11}$$

where m_2 and m_0 are the masses of the spin-2 and spin-0 states discussed in section 2.2.2. The integration constants have been chosen in such a way that the $1/r$ singularity in the origin is absent. If λ and ξ are numbers of order one, $m_2 \approx m_0 \approx m_P$ and the corrections can be neglected at distances larger than the Planck length. Effectively, L_4 gives rise to contact interactions.

There are however some significant differences between gravity and the chiral model. The most striking one is the appearance of the overall factor $\sqrt{g}$ which is necessary to ensure diffeomorphism invariance of the measure. The second, related, difference is the cosmological term L_0, which contains the field but no derivatives. This gives rise to non-derivative interactions, which are absent in the chiral model. A third and less evident difference lies in the fact that the curvature tensor also involves terms of the form $\partial\Gamma = \partial(g^{-1}\partial g)$. When the derivative acts on g^{-1} it produces further terms of the form $(g^{-1}\partial g)^2$, but there are also terms $(g^{-1}\partial^2 g)$. In the expansion around flat spacetime, the curvature squared terms contain $(g^{-1}\partial g)^4 \sim (\partial h)^4$, which are the analogs of the terms in L_4 in (4.3) and correspond to vertices with at least four legs, but also terms $(g^{-1}\partial^2 g)^2 \sim (\Box h)^2$. As we have discussed in section 2.2.2, propagators with four derivatives would be problematic because they lead to physical ghosts. From the point of view of perturbative EFT, the correct way to treat these terms is to not include them in the propagator but rather to treat them as perturbations. (In the power counting argument given above it was indeed assumed that the propagator decays as $1/q^2$).

An alternative approach is to use the freedom of redefining the field. As discussed in section 3.5, in four dimensions and with zero cosmological constant, the only curvature squared term that survives on-shell is a total derivative. It has been shown that in the absence of a cosmological constant, or on a background of constant curvature, it is possible also in other dimensions, by means of field redefinitions, to remove the higher-derivative modifications of the propagator [148]. In this way one modifies the vertices, but unitarity is guaranteed.

4.5.4 *The leading corrections to the Newtonian potential*

To substantiate the general picture outlined in the previous section, it would be desirable to have an explicit calculation of an observable quantity where such quantum gravity effects are present. Such a calculation has been proposed by Donoghue [144]: it is the leading quantum correction to the Newtonian potential between massive non-relativistic particles. Here we outline the main points of this calculation.

We begin from the tree-level calculation, which reproduces the Newtonian potential. For simplicity the two gravitating particles will be assumed to have spin-0. The general idea has already been mentioned in section 2.1.1. The Newtonian potential can be seen as arising from the exchange of a graviton, as in the following Feynman diagram:

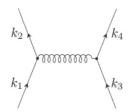

Fig. 4.2

The spring-like line is the propagator of the canonically normalized field $\phi_{\mu\nu}$, related to the fluctuation of the metric by Eq. (2.18). If we choose the Feynman-de Donder gauge $\alpha = \beta = 1$, its propagator is given by

$$D_{\mu\nu\rho\sigma}(q) = \frac{1}{2} \frac{\eta_{\mu\rho}\eta_{\nu\sigma} + \eta_{\mu\sigma}\eta_{\nu\rho} - \eta_{\mu\nu}\eta_{\rho\sigma}}{-q^2} \tag{4.12}$$

The vertex comes from the scalar action

$$\int d^4x \sqrt{|g|} \left(-\frac{1}{2} \sqrt{g} g^{\mu\nu} \partial_\mu \phi \partial_\nu \phi - \frac{1}{2} m^2 \phi^2 \right) . \tag{4.13}$$

When expanded around flat spacetime to first order in $h_{\mu\nu}$ and second order in ϕ, it gives the three-point vertex

$$\frac{1}{2} \int d^4x \, h^{\mu\nu} T_{\mu\nu} = \frac{1}{2} \int d^4x \, 2\kappa \, \phi^{\mu\nu} \left(\partial_\mu \phi \partial_\nu \phi - \frac{1}{2} \eta_{\mu\nu} \left((\partial\phi)^2 + m^2 \right) \right) . \tag{4.14}$$

In momentum space, calling k_1^μ and k_2^ν the momenta of the scalar particles (k_1 ingoing and k_2 outgoing, and $q = k_2 - k_1$), the vertex on the left in Fig. 4.2 reads

$$V^{\mu\nu}(k_1, k_2) = 2\kappa \left(k_1^{(\mu} k_2^{\nu)} - \frac{1}{2} \eta^{\mu\nu}(k_1 \cdot k_2 + m^2) \right) . \tag{4.15}$$

We assume that the initial particles are at rest and that the outgoing particles have three-momenta $\mathbf{p}$ and $-\mathbf{p}$, with $|\mathbf{p}| \ll m_1, m_2$. Performing the contractions, the amplitude in the diagram is, in the non-relativistic limit,

$$V^{\mu\nu}(k_1, k_2) D_{\mu\nu\rho\sigma}(k_2 - k_1) V^{\rho\sigma}(k_3, k_4) = -\frac{2\kappa^2 m_1^2 m_2^2}{q^2} = -\frac{16\pi G m_1^2 m_2^2}{\mathbf{p}^2} .$$

We can interpret this amplitude as the Fourier transform of a scattering potential in the Born approximation. To reconstruct the scattering potential one has to divide the amplitude by $2m_1 \times 2m_2$ to account for the difference between the relativistic and non-relativistic normalization of the states. This produces the coordinate-space potential[1]

$$V(r) = -4\pi G m_1 m_2 \int \frac{d^3\mathbf{p}}{(2\pi)^3} \frac{e^{i\mathbf{p} \cdot \mathbf{r}}}{\mathbf{p}^2} = -\frac{G m_1 m_2}{r} \tag{4.16}$$

Loop corrections will modify this potential. On dimensional grounds, the leading corrections must have the form

$$V(r) = -\frac{G m_1 m_2}{r} \left[1 + a \frac{G(m_1 + m_2)}{rc^2} + b \frac{G\hbar}{r^2 c^3} + \cdots \right] \tag{4.17}$$

The second term in the bracket does not contain factors of $\hbar$ and is therefore a classical effect. We shall discuss it further below. The third term is linear in $\hbar$ and therefore represents the leading quantum correction to the potential. One would like to calculate the coefficients a and b. In perturbation theory at one loop, one has to evaluate several diagrams. First one has the vacuum polarization diagrams shown in Fig. 4.3. These were considered originally in [150]. Then there are the vertex corrections shown in Fig. 4.4 and the one-particle irreducible diagrams shown in Figs. (4.5–4.8)

Fig. 4.3   Vacuum polarization diagrams.

These are all divergent, and one may worry about the renormalization ambiguities. As a matter of fact, the terms that enter in the evaluation of the coefficients

[1]An early historical reference for this calculation is [149].

Fig. 4.4   Vertex corrections.

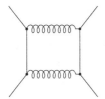

Fig. 4.5   Box diagram.

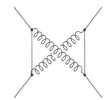

Fig. 4.6   Cross-box diagram.

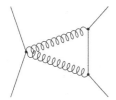

Fig. 4.7   Triangle diagram.

Fig. 4.8   Seagull diagram.

a and b are completely immune from these ambiguities. The Fourier transforms of the corrections to the potential in (4.17) are

$$\int \frac{d^3\mathbf{p}}{(2\pi)^3} \frac{1}{|\mathbf{p}|} e^{i\mathbf{p}\cdot\mathbf{r}} = \frac{1}{2\pi^2 r^2} \qquad (4.18)$$

$$\int \frac{d^3\mathbf{p}}{(2\pi)^3} \log\left(\frac{\mathbf{p}^2}{\mu^2}\right) e^{i\mathbf{p}\cdot\mathbf{r}} = -\frac{1}{2\pi^2 r^3} \qquad (4.19)$$

These momentum-space amplitudes are non-analytic in $\mathbf{p}$ and are clearly distinct from contributions of local counterterms, that give analytic corrections to the amplitude. For example, as mentioned in the end of the preceding section, a curvature-squared perturbation with coupling ℓ generates a "two-point vertex" proportional to ℓq^4. Insertion of such a vertex in the diagram (4.2) would produce an amplitude that is independent of momentum transfer:

$$\frac{1}{-q^2} \ell q^4 \frac{1}{-q^2} \approx \ell$$

Following the preceding reasoning, this generates a scattering potential

$$\int \frac{d^3\mathbf{p}}{(2\pi)^3} \ell e^{i\mathbf{p}\cdot\mathbf{r}} = \ell\delta(r) . \tag{4.20}$$

From the point of view of the low-energy EFT, the higher-derivative, UV divergent terms appear as contact interactions, whereas the terms we are interested in originate from the low energy part of the momentum integrations. It is therefore possible to neatly disentangle the UV effects, which is subject to renormalization ambiguities, from the IR effects, which are not.

In order to calculate the coefficients a and b one has to isolate the terms proportional to $\frac{1}{|\mathbf{P}|}$ and $\log\left(\frac{\mathbf{P}^2}{\mu^2}\right)$ in the diagrams listed above. The actual calculation of the coefficients took a while, with several papers making conflicting claims [151–154]. In the end agreement was established [155, 156]. The following table gives the contributions of each type of diagram:

diagram	contribution to a	contribution to πb
vacuum polarization	0	$\frac{43}{30}$
vertex corrections	-1	$\frac{5}{3} - \frac{26}{3}$
box and crossbox	0	$\frac{23}{3} + 8$
triangle	4	-28
seagull	0	22
total	3	$\frac{41}{10}$

The final result for the leading quantum corrections to the scattering potential is therefore

$$V(r) = -\frac{Gm_1m_2}{r}\left[1 + 3\frac{G(m_1 + m_2)}{rc^2} + \frac{41}{10\pi}\frac{G\hbar}{r^2c^3} + \cdots\right] . \tag{4.21}$$

Several comments are in order at this point. The second term in the square bracket is related to general relativistic corrections to the metric. It had been computed originally in [157]. One can show that the same vacuum polarization and vertex correction diagrams considered above generate non-analytic terms in the form factors of the energy-momentum tensor, and via the gravitational field equations these give rise to modifications of the metric. Such modifications reproduce the leading terms of the expansion of the Schwarzschild metric, written in the harmonic gauge [158]. Similar diagrams involving photons generate the leading terms of the Reissner-Nordstrøm metric [159]. The fact that loop diagrams reproduce the classical general relativistic correction to the Newtonian potential provides a counterexample to the general belief that the expansion in loops is also an expansion in powers of $\hbar$. Another counterexample, and a detailed explanation of the reasons behind the failure of the standard argument, has been given in [160].

The second term in the square bracket is truly of quantum origin. We have already emphasized that it is completely unaffected by renormalization ambiguities. Here we note that the coefficient b is also unaffected by higher loop corrections. For example, the two-loop correction must be proportional to $G^2\hbar^2$ and therefore gives rise to a term in the square bracket in (4.17) proportional to

$$\frac{G^2\hbar^2}{c^6 r^4}.$$

Higher loops give contributions that fall off even faster with distance. The coefficient b will be different in the presence of other massless particles. The most interesting case is that of the photon. The quantum corrections to the gravitational and electrostatic potential in scalar QED have been considered in [161, 162].

The calculation of the quantum-corrected Newtonian potential has been performed also with the modern unitarity-based methods mentioned in section 4.4, confirming the result (4.21) [163]. In principle it should be possible to derive it from the covariant effective action, by methods similar to those described in chapter 3. Unfortunately it has not yet been possible to write all the non-local terms that are needed. There is however a partial check. It is possible to compute the part of the effective action that contains up to two powers of curvature (and any number of derivatives). This part of the effective action for gravity contains the following finite terms [164]

$$\Gamma \sim \frac{1}{32\pi^2} \int d^4x \sqrt{g} \left[\frac{1}{60} R \log\left(\frac{-\Box}{\mu^2}\right) R + \frac{7}{10} R_{\mu\nu} \log\left(\frac{-\Box}{\mu^2}\right) R^{\mu\nu} \right], \qquad (4.22)$$

which accounts for the vacuum polarization generated by gravitons and ghosts.[2] It can be shown that these terms would lead to the potential [164]

$$V(r) = -\frac{Gm_1 m_2}{r} \left[1 + \frac{43}{30\pi} \frac{G\hbar}{r^2 c^3} + \ldots \right]. \qquad (4.23)$$

We see from the preceding table that this agrees exactly with the vacuum polarization contribution to the scattering potential. To completely reproduce the scattering potential (4.21) one would need also the terms in the effective action that are cubic and quartic in curvature. The form of these terms is presently not known.

Finally let us consider the order of magnitude of the corrections. The following table gives the values of the dimensionless corrections for two values of r: at the surface of the Sun, and at the Schwarzschild radius, assuming the Sun collapsed to a black hole.

$M = M_\odot$	$\frac{GM_\odot}{rc^2}$	$\frac{G\hbar}{r^2 c^3}$
$r = R_\odot$	2×10^{-6}	5×10^{-88}
$r = r_{S\odot}$	0.5	3×10^{-77}

---

[2] A similar calculation involving also the dynamics of the scalar is discussed in [165, 166].

The classical general relativistic correction is small but detectable in the former case, and of order one in the latter. By contrast the quantum correction is extremely small in both cases. To have a sizable quantum correction one would have to go near a Planck mass black hole. One can draw from this the standard negative conclusion about the impossibility of detecting quantum gravitational effects, but it is also possible to view this result positively: the smallness of the quantum correction means that the predictions of the effective QFT of gravity agree with the predictions of classical GR. To the extent that GR is tested experimentally, this EFT also is.

One can also draw from here a more general conclusion: it is not true, as is very often stated, that "in spite of many efforts we still do not have a quantum theory of gravity" or that "there is a fundamental clash between gravity and quantum physics". The EFT of gravity is a perfectly well-defined and predictive quantum field theory. Its status is comparable to that of the standard model, which in the absence of "new physics" is also expected to break down at some scale.

The smallness of the quantum effects is the result of the wide separation between the scales where observations are made and the "scale of new physics", which in this case is the Planck scale. In no other EFT is the separation of scales so large, and the expansion parameter so small. In this limited sense one could paradoxically say that this EFT of gravity is *the best* perturbative QFT. It certainly has a very broad range of applicability.

Of course, this EFT does not solve any of the problems that are usually given as motivations for research in quantum gravity: UV problems (the EFT methods break down at the Planck scale), IR problems (the EFT does not seem to say anything about the cosmological constant) and strong field problems (gravitational singularities). It is quite possible that to solve these issues it will be necessary to abandon standard QFT. The rest of this book is devoted to asymptotic safety, which is an attempt to construct a complete theory of gravity while remaining within the framework of QFT.

Chapter 5

Interlude: Technical developments

In this and in the next chapter we will work our way towards the discussion of asymptotic safety in gravity. This requires the use of additional techniques that have not been necessary so far. Such techniques have wider applicability, and it is pedagogically useful to introduce them in the simpler and more familiar setting of one-loop QFT in curved spacetime, introduced in Chapter 3. As a concrete application, we shall use these techniques to define the EA in a more general gauge than the one used in chapter 3, and prove independence of the gauge parameters on shell. The calculation will be continued in section 7.3.3, where the one-loop divergences will be obtained.

5.1 York decomposition

In electromagnetism and more generally in gauge theories, it is often convenient to decompose the gauge potential into its longitudinal and transverse parts:

$$A_\mu = A_\mu^T + \nabla_\mu \phi ; \qquad \nabla^\mu A_\mu^T = 0 . \tag{5.1}$$

The longitudinal part $\nabla_\mu \phi$, which is a spin-0 degree of freedom, is pure gauge and hence unphysical. The transverse part A_μ^T carries spin-1 and contains the physical degrees of freedom. We note that the constant mode of ϕ gives no contribution to A_μ and has therefore to be removed from the list of the degrees of freedom.

One can change variables in the functional integral, from the original degree of freedom A_μ to A_μ^T and ϕ. In order to compute the functional Jacobian we proceed as follows. As in (3.11), we define the measure such that for every field ψ the Gaussian integral is normalized to one:

$$\int (d\psi) e^{-\int dx \sqrt{g} \psi^2} = 1 , \tag{5.2}$$

where we are assuming Euclidean signature and for generality we work in a curved background metric g. This normalization depends on the choice of an inner product on the space of the fields, that has been discussed for different reasons in section 3.4. Let us apply this formula to the field A_μ. Applying the decomposition (5.1) and assuming that integrations by parts do not leave any boundary term, we have

$$\int dx \sqrt{g} A_\mu A^\mu = \int dx \sqrt{g} A_\mu^T A^{T\mu} + \int dx \sqrt{g} \, \phi(-\nabla^2)\phi \ . \tag{5.3}$$

If we define the Jacobian J by $(dA_\mu) = J(dA_\mu^T)(d\phi)$, then the integral becomes

$$1 = J \int (dA_\mu^T) e^{-\int dx \sqrt{g} A_\mu^T A^{T\mu}} \int (d\phi) e^{-\int dx \sqrt{g} \phi(-\nabla^2)\phi} = J(\text{det}'_\phi(-\nabla^2))^{-1/2} \ , \tag{5.4}$$

where in the last step we have used (5.2) for A_μ^T. Thus

$$J = (\text{det}'_\phi(-\nabla^2))^{1/2} \ . \tag{5.5}$$

The prime means that the zero mode has to be removed in the calculation of the determinant. This is necessary to make the calculation meaningful, but it is also dictated by the previous observation that constant ϕ gives no contribution to A_μ.

We note that if A_μ has the normal canonical dimension of mass (in four dimensions), then ϕ is dimensionless. Furthermore, ϕ naturally comes with a higher derivative kinetic term. To see this we observe first that the ordinary Maxwell term is independent of ϕ and that the dynamics of ϕ is entirely contained in the gauge fixing term. A Lorentz gauge fixing term $(\nabla^\mu A_\mu)^2$, which is independent of A_μ^T, leads to $\phi(-\nabla^2)^2\phi$. One can avoid these non-standard features by redefining the field $\hat{\phi} = \sqrt{-\nabla^2}\phi$.

This field has the normal canonical dimension and the Lorentz gauge condition gives rise to a normal second order kinetic term. (We note that such nonlocal redefinitions are generally not allowed in the case of physical degrees of freedom.) The Jacobian for this transformation is

$$(d\phi) = \det(-\nabla^2)^{-1/2}(d\hat{\phi}) \ , \tag{5.6}$$

so that the Jacobian of the transformation from A_μ to $(A_\mu^T, \hat{\phi})$ is one.

There is a close analog of this discussion in the case of gravity. In the background field method one has a background metric $\bar{g}$, and a fluctuation field $h_{\mu\nu}$ which is an ordinary symmetric tensor. The analog of (5.1) is the York decomposition [167]. First one can split algebraically

$$h_{\mu\nu} = h_{\mu\nu}^T + \frac{1}{d}\bar{g}_{\mu\nu} h \ , \tag{5.7}$$

where $h = \bar{g}^{\mu\nu} h_{\mu\nu}$ and $h_{\mu\nu}^T$ is tracefree: $\bar{g}^{\mu\nu} h_{\mu\nu}^T = 0$. In flat space one can further decompose the symmetric traceless tensor $h_{\mu\nu}^T$ into irreducible representations of the Lorentz group with spins 0, 1 and 2, as we did in section 2.1.5. In contrast to (5.7), this decomposition uses differential conditions. In curved spacetime it is more common to use the York decomposition:

$$h_{\mu\nu} = h_{\mu\nu}^{TT} + \bar{\nabla}_\mu \xi_\nu + \bar{\nabla}_\nu \xi_\mu + \bar{\nabla}_\mu \bar{\nabla}_\nu \sigma - \frac{1}{d}\bar{g}_{\mu\nu}\bar{\nabla}^2\sigma + \frac{1}{d}\bar{g}_{\mu\nu}h, \tag{5.8}$$

where $h_{\mu\nu}^{TT}$ is transverse and traceless and ξ_μ is transverse:

$$\nabla^\mu h_{\mu\nu}^{TT} = 0 \ , \qquad \bar{g}^{\mu\nu} h_{\mu\nu}^{TT} = 0 \ , \qquad \bar{\nabla}^\mu \xi_\mu = 0 \ . \tag{5.9}$$

Recalling the discussion of section 2.1.3, it is clear that $h_{\mu\nu}^{TT}$ is the spin-2 degree of freedom, whereas the transverse vector ξ_μ has spin-1 and σ and h have spin-0.

Now consider an infinitesimal diffeomorphism ϵ^μ. Using the background $\bar{g}$ we can decompose the transformation parameter ϵ^μ in its longitudinal and transverse parts:

$$\epsilon^\mu = \epsilon^{T\mu} + \bar{\nabla}_\mu \frac{1}{\sqrt{-\bar{\nabla}^2}} \psi ; \qquad \bar{\nabla}_\mu \epsilon^{T\mu} = 0 . \tag{5.10}$$

The inverse square root of the background Laplacian has been inserted conventionally in the definition of ψ so that it has the same dimension as ϵ^μ. We can then calculate the separate transformation properties of the York-decomposed metric under longitudinal and transverse infinitesimal diffeomorphisms. We have

$$\delta_{\epsilon^T}\xi^\mu = \epsilon^{T\mu} ; \qquad \delta_\psi h = -2\sqrt{-\bar{\nabla}^2}\psi ; \qquad \delta_\psi \sigma = \frac{2}{\sqrt{-\bar{\nabla}^2}}\psi , \tag{5.11}$$

all other transformations being zero. Note that σ and h are gauge-variant but the scalar combination

$$s = h - \bar{\nabla}^2 \sigma \tag{5.12}$$

is invariant.

To establish the connection with the spin projectors defined in section 2.1.5, we specialize $\bar{g}_{\mu\nu} = \delta_{\mu\nu}$ (or $\eta_{\mu\nu}$). Then the York decomposition can be written in momentum space:

$$h_{\mu\nu} = h_{\mu\nu}^{TT} + i(q_\mu \xi_\nu + q_\nu \xi_\mu) - q_\mu q_\nu \sigma + \frac{1}{d}\delta_{\mu\nu}q^2\sigma + \frac{1}{d}\delta_{\mu\nu}h. \tag{5.13}$$

It is easy to check that the first two terms correspond exactly to the spin-2 and spin-1 fields defined by the respective projectors:

$$P^{(2)}{}_{\mu\nu}{}^{\rho\sigma}h_{\rho\sigma} = h_{\mu\nu}^{TT} ; \tag{5.14}$$

$$P^{(1)}{}_{\mu\nu}{}^{\rho\sigma}h_{\rho\sigma} = i(q_\mu \xi_\nu + q_\nu \xi_\mu) . \tag{5.15}$$

On the other hand, the remaining three terms can be rewritten in the form $\frac{1}{d}(T_{\mu\nu}s + L_{\mu\nu}w)$, where

$$s = h + q^2\sigma$$
$$w = h - (d-1)q^2\sigma . \tag{5.16}$$

Then one finds that s and w are the degrees of freedom defined by the remaining two projectors:

$$P^{(ss)}{}_{\mu\nu}{}^{\rho\sigma}h_{\rho\sigma} = \frac{1}{d}T_{\mu\nu}s ; \tag{5.17}$$

$$P^{(ww)}{}_{\mu\nu}{}^{\rho\sigma}h_{\rho\sigma} = \frac{1}{d}L_{\mu\nu}w . \tag{5.18}$$

We see that, aside from the trivial rescaling by $\sqrt{2\kappa}$, the York decomposition differs from the decomposition (2.56) defined by the spin projectors only by a linear transformation in the spin-0 sector.

To find the functional Jacobian of the transformation $h_{\mu\nu} \to (h_{\mu\nu}^{TT}, \xi_\mu, \sigma, h)$ we use the relation

$$1 = \int (dh_{\mu\nu}) e^{-\mathcal{G}(h,h)} , \tag{5.19}$$

where $\mathcal{G}$ is an inner product in the space of symmetric two-tensors. If we require $\mathcal{G}$ to be ultralocal, the most general form is (3.137). Then, assuming that $\bar{g}$ is Einstein (*i.e.* that it satisfies Eq. (3.129)), we calculate

$$\begin{aligned}
\mathcal{G}(h,h) &= \int dx \sqrt{\bar{g}} \left(h_{\mu\nu} h^{\mu\nu} + \frac{a}{2} h^2 \right) \\
&= \int dx \sqrt{\bar{g}} \Big[h^{TT}{}_{\mu\nu} h^{TT\mu\nu} + 2\xi_\mu \left(-\bar{\nabla}^2 - \frac{\bar{R}}{d} \right) \xi^\mu \\
&\quad + \frac{d-1}{d} \sigma(-\bar{\nabla}^2) \left(-\bar{\nabla}^2 - \frac{\bar{R}}{d-1} \right) \sigma + \left(\frac{1}{d} + \frac{a}{2} \right) h^2 \Big] .
\end{aligned} \tag{5.20}$$

(If the background metric was not Einstein, there would be a mixing term between ξ_μ and σ.) Then, proceeding as before, we find

$$J = \left(\det_\xi \left(-\bar{\nabla}^2 - \frac{\bar{R}}{d} \right) \right)^{1/2} (\det'_\sigma(-\bar{\nabla}^2))^{1/2} \left(\det_\sigma \left(-\bar{\nabla}^2 - \frac{\bar{R}}{d-1} \right) \right)^{1/2} . \tag{5.21}$$

The meaning of the prime in this formula is analogous to the one of the prime in (5.5): a constant σ gives no contribution to $h_{\mu\nu}$. For some background metrics there can be additional modes of this type that have to be removed from the spectrum, for example, if ξ_μ is a Killing vector. We shall discuss this point later on.

Notice that the Jacobian does not depend on the free parameter a in the definition of the metric $\mathcal{G}$. This is because a enters only in the square of the trace, and the Gaussian integral over the trace does not contain a kinetic term, so it can just be absorbed in the overall normalization.

As in the case of electromagnetism, the fields ξ_μ and σ have nonstandard dimensions. They can be redefined as follows:

$$\hat{\xi}_\mu = \sqrt{-\bar{\nabla}^2 - \frac{\bar{R}}{d}} \, \xi_\mu \tag{5.22}$$

$$\hat{\sigma} = \sqrt{-\bar{\nabla}^2} \sqrt{-\bar{\nabla}^2 - \frac{\bar{R}}{d-1}} \, \sigma \tag{5.23}$$

The Jacobian of this transformation exactly cancels the one of the York decomposition, so that the transformation $h_{\mu\nu} \to (h_{\mu\nu}^{TT}, \hat{\xi}, \hat{\sigma}, h)$ has unit Jacobian.

We note the transformation properties of the redefined variables:

$$\delta_{\epsilon^T} \hat{\xi}_\mu = \sqrt{-\bar{\nabla}^2 - \frac{\bar{R}}{d}} \, \epsilon_\mu^T ; \qquad \delta_\psi \hat{\sigma} = 2\sqrt{-\bar{\nabla}^2 - \frac{\bar{R}}{d-1}} \, \psi . \tag{5.24}$$

Finally we record that on a maximally symmetric space, using equations (3.176), the following relations hold:

$$\int dx\sqrt{\bar{g}}\,h_{\mu\nu}h^{\mu\nu} = \int dx\sqrt{\bar{g}}\Big[h^{TT}{}_{\mu\nu}h^{TT\mu\nu} + 2\xi_\mu\left(-\bar{\nabla}^2 - \frac{\bar{R}}{d}\right)\xi^\mu$$
$$+\frac{d-1}{d}\sigma(-\bar{\nabla}^2)\left(-\bar{\nabla}^2 - \frac{\bar{R}}{d-1}\right)\sigma + \frac{1}{d}h^2\Big]. \tag{5.25}$$

$$\int dx\sqrt{\bar{g}}\,h_{\mu\nu}\bar{\nabla}^2 h^{\mu\nu} = \int dx\sqrt{\bar{g}}\Big[h^{TT}{}_{\mu\nu}\bar{\nabla}^2 h^{TT\mu\nu} + 2\hat{\xi}_\mu\left(\bar{\nabla}^2 + \frac{d+1}{d(d-1)}\bar{R}\right)\hat{\xi}^\mu$$
$$+\frac{d-1}{d}\hat{\sigma}\left(\bar{\nabla}^2 + \frac{2\bar{R}}{d-1}\right)\hat{\sigma} + \frac{1}{d}h\bar{\nabla}^2 h\Big], \tag{5.26}$$

$$\int dx\sqrt{\bar{g}}\,h_{\mu\nu}\nabla^\mu\nabla_\rho h^{\rho\nu} = \int dx\sqrt{\bar{g}}\Big[\hat{\xi}_\mu\left(\bar{\nabla}^2 + \frac{\bar{R}}{d}\right)\hat{\xi}^\mu + \frac{(d-1)^2}{d^2}\hat{\sigma}\left(\bar{\nabla}^2 + \frac{\bar{R}}{d-1}\right)\hat{\sigma}$$
$$+\frac{2(d-1)}{d^2}h\sqrt{-\bar{\nabla}^2}\sqrt{-\bar{\nabla}^2 - \frac{\bar{R}}{d-1}}\hat{\sigma} + \frac{1}{d^2}h\bar{\nabla}^2 h\Big]. \tag{5.27}$$

The first of these relations holds without modification also on a generic Einstein space; the second and third would be modified by mixing terms.

5.2 The Wick rotation revisited

We return here to the definition of the Euclidean continuation of gravity. In chapter 3 we uncritically assumed that it is possible to perform an analytic continuation to imaginary time in a way that parallels the standard treatment in flat space QFT. We shall now discuss the difficulties of that procedure and discuss a better one.

Defining the Euclidean continuation of GR as an analytic continuation of the time coordinate is at least unnatural, since time has no physical meaning in GR. If one tries to do that, one immediately finds that the result depends very strongly on the coordinate system. Thus for example beginning from the de Sitter metric, written in three different forms: the form with flat spatial sections

$$ds^2 = -dt^2 + H^{-2}e^{Ht}\left(dr^2 + r^2(d\theta^2 + \sin^2\theta d\varphi^2)\right) \tag{5.28}$$

or the form with positively curved spatial sections

$$ds^2 = -dt^2 + H^{-2}\cosh^2(Ht)\left(\frac{dr^2}{1-r^2} + r^2(d\theta^2 + \sin^2\theta d\varphi^2)\right) \tag{5.29}$$

or the form with negatively curved spatial sections

$$ds^2 = -dt^2 + H^{-2}\sinh^2(Ht)\left(\frac{dr^2}{1+r^2} + r^2(d\theta^2 + \sin^2\theta d\varphi^2)\right) \tag{5.30}$$

the prescription $t \to -it$ leads to a metric that is either complex, or positive definite, or again Lorentzian but with opposite signature. It is generally assumed that the "correct" Euclidean de Sitter metric is the second, because it is the unique maximally symmetric space with positive curvature, but this argument could not be used for more general metrics.

Another fact that should be cause of concern is the following [168, 169]. In flat spacetime, the sense of the Wick rotation is fixed by the requirement that the analytic continuation of the Feynman propagator of a free particle should not cross the poles in the complex energy plane. This is related to Feynman's "$i\epsilon$" prescription, which is a way to incorporate the notion of causality in the two-point function. Furthermore, the Euclidean continuation of any correlation functions must satisfy Osterwalder-Schrader positivity, which is again a consequence of causality. No such restrictions from causality seem to limit the analytic continuation of a time coordinate in a generic Lorentzian manifold.

Third, if we allow the Euclidean functional integral for gravity to include a sum over all Euclidean topologies, we are confronted with the fact that in four dimensions even the classification of the topologies is impossible. This is a major challenge to the definition of a functional integral, over and above the usual functional analytic issues.

To avoid such issues, in section 3.1.1 it was assumed that spacetime manifold is a product of the form $\mathbf{R} \times \Sigma$, and that this defines a preferred time direction on which the Wick rotation is to be performed. This, however, has the effect of restricting the class of Lorentzian metrics that can be analytically continued.

A better procedure is not to think of the Wick rotation as an analytic continuation of a coordinate but rather of the metric. One may start by recalling that every manifold admits a Riemannian (Euclidean) metric but that there are topological restrictions for the existence of Lorentzian metrics, namely there must exist a nowhere zero vectorfield [170]. Without loss of generality, such a vectorfield can be unit-normalized. Then, a Lorentzian metric $g_{(L)\mu\nu}$ can be constructed starting from a Euclidean metric $g_{(E)\mu\nu}$ and a nonvanishing unit vector field X by the formula:

$$g_{(L)\mu\nu} = g_{(E)\mu\nu} - 2X_\mu X_\nu . \tag{5.31}$$

In the Lorentzian metric, $g_{(L)\mu\nu}X^\mu X^\nu = -1$, so X is a unit timelike vectorfield. It is clear that the same formula can be used to construct a Euclidean metric out of a given Lorentzian metric and a unit timelike vectorfield.

One would like to see this as a continuous deformation. This is provided by the formula:

$$g_{\mu\nu}^{(t)} = g_{(L)\mu\nu} + 2tX_\mu X_\nu , \tag{5.32}$$

where t varies between 0 and 1. Clearly $g^{(0)} = g_{(L)}$ and $g^{(1)} = g_{(E)}$. Let us see how this procedure reproduces the Wick rotation in flat spacetime. We have $g_{\mu\nu}^{(0)} = g_{(L)\mu\nu} = \eta_{\mu\nu}$, so the interpolating metric is $g_{\mu\nu}^{(t)} = \text{diag}(-1 + 2t, 1, 1, 1)$, and $g_{(E)\mu\nu} \equiv g_{\mu\nu}^{(1)} = \delta_{\mu\nu}$. The volume element $\sqrt{\det g^{(t)}}$ is $\sqrt{\det(\eta_{\mu\nu})} = i$ for $t = 0$ and $\sqrt{\det(\delta_{\mu\nu})} = 1$ for $t = 1$. Consider for example the one-parameter family of actions

$$S^{(t)}(\phi) = \frac{1}{2} \int dx \sqrt{\det g^{(t)}} g^{(t)\mu\nu} \partial_\mu \phi \partial_\nu \phi . \tag{5.33}$$

For $t = 0$ this action is imaginary because of the measure. The correct Lorentzian action of a scalar field, giving rise to a positive definite Hamiltonian, is

$$S_L = iS^{(0)} = -\frac{1}{2} \int dx \sqrt{-\det g_{(L)}} g_{(L)}^{\mu\nu} \partial_\mu \phi \partial_\nu \phi \ .$$

In the Lorentzian functional integral one has the weight factor $e^{iS_L} = e^{-S^{(0)}}$. Deforming the metric from $t = 0$ to $t = 1$ one arrives continuously at $e^{-S^{(1)}}$, where

$$S^{(1)} = S_E = \frac{1}{2} \int dx \sqrt{\det g_{(E)}} g_{(E)}^{\mu\nu} \partial_\mu \phi \partial_\nu \phi$$

is the correct positive definite action of a Euclidean scalar field. One can similarly check that

$$S^{(t)} = \frac{1}{4} \int dx \sqrt{\det g^{(t)}} g^{(t)\mu\nu} g^{(t)\rho\sigma} F_{\mu\rho} F_{\nu\sigma} \tag{5.34}$$

and

$$S^{(t)} = \frac{1}{2\kappa^2} \int dx \sqrt{\det g^{(t)}} (2\Lambda - R(g^{(t)})) \tag{5.35}$$

interpolate between Lorentzian and Euclidean path integrals for electromagnetism and gravity, always with the identifications $S_L = iS^{(0)}$ and $S_E = S^{(1)}$. With this definition of the functional integral, the "i" in the exponent comes from taking the square root of the determinant of a metric with Lorentzian signature.

There is still one subtlety, however: if t is treated as a real parameter in the interval $[0, 1]$, then for $t = 1/2$ the metric would become degenerate. To avoid this, one has to allow t to describe a path in the complex plane. The question of the contour then arises: does the path pass above or below the point $t = 1/2$? We note that the propagator constructed with this metric

$$\frac{i}{-g_{\mu\nu}^{(t)} p^\mu p^\nu - m^2} = \frac{i}{E^2 - \vec{p}^2 - m^2 - 2tE^2} \tag{5.36}$$

coincides with the causal (Feynman) propagator,

$$\Delta^F = \frac{i}{E^2 - \vec{p}^2 - m^2 + i\epsilon} \tag{5.37}$$

if

$$t = -\frac{i\epsilon}{2E^2} \ . \tag{5.38}$$

We see that the usual prescription for the choice of integration contour in the definition of the propagator can be interpreted naturally as an incipient complexification of the metric. After allowing $\mathrm{Re}(t)$ to grow from 0 to 1 and letting $\mathrm{Im}(t)$ go back to zero, and taking into account the factor i from the volume element, the propagator takes the Euclidean form

$$\frac{-1}{-g_{\mu\nu}^{(1)} p^\mu p^\nu - m^2} = \frac{1}{E^2 + \vec{p}^2 + m^2} \ . \tag{5.39}$$

The procedure outlined here clearly avoids the first of the three issues mentioned above, because it is manifestly independent of the coordinate system. Concerning the second issue, we see that the sense of the Euclidean continuation is not arbitrary but rather is dictated by arguments of causality. There is in general no notion of reflection positivity in curved spacetime because generically there is no isometry that can serve the function of reflection. At least on static spacetimes, where such a reflection exists, a suitable generalization of reflection positivity holds [171].

Finally concerning the very broad issue of what topologies should be included in the functional integral, we have already noted that the existence of a Lorentzian metric is a rather strong restriction on the topology. This should be welcome, since a sum over all topologies is very hard to imagine. There is strong evidence from lattice calculations that a restriction of this type may be sufficient to define a gravitational functional integral. In CDT, the sum is over discrete triangulated spacetimes that admit a Lorentzian metric and in contrast to generic Dynamical Triangulations, in CDT there exists a phase that closely resembles an extended four-dimensional spacetime [172].

To close this section we return to the question of the Euclidean continuation of the de Sitter metric. As is clear from the preceding discussion, the Euclidean continuation depends on a choice of a unit timelike vectorfield and is therefore not unique. Choosing the unit vectorfield ∂_t in the coordinates where the metric has the forms (5.28), (5.29) or (5.30) corresponds simply to changing the sign of the term dt^2. Unlike the imaginary rotation of t, this generates three Riemannian metrics, but now the problem is that these either are not solutions of Einstein's equations (cases (5.29) and (5.30)) or are solutions with opposite value of the cosmological constant (case (5.28)). It is clearly desirable to define Euclidean de Sitter space as a solution of the Euclidean Einstein equations with the same value of the cosmological constant as the Lorentzian one. For this purpose let us go to a coordinate system where the metric is static. De Sitter space can be embedded in a five-dimensional Minkowski space with metric $ds^2 = -(dz^0)^2 + (dz^1)^2 + (dz^2)^2 + (dz^3)^2 + (dz^4)^2$ by the equation

$$-(z^0)^2 + (z^1)^2 + (z^2)^2 + (z^3)^2 + (z^4)^2 = r^2 \ .$$

One can choose coordinates $\tau, \chi, \theta, \varphi$ on de Sitter space, defined by

$$
\begin{aligned}
z^1 &= r \, \sin \chi \sin \theta \sin \varphi \ , \\
z^2 &= r \, \sin \chi \sin \theta \cos \varphi \ , \\
z^3 &= r \, \sin \chi \cos \theta \ , \\
z^4 &= r \, \cos \chi \cosh \tau, \\
z^0 &= r \, \cos \chi \sinh \tau.
\end{aligned}
\tag{5.40}
$$

This embedding gives rise to the metric

$$ds^2 = r^2 \left(-\cos^2 \chi d\tau^2 + d\chi^2 + \sin^2 \chi (d\theta^2 + \sin^2 \theta d\varphi^2) \right) \ . \tag{5.41}$$

The one-form $X = r \cos \chi d\tau$ has norm -1, and can be used in (5.31) to generate the Euclidean metric

$$ds^2 = r^2 \left(\cos^2 \chi d\tau^2 + d\chi^2 + \sin^2 \chi (d\theta^2 + \sin^2 \theta d\varphi^2) \right) . \tag{5.42}$$

This is the standard metric on the 4-sphere, embedded in the standard way in a five-dimensional Euclidean space. The coordinates are now defined by

$$
\begin{aligned}
z^1 &= r \sin \chi \sin \theta \sin \varphi , \\
z^2 &= r \sin \chi \sin \theta \cos \varphi , \\
z^3 &= r \sin \chi \cos \theta , \\
z^4 &= r \cos \chi \cos \tau, \\
z^5 &= r \cos \chi \sin \tau .
\end{aligned}
\tag{5.43}
$$

These embeddings guarantee that the metrics are solutions of Einstein's equations with cosmological constant $\Lambda = 3/r^2$ (the curvature scalar is $R = 12/r^2$). It is worth stressing that the role of the static coordinates in the preceding construction is just to make the form of the appropriate vectorfield easy to guess. One can derive the Euclidean metric in any other coordinate system by just transforming the objects given above.

In a similar way one can obtain a Euclidean version of Anti-de Sitter space. We start from the embedding

$$-(z^0)^2 + (z^1)^2 + (z^2)^2 + (z^3)^2 - (z^4)^2 = -r^2 .$$

in the flat space with metric $ds^2 = -(dz^0)^2 + (dz^1)^2 + (dz^2)^2 + (dz^3)^2 - (dz^4)^2$. One can choose coordinates $\tau, \chi, \theta, \varphi$ on Anti-de Sitter space, defined by

$$
\begin{aligned}
z^1 &= r \sinh \chi \sin \theta \sin \varphi , \\
z^2 &= r \sinh \chi \sin \theta \cos \varphi , \\
z^3 &= r \sinh \chi \cos \theta , \\
z^4 &= r \cosh \chi \sin \tau, \\
z^0 &= r \cosh \chi \cos \tau .
\end{aligned}
\tag{5.44}
$$

This embedding gives rise to the metric

$$ds^2 = r^2 \left(- \cosh^2 \chi d\tau^2 + d\chi^2 + \sinh^2 \chi (d\theta^2 + \sin^2 \theta d\varphi^2) \right) . \tag{5.45}$$

The one-form $X = r \cosh \chi d\tau$ has norm -1, and can be used in (5.31) to generate the Euclidean metric

$$ds^2 = r^2 \left(\cosh^2 \chi d\tau^2 + d\chi^2 + \sinh^2 \chi (d\theta^2 + \sin^2 \theta d\varphi^2) \right) . \tag{5.46}$$

This is the standard metric on the 4-four-dimensional one-sheeted hyperboloid, which is embedded in a five-dimensional Minkowski space with metric $ds^2 = (dz^1)^2 + (dz^2)^2 + (dz^3)^2 + (dz^4)^2 - (dz^5)^2$ by the condition

$$(z^1)^2 + (z^2)^2 + (z^3)^2 + (z^4)^2 - (z^5)^2 = -r^2 .$$

The coordinates are defined by

$$
\begin{aligned}
z^1 &= r \sinh \chi \sin \theta \sin \varphi , \\
z^2 &= r \sinh \chi \sin \theta \cos \varphi , \\
z^3 &= r \sinh \chi \cos \theta , \\
z^4 &= r \cosh \chi \cosh \tau, \\
z^5 &= r \cosh \chi \sinh \tau .
\end{aligned} \tag{5.47}
$$

The curvature scalar of this space is $R = -12/r^2$, and therefore it is a solution of Einstein's equations with cosmological constant $\Lambda = -3/r^2$.

5.3 Some Laplace-type operators

On a Riemannian manifold it may seem natural to define the Laplacian on any type of tensor field to be just $-\nabla^2$, where ∇ is the Levi-Civita connection. In fact this is one of many possibilities (it is sometimes called the Bochner Laplacian), and it is not the one that has the most convenient properties.

One class of tensors on which one can define a natural Laplacian are the differential forms (totally antisymmetric covariant tensors). Recall the differential d mapping p-forms to $(p+1)$-forms by[1]

$$
(d\omega)_{\mu_1 \ldots \mu_{p+1}} = (p+1)\partial_{[\mu_1} \omega_{\mu_2 \ldots \mu_{p+1}]} . \tag{5.48}
$$

With a Riemannian metric g one can define the inner product on p-forms

$$
(\alpha, \beta) = \frac{1}{p!} \int dx \sqrt{g} g^{\mu_1 \nu_1} \ldots g^{\mu_p \nu_p} \alpha_{\mu_1 \ldots \mu_p} \beta_{\nu_1 \ldots \nu_p} . \tag{5.49}
$$

The adjoint of d with respect to this inner product is called the co-differential and denoted δ:

$$
(d\alpha, \beta) = (\alpha, \delta\beta) . \tag{5.50}
$$

A straightforward calculation in coordinates shows that

$$
(\delta\beta)^{\mu_1 \ldots \mu_{p-1}} = -\frac{1}{\sqrt{g}} \partial_\lambda \left(\sqrt{\det g} \beta^{\lambda \mu_1 \ldots \mu_{p-1}} \right) , \tag{5.51}
$$

or, using the Levi-Civita connection of g,

$$
(\delta\beta)_{\mu_1 \ldots \mu_{p-1}} = -\nabla^\lambda \beta_{\lambda \mu_1 \ldots \mu_{p-1}} . \tag{5.52}
$$

The Laplacian on p-forms is defined by

$$
\Delta = d\delta + \delta d . \tag{5.53}
$$

In particular on functions (zero-forms)

$$
\Delta f = -\frac{1}{\sqrt{g}} \partial_\lambda (\sqrt{g} g^{\lambda\rho} \partial_\rho f) = -\nabla^2 f . \tag{5.54}
$$

---

[1]Square brackets around indices denote antisymmetrization with weight one, round brackets symmetrization.

This is called the Laplace-Beltrami operator. On one-forms

$$\Delta\omega_\mu = -\nabla_\lambda\nabla^\lambda\omega_\mu + R_\mu{}^\nu\omega_\nu \;, \tag{5.55}$$

and on two-forms

$$\Delta\omega_{\mu\nu} = -\nabla_\lambda\nabla^\lambda\omega_{\mu\nu} + R_\mu{}^\rho\omega_{\rho\nu} + R_\nu{}^\rho\omega_{\mu\rho} - 2R_\mu{}^\rho{}_\nu{}^\sigma\omega_{\rho\sigma} \;. \tag{5.56}$$

A form that is in the kernel of d is said to be closed, a form that is in the kernel of δ is said to be co-closed and a form that is in the kernel of Δ is said to be harmonic. The Hodge theorem states that on a compact manifold without boundary every p-form can be decomposed into the sum of a harmonic, a closed and a co-closed form, and these are orthogonal with respect to the inner product (5.49).

The operators d and δ satisfy $d^2 = 0$ and $\delta^2 = 0$. It then follows trivially from the definition (5.53) that they intertwine with the Laplacian:

$$d\Delta = \Delta d \;,$$
$$\delta\Delta = \Delta\delta \;. \tag{5.57}$$

(Note that if Δ on the l.h.s acts on p-forms, the one on the r.h.s. acts on $(p+1)$-forms or $(p-1)$-forms respectively.)

To some extent, one can generalize this construction to any tensor. The Lichnerowicz Laplacian acting on covariant p-tensors (without any symmetry properties) is defined by [173]

$$(\Delta_L T)_{\mu_1\ldots\mu_p} = -\nabla^2 T_{\mu_1\ldots\mu_p} + \sum_i R_{\mu_i}{}^\rho T_{\mu_1\ldots\rho\ldots\mu_p} - \sum_{i\neq j} R_{\mu_i}{}^\rho{}_{\mu_j}{}^\sigma T_{\mu_1\ldots\rho\ldots\sigma\ldots\mu_p} \tag{5.58}$$

where ρ and σ are in positions i and j respectively. Since $\nabla_\rho g_{\mu\nu} = 0$, one can freely raise and lower indices in the above definition and obtain Lichnerowicz Laplacians acting on tensors of any type.

The Lichnerowicz Laplacian preserves the type and the symmetries of the tensor it acts on, it is self-adjoint and it commutes with contractions. It coincides with (5.53) when acting on totally antisymmetric tensors. On covariant symmetric two-tensors it is given by

$$(\Delta_{L2}T)_{\mu\nu} = -\nabla^2 T_{\mu\nu} + R_\mu{}^\rho T_{\rho\nu} + R_\nu{}^\rho T_{\rho\mu} - 2R_\mu{}^\rho{}_\nu{}^\sigma T_{\rho\sigma} \;. \tag{5.59}$$

This is also the only case besides the Laplacian on forms that we shall need in the following. Its special significance for the theory of gravity comes from the observation that Eq. (3.122) for the first variation of the Ricci tensor can be written in the form

$$R^{(1)}_{\mu\nu} = \frac{1}{2}\Delta_{L2}h_{\mu\nu} + \nabla_\mu\left(\nabla_\rho h^\rho{}_\nu - \frac{1}{2}\nabla_\nu h\right) + \nabla_\nu\left(\nabla_\rho h^\rho{}_\mu - \frac{1}{2}\nabla_\mu h\right) \;. \tag{5.60}$$

Thus, when the fluctuation satisfies the de Donder condition, the Lichnerowicz Laplacian gives twice the variation of the Ricci tensor.

Recall that a metric is said to be Einstein if $R_{\mu\nu}$ is a constant multiple of $g_{\mu\nu}$ (this is the same as Einstein's equation with a cosmological constant). This implies that $R_{\mu\nu}$ is covariantly constant and that R is constant. The Lichnerowicz Laplacians have the useful property that on Einstein spaces they intertwine covariant derivatives:

$$\Delta_{L1}\nabla_\mu\phi = \nabla_\mu\Delta_{L0}\phi \tag{5.61}$$

$$\nabla_\mu\Delta_{L1}\xi^\mu = \Delta_{L0}\nabla_\mu\xi^\mu \tag{5.62}$$

$$\Delta_{L2}(\nabla_\mu\nabla_\nu\phi) = \nabla_\mu\nabla_\nu\Delta_{L0}\phi \,, \tag{5.63}$$

$$\Delta_{L2}(\nabla_\mu\xi_\nu + \nabla_\nu\xi_\mu) = \nabla_\mu\Delta_{L1}\xi_\nu + \nabla_\nu\Delta_{L1}\xi_\mu, \tag{5.64}$$

$$\Delta_{L2}g_{\mu\nu}\phi = g_{\mu\nu}\Delta_{L0}\phi \,. \tag{5.65}$$

5.4   Quantum GR in general gauges

Our goal in chapter 3 has been to arrive as quickly as possible at the main results concerning the non-renormalizability of GR, and to this end we have restricted our attention to the Feynman-de Donder gauge $\alpha = \beta = 1$ (in four dimensions). Technically this choice is convenient because in this case, and only in this case, the kinetic operators of gravitons and ghosts is a minimal Laplacian of the form $-\nabla^2\mathbf{1} +$ **E**. In other gauges the operator contains non-minimal terms such as $\nabla^\mu\nabla^\nu h_{\mu\nu}$. Such situations can be dealt with using the York decomposition. We shall use these methods to prove the gauge-independence of the one-loop effective action in GR, and in particular of its divergent part. For a more general but formal proof of this fact at the level of path integral, see [174].

5.4.1   *York-decomposed Hessian*

In this section we shall further manipulate the second variation of the Hilbert action, Eq. (3.126). The starting observation is that the Riemann term and the term $-\bar\nabla^2$ occur in the same combination in which they appear in the Lichnerowicz Laplacian. We can therefore rewrite (3.126) in the form

$$S^{(2)}(h;\bar g) = \frac{1}{4\kappa^2}\int d^d x \sqrt{\bar g}\left[\frac{1}{2}h_{\mu\nu}\Delta_{L2}h^{\mu\nu} + h_{\mu\nu}\bar\nabla^\mu\bar\nabla^\rho h_\rho{}^\nu - h\bar\nabla^\mu\bar\nabla^\nu h_{\mu\nu} - \frac{1}{2}h\Delta_{L0}h\right.$$
$$\left. + \left(\frac{\Lambda}{d-2} - \frac{1}{4}E\right)(h^2 - 2h_{\mu\nu}h^{\mu\nu}) + hE^{\mu\nu}h_{\mu\nu} - 2h_{\mu\nu}E^{\nu\rho}h^\mu{}_\rho\right], \tag{5.66}$$

where we defined

$$E^{\mu\nu} = \bar R^{\mu\nu} - \frac{2\Lambda}{d-2}\bar g^{\mu\nu} \,; \qquad E = E^\mu_\mu = \bar R - \frac{2d\Lambda}{d-2} \,. \tag{5.67}$$

All explicit appearances of the Ricci tensor have been rewritten in terms of E. We have thus isolated in the second line three terms that are proportional to the equations of motion. To proceed further we assume that the metric $\bar g_{\mu\nu}$ is Einstein:

$$\bar R_{\mu\nu} = \frac{\bar R}{d}\bar g_{\mu\nu} \,, \tag{5.68}$$

but we leave the constant $\bar{R}$ as a free parameter. This ansatz nearly solves the equations of motion $E^{\mu\nu} = 0$. More precisely, it solves all of them except for the trace equation $E = 0$. We will refer to this as an "almost on-shell" background. The Einstein condition allows us to further simplify the Hessian:

$$S^{(2)}(h;\bar{g}) = \frac{1}{4\kappa^2}\int d^d x \sqrt{\bar{g}}\left[\frac{1}{2}h_{\mu\nu}\Delta_{L2}h^{\mu\nu} + h_{\mu\nu}\bar{\nabla}^\mu\bar{\nabla}^\rho h_\rho{}^\nu - h\bar{\nabla}^\mu\bar{\nabla}^\nu h^{\mu\nu} - \frac{1}{2}h\Delta_{L0}h\right.$$

$$\left. + \left(\frac{\Lambda}{d-2} - \frac{d-4}{4d}E\right)(h^2 - 2h_{\mu\nu}h^{\mu\nu})\right]. \tag{5.69}$$

This Hessian is such that we can usefully apply the York decomposition. Using the properties (5.61–5.65) one has the following intermediate results:[2]

$$\Delta_{L2}h_{\mu\nu} = \Delta_{L2}h_{\mu\nu}^{TT} + \bar{\nabla}_\mu\Delta_{L1}\xi_\nu + \bar{\nabla}_\nu\Delta_{L1}\xi_\mu$$

$$+ \left(\bar{\nabla}_\mu\bar{\nabla}_\nu + \frac{1}{d}\bar{g}_{\mu\nu}\Delta_{L0}\right)\Delta_{L0}\sigma + \frac{1}{d}\bar{g}_{\mu\nu}\Delta_{L0}h , \tag{5.70}$$

$$\int d^d x \sqrt{\bar{g}}\, h_{\mu\nu}\Delta_{L2}h^{\mu\nu} = \int d^d x \sqrt{\bar{g}}\left[h_{\mu\nu}^{TT}\Delta_{L2}h^{TT\mu\nu} + 2\xi_\mu\Delta_{L1}\left(\Delta_{L1} - \frac{2\bar{R}}{d}\right)\xi^\mu\right.$$

$$\left. + \frac{d-1}{d}\sigma\Delta_{L0}^2\left(\Delta_{L0} - \frac{\bar{R}}{d-1}\right)\sigma + \frac{1}{d}h\Delta_{L0}h\right], \tag{5.71}$$

$$\bar{\nabla}^\nu h_{\mu\nu} = \left(\bar{\nabla}^2 + \frac{\bar{R}}{d}\right)\xi_\nu + \frac{d-1}{d}\bar{\nabla}_\nu\left(\bar{\nabla}^2 + \frac{\bar{R}}{d-1}\right)\sigma + \frac{1}{d}\bar{\nabla}_\nu h \tag{5.72}$$

$$\bar{\nabla}^\mu\bar{\nabla}^\nu h_{\mu\nu} = \frac{d-1}{d}\bar{\nabla}^2\left(\bar{\nabla}^2 + \frac{\bar{R}}{d-1}\right)\sigma + \frac{1}{d}\bar{\nabla}^2 h . \tag{5.73}$$

(We recall $\Delta_{L0} = -\bar{\nabla}^2$.)

Using these formulae, the quadratic part of the Hilbert action can be written

$$S^{(2)}(h;\bar{g}) = \frac{1}{4\kappa^2}\int d^d x \sqrt{\bar{g}}\left\{\frac{1}{2}h_{\mu\nu}^{TT}\left(\Delta_{L2} - \frac{4\Lambda}{d-2} + \frac{d-4}{d}E\right)h^{TT\mu\nu}\right.$$

$$+ \frac{d-2}{d}E\hat{\xi}_\mu\hat{\xi}^\mu - \frac{(d-1)(d-2)}{2d^2}\left[\hat{\sigma}(\Delta_{L0} - E)\hat{\sigma}\right.$$

$$\left.\left. + 2\hat{\sigma}\sqrt{\Delta_{L0}}\sqrt{\Delta_{L0} - \frac{\bar{R}}{d-1}}h + h\left(\Delta_{L0} - \frac{\bar{R}}{d-1} + \frac{d-2}{2(d-1)}E\right)h\right]\right\}. \tag{5.74}$$

Now we observe that the terms in the last three lines that do not contain E form a perfect square, which can be rewritten in terms of the gauge-invariant scalar variable

---

[2]We note here another virtue of the Lichnerowicz Laplacian over the Bochner Laplacian: in (5.26), in order to eliminate the mixed terms, it was necessary to assume that the background is maximally symmetric. In (5.71) it is enough to assume the Einstein condition.

$s = h + \Delta_{L0}\sigma$. We have

$$S^{(2)}(h;\bar{g}) = \frac{1}{4\kappa^2}\int d^d x \sqrt{\bar{g}}\left[\frac{1}{2}h_{\mu\nu}^{TT}\left(\Delta_{L2} - \frac{4\Lambda}{d-2} + \frac{d-4}{d}E\right)h^{TT\mu\nu}\right.$$
$$-\frac{(d-1)(d-2)}{2d^2}s\left(\Delta_{L0} - \frac{2d\Lambda}{(d-1)(d-2)} - \frac{E}{d-1}\right)s$$
$$\left.+\frac{d-2}{d}E\hat{\xi}_\mu\hat{\xi}^\mu + \frac{(d-1)(d-2)}{2d^2}E\hat{\sigma}^2 - \frac{(d-2)^2}{4d^2}Eh^2\right].\qquad(5.75)$$

We have rewritten $\bar{R}$ in terms of Λ and E, in such a way that the on-shell Hessian can be simply obtained by suppressing all terms proportional to E. We see that on-shell the Hessian depends only on the gauge-invariant variables $h_{\mu\nu}^{TT}$ and s. In particular in four dimensions we have

$$S^{(2)}_{\text{on-shell}}(h;\bar{g}) = \frac{1}{4\kappa^2}\int d^4 x \sqrt{\bar{g}}\left[\frac{1}{2}h_{\mu\nu}^{TT}\left(\Delta_{L2} - 2\Lambda\right)h^{TT\mu\nu} - \frac{3}{16}s\left(\Delta_{L0} - \frac{4}{3}\Lambda\right)s\right].$$
$$(5.76)$$

Also, note that in (5.75) the variables s, $\hat{\sigma}$ and h appear simultaneously, but only two of these can be regarded as independent. We defer this choice until later, when we will write the gauge-fixing term, but we note right away that it is clearly desirable to keep s as one of the independent variables.

These formulae hold for a generic Einstein background. In the literature one often encounters calculations where the background is maximally symmetric. In this case, using (3.176), the Lichnerowicz Laplacian on symmetric tensors is

$$\Delta_{L2} = -\bar{\nabla}^2 + \frac{2\bar{R}}{d-1},\qquad(5.77)$$

so the operator acting on the TT fields can be rewritten in the form

$$\Delta_{L2} - \frac{4\Lambda}{d-2} + \frac{d-4}{d}E = -\bar{\nabla}^2 + \frac{4\Lambda}{(d-1)(d-2)} + \frac{d^2 - 3d + 4}{d(d-1)}E\qquad(5.78)$$

or equivalently

$$-\bar{\nabla}^2 + \frac{d^2 - 3d + 4}{d(d-1)}\bar{R} - 2\Lambda.\qquad(5.79)$$

5.4.2   *The conformal factor problem*

The first term in (5.76) is positive, but the second is not. Normally a negative Euclidean kinetic term would be related to negative energy in the Minkowskian theory, but there is no such pathology in Einstein's theory, at least when one expands around flat space [175]. This is the same issue that we encountered already in the discussion of the Fierz-Pauli action, see Eq. (2.10), but is not just a problem of the linearized theory: the full Euclidean Hilbert action (3.125) is unbounded from below [176]. To see this consider a conformal tranformation of the metric $g_{\mu\nu} \to \Omega^2 g_{\mu\nu}$. The Ricci scalar transforms as

$$R \to \Omega^{-2} \left[R - 2(d-1)\Omega^{-1}\nabla^2\Omega - (d-1)(d-4)(\Omega^{-1}\nabla\Omega)^2 \right] .$$

In $d = 4$ only the first term is present and since ∇^2 has negative spectrum, one sees that R can be made arbitrarily positive and hence the Euclidean Hilbert action is unbounded from below.

There are various possible attitudes one can take in this respect. The pragmatic solution of the Cambridge school is that the integration over the conformal degree of freedom should be rotated in the complex plane so as to make the integral (formally) convergent [176]. A more detailed analysis of the issue, at least at one loop, has been made by Mazur and Mottola [177]. They make two essential observations. The first is that there should not be a kinetic term for s in the linearized action, because there is no propagating scalar degree of freedom in the theory. The appearance of such a term in (5.76) is the result of not having taken properly into account the integration measure in the definition of the path integral. More explicitly, we have to take into account the Jacobians that arise due to the change of variables we have performed to arrive at (5.76). We recall that the change of variable from $h_{\mu\nu}$ to $(h_{\mu\nu}^{TT}, \hat{\xi}_\mu, \hat{\sigma}, h)$ has unit Jacobian. The additional scalar change of variable from $\hat{\sigma}$ and h to s and h has Jacobian

$$\frac{\sqrt{\mathrm{Det}\left(-\bar{\nabla}^2 - \frac{\bar{R}}{d-1}\right)}}{\sqrt{\mathrm{Det}\left(-\bar{\nabla}^2\right)}} \tag{5.80}$$

Going on-shell, in $d = 4$, this is

$$\frac{\sqrt{\mathrm{Det}\left(-\bar{\nabla}^2 - \frac{4}{3}\Lambda\right)}}{\sqrt{\mathrm{Det}\left(-\bar{\nabla}^2\right)}} \tag{5.81}$$

Now we note that the numerator exactly cancels the determinant that comes from the Gaussian integration over s.

Alternatively, one could remove the numerator in the Jacobian by a further transformation to a new variable

$$S = \sqrt{\left(-\bar{\nabla}^2 - \frac{4}{3}\Lambda\right)}\, s$$

in terms of which (5.76) becomes

$$S^{(2)}(h;\bar{g}) = \frac{1}{4\kappa^2} \int d^4x \sqrt{\bar{g}} \left[\frac{1}{2} h_{\mu\nu}^{TT} \left(\Delta_{L2} - 2\Lambda\right) h^{TT\mu\nu} - \frac{3}{16} S^2 \right] .$$

Such non-local redefinitions, transforming a kinetic term into a mass term, are generally not allowed for physical degrees of freedom but this procedure may be justified in this case due to the fact that s does not propagate. There is then no longer a kinetic term for a scalar degree of freedom, only a "potential" term, whose variation gives the equation $S = 0$. The action now correctly reflects the absence of propagating scalars in GR, but the potential still has the wrong sign for Euclidean space.

The second observation, that is meant to correct this issue, is that we still have the freedom in the choice of the DeWitt metric in the space of symmetric tensors, Eq. (3.137). We have already used this metric in (5.19), (5.20) to calculate the functional Jacobian associated to the York decomposition, and we have seen that the Jacobian is actually independent of the free parameter a. However, if one looks at the last term in (5.20), one notices that it is positive if $a > -2/d$, negative if $a < -2/d$ (the DeWitt metric is degenerate when $a = -2/d$). Mazur and Mottola then argue that in the case $a < -2/d$ the analytic continuation of the trace mode has to be performed in the opposite way of the tracefree modes. This corrects the wrong sign of the operator appearing in the Euclidean Hessian. The final outcome of this analysis is thus equivalent in practice to the Cambridge prescription.

Some further discussion of the correct action to be used in Euclidean quantum gravity can be found in [177, 178]. At a more elementary level, we observe that already the addition of terms quadratic in curvature would correct the problem of the unboundedness of the action. It appears that the issue is strictly limited to the perturbative treatment of GR. Later we shall see that the unboundedness of the action is easily circumvented in the formulation of a functional RG equation for gravity, even when one considers only the Hilbert action. For the rest of this section we shall treat the second term in (5.76) with the Cambridge prescription.

5.4.3   *Gauge fixed Hessian*

Now let us consider a standard gauge fixing term

$$S_{GF} = \frac{1}{4\kappa^2\alpha} \int dx \sqrt{\bar{g}} \, \bar{g}^{\mu\nu} F_\mu F_\nu \tag{5.82}$$

with

$$F_\mu = \bar{\nabla}_\rho h^\rho{}_\mu - \frac{\beta+1}{d} \bar{\nabla}_\mu h \ . \tag{5.83}$$

Using the York decomposition,

$$F_\mu = -\left(\Delta_{L1} - \frac{2\bar{R}}{d}\right)\xi_\mu - \nabla_\mu\left(\frac{d-1}{d}\left(\Delta_{L0} - \frac{\bar{R}}{d-1}\right)\sigma + \frac{\beta}{d}h\right) \tag{5.84}$$

where, using the Einstein condition,

$$\Delta_{L1} = -\bar{\nabla}^2 + \frac{\bar{R}}{d} \ . \tag{5.85}$$

We see that a specific combination of scalar degrees of freedom appears in this formula. It is sometimes convenient to reparameterize the scalar sector in terms of the variable s introduced above, and this new degree of freedom:

$$\chi = \frac{((d-1)\Delta_{L0} - \bar{R})\sigma + \beta h}{(d-1-\beta)\Delta_{L0} - \bar{R}} \ . \tag{5.86}$$

The gauge fixing condition then reads

$$F_\mu = - \left(\Delta_{L1} - \frac{2\bar{R}}{d} \right) \xi_\mu - \frac{d - 1 - \beta}{d} \nabla_\mu \left(\Delta_{L0} - \frac{\bar{R}}{d - 1 - \beta} \right) \chi \, . \tag{5.87}$$

Using the variables $\hat{\xi}_\mu$ and

$$\hat{\chi} = \sqrt{\Delta_{L0}} \sqrt{\Delta_{L0} - \frac{\bar{R}}{d - 1 - \beta} \chi} \, , \tag{5.88}$$

the gauge fixing action can be written in the form

$$S_{GF} = \frac{1}{4\kappa^2 \alpha} \int dx \sqrt{\bar{g}} \left[\hat{\xi}_\mu \left(\Delta_{L1} - \frac{2\bar{R}}{d} \right) \hat{\xi}^\mu + \frac{(d - 1 - \beta)^2}{d^2} \hat{\chi} \left(\Delta_{L0} - \frac{\bar{R}}{d - 1 - \beta} \right) \hat{\chi} \right] . \tag{5.89}$$

Using (5.86) we see that under the transformation (5.11) the variable χ transforms in the same way as σ:

$$\delta_\psi \chi = \frac{2}{\sqrt{-\bar{\nabla}^2}} \psi \tag{5.90}$$

and $\hat{\chi}$ transforms as $\hat{\sigma}$ in (5.24). Thus $\hat{\xi}$ and $\hat{\chi}$ can be viewed as the gauge degrees of freedom. The ghost action for this gauge fixing contains a non-minimal operator

$$S_{gh} = - \int dx \sqrt{\bar{g}} \, \bar{C}^\mu \left(\delta_\mu^\nu \bar{\nabla}^2 + \left(1 - 2\frac{\beta + 1}{d} \right) \bar{\nabla}_\mu \bar{\nabla}^\nu + \bar{R}_\mu{}^\nu \right) C_\nu \, . \tag{5.91}$$

Decomposing the ghost into transverse and longitudinal parts

$$C_\nu = C_\nu^T + \nabla_\nu \frac{1}{\sqrt{-\bar{\nabla}^2}} C^L \tag{5.92}$$

and likewise for $\bar{C}$, the ghost action splits in two terms

$$S_{gh} = \int dx \sqrt{\bar{g}} \left[\bar{C}^{T\mu} \left(\Delta_{L1} - \frac{2\bar{R}}{d} \right) C_\mu^T + 2\frac{d - 1 - \beta}{d} \bar{C}^L \left(\Delta_{L0} - \frac{\bar{R}}{d - 1 - \beta} \right) C^L \right] . \tag{5.93}$$

We note for future reference that this change of variables has unit Jacobian.

As noted in section 5.4.1, Eq. (5.75) contains the three variables $\hat{\sigma}$, h and s, and in this section we have introduced a fourth scalar variable $\hat{\chi}$. Only two of these can be regarded as independent, and one has to choose which ones.

In much of the literature on asymptotic safety, reviewed in Chapter 8, one needs to work off-shell and the original York variables $\hat{\sigma}$ and h are used. In this case there is no advantage in doing further redefinitions and the symbols s and χ should be interpreted as particular linear combinations of the variables $\hat{\sigma}$ and h. On the other hand, $\hat{\sigma}$ and h appear in the gauge fixing condition only in the combination $\hat{\chi}$, and when one goes on-shell (by setting $E = 0$) they appear in (5.75) only in the combination s. It is then convenient to think of s and $\hat{\chi}$ as the independent scalar degrees of freedom and the full quadratic action, including the gauge fixing term, becomes diagonal in these variables. One could insist on this choice also off-shell, but then $\hat{\chi}$ would mix with s through the last two terms in (5.75) and the advantage of performing this redefinition would be lost.

We remark that in general one is not allowed to perform nonlocal field redefinitions such as (5.22), (5.23), (5.86), (5.88) on quantum fields that have physical asymptotic states. By such transformations one could turn a kinetic term into a mass term, for example. There is no such constraint on gauge variables such as ξ_μ, σ, χ etc.

We report for future reference the form of the full quadratic gauge-fixed action in the general α-β gauge, in terms of the York variables, on a maximally symmetric background:

$$
\begin{aligned}
S^{(2)} + S_{GF} = \frac{1}{4\kappa^2} \int d^d x \sqrt{\bar{g}} \Bigg[&\frac{1}{2} h_{\mu\nu}^{TT} \left(-\bar{\nabla}^2 + \frac{d^2 - 3d + 4}{d(d-1)} \bar{R} - 2\Lambda \right) h^{TT\mu\nu} \\
&+ \frac{1}{\alpha} \hat{\xi}_\mu \left(-\bar{\nabla}^2 + \frac{(d-2)\alpha - 1}{d} \bar{R} - 2\alpha\Lambda \right) \hat{\xi}^\mu \\
&- \frac{(d-1)((d-2)\alpha - 2(d-1))}{2d^2\alpha} \hat{\sigma} \left(-\bar{\nabla}^2 + \frac{(\alpha(d-2)-2)\bar{R} - 2d\alpha\Lambda}{2(d-1) - \alpha(d-2)} \right) \hat{\sigma} \\
&- \frac{(d-1)((d-2)\alpha - 2\beta)}{d^2\alpha} h \sqrt{-\bar{\nabla}^2} \sqrt{-\bar{\nabla}^2 - \frac{\bar{R}}{d-1}} \hat{\sigma} \qquad (5.94) \\
&- \frac{(d-1)(d-2)\alpha - 2\beta^2}{2d^2\alpha} h \left(-\bar{\nabla}^2 - \frac{(d-2)\alpha((d-4)\bar{R} - 2d\Lambda)}{2((d-1)(d-2)\alpha - 2\beta^2)} \right) h \Bigg] .
\end{aligned}
$$

The Lichnerowicz Laplacians have been rewritten in terms of Bochner Laplacians. This form can also be obtained directly using (5.25)–(5.27).

It is instructive to check this against the gauge-fixed Hessian in Minkowski space (2.71). To this end we first go from the variable $h_{\mu\nu}$ to the canonical variable $\phi_{\mu\nu}$ by means of the redefinition (2.18). The variable ϕ has a York decomposition that is identical to (5.8). This eliminates the prefactor $1/4\kappa^2$ from (5.94). We choose the de Donder gauge $\beta = d/2 - 1$ and we choose $\bar{g}$ to be flat (this is a special case of maximally symmetric space). The Hessian becomes

$$
\begin{aligned}
S^{(2)} + S_{GF} = \int d^d x \Bigg[&\frac{1}{2} h_{\mu\nu}^{TT} \left(-\partial^2 \right) h^{TT\mu\nu} + \frac{1}{\alpha} \hat{\xi}_\mu \left(-\partial^2 \right) \hat{\xi}^\mu \\
&- \frac{(d-1)((d-2)\alpha - 2(d-1))}{2d^2\alpha} \hat{\sigma} \left(-\partial^2 \right) \hat{\sigma} \\
&- \frac{(d-1)(d-2)(\alpha - 1)}{d^2\alpha} \hat{\sigma} \left(-\partial^2 \right) h \\
&- \frac{(d-2)(2(d-1)\alpha - (d-2))}{4d^2\alpha} h \left(-\partial^2 \right) h \Bigg] .
\end{aligned}
$$

Finally we go to the variables s and w by performing the linear transformation

$$
h = \frac{1}{d}((d-1)s + w) , \qquad \hat{\sigma} = \frac{1}{d}(s + w) , \qquad (5.95)
$$

and transform to momentum space. In this way (5.94) becomes

$$S^{(2)} + S_{GF} = \frac{1}{2} \int d^d q \, q^2 \left[\phi_{\mu\nu}^{TT}(-q) \phi^{TT\mu\nu}(q) + \frac{2}{\alpha} \hat{\xi}_\mu(-q) \hat{\xi}^\mu(q) \right.$$

$$- \frac{(d-1)(2\alpha(d-2) - (d-1))}{2d^2\alpha} s(-q)s(q)$$

$$\left. - \frac{(d-1)}{d^2\alpha} s(-q)w(q) + \frac{1}{2d^2\alpha} w(-q)w(q) \right] . \tag{5.96}$$

On the other hand, using the decomposition (2.56) and the spin projectors (2.47)–(2.52), we obtain the following relations:

$$\phi_{\mu\nu}(-q) P^{(2)\mu\nu\rho\sigma} \phi_{\rho\sigma}(q) = \phi_{\mu\nu}^{TT}(-q) \phi^{TT\mu\nu}(q) ,$$

$$\phi_{\mu\nu}(-q) P^{(1)\mu\nu\rho\sigma} \phi_{\rho\sigma}(q) = 2 \hat{\xi}_\mu(-q) \hat{\xi}^\mu(q) ,$$

$$\phi_{\mu\nu}(-q) P^{(0ss)\mu\nu\rho\sigma} \phi_{\rho\sigma}(q) = \frac{d-1}{d^2} s(-q)s(q) ,$$

$$\phi_{\mu\nu}(-q) P^{(0sw)\mu\nu\rho\sigma} \phi_{\rho\sigma}(q) = \frac{\sqrt{d-1}}{d^2} s(-q)w(q) , \tag{5.97}$$

$$\phi_{\mu\nu}(-q) P^{(0ws)\mu\nu\rho\sigma} \phi_{\rho\sigma}(q) = \frac{\sqrt{d-1}}{d^2} w(-q)s(q) ,$$

$$\phi_{\mu\nu}(-q) P^{(0ww)\mu\nu\rho\sigma} \phi_{\rho\sigma}(q) = \frac{1}{d^2} w(-q)w(q) .$$

Using these in (2.71) one arrives again at (5.96), up to an overall sign that is due to the use of the Euclidean signature in this chapter and Minkowskian signature in chapter 2.

5.4.4 *Gauge invariance of the one-loop effective action on-shell*

In this section we restrict ourselves to four dimensions. We consider the linearized action written in terms of the variables $h_{\mu\nu}^{TT}$, $\hat{\xi}$, s and $\hat{\chi}$. On shell, the gauge fixed action is given by (5.76) plus (5.89), and the ghost action is given by (5.93):

$$\frac{1}{4\kappa^2} \int d^4 x \sqrt{\bar{g}} \left\{ \frac{1}{2} h_{\mu\nu}^{TT} \Delta_2 h^{TT\mu\nu} - \frac{3}{16} s \Delta_0 s + \frac{1}{\alpha} \left[\hat{\xi}_\mu \Delta_1 \hat{\xi}^\mu + \frac{(3-\beta)^2}{16} \hat{\chi} \Delta_\beta \hat{\chi} \right] \right\}$$

$$+ \int dx \sqrt{\bar{g}} \left[\bar{C}^{T\mu} \Delta_1 C_\mu^T + \frac{3-\beta}{2} \bar{C}^L \Delta_\beta C^L \right] , \tag{5.98}$$

where we have introduced the following notation for operators in $d = 4$, on-shell:

$$\Delta_2 = \Delta_{L2} - 2\Lambda = -\bar{\nabla}^2 + \frac{2}{3}\Lambda ,$$

$$\Delta_1 = \Delta_{L1} - \frac{\bar{R}}{2} = -\bar{\nabla}^2 - \Lambda ,$$

$$\Delta_0 = \Delta_{L0} - \frac{\bar{R}}{3} = -\bar{\nabla}^2 - \frac{4}{3}\Lambda ,$$

$$\Delta_\beta = \Delta_{L0} - \frac{\bar{R}}{3-\beta} = -\bar{\nabla}^2 - \frac{4}{3-\beta}\Lambda .$$

The virtue of the changes of variables we have made is the perfect separation of the physical from the gauge degrees of freedom, on-shell: the gauge-invariant variables h^{TT} and s only appear in the linearization of the Hilbert action, whereas the gauge degrees of freedom $\hat{\xi}_\mu$ and $\hat{\chi}$ only appear in the gauge fixing term.

Then we have the following contributions to the one-loop partition function. The spin-2 graviton $h^{TT}_{\mu\nu}$ and the scalar s contribute

$$\mathrm{Det}\Delta_2^{-1/2}\mathrm{Det}\Delta_0^{-1/2} \,,$$

the fields $\hat{\xi}$ and $\hat{\chi}$ contribute

$$\mathrm{Det}\Delta_1^{-1/2}\mathrm{Det}\Delta_\beta^{-1/2} \,,$$

the ghosts give

$$\mathrm{Det}\Delta_1 \, \mathrm{Det}\Delta_\beta \,.$$

Finally we have to consider the Jacobian determinants. As already discussed, the changes of variable $h_{\mu\nu} \to (h^{TT}_{\mu\nu}, \hat{\xi}_\mu, \hat{\sigma}, h)$ and $(\bar{C}_\mu, C_\mu) \to (\bar{C}^T_\mu, \bar{C}^L, C^T_\mu, C^L)$ have unit Jacobians. The remaining change of variable $(\hat{\sigma}, h) \to (s, \hat{\chi})$ can be decomposed in $(\hat{\sigma}, h) \to (\sigma, h)$, which from (5.23) is seen to have Jacobian $\mathrm{Det}(-\bar{\nabla}^2)^{1/2}\mathrm{Det}\Delta_0^{1/2}$, $(\sigma, h) \to (s, \chi)$ which from (5.86) is seen to have unit Jacobian and $(s, \chi) \to (s, \hat{\chi})$, which from (5.88) is seen to have Jacobian $\mathrm{Det}(-\bar{\nabla}^2)^{-1/2}\mathrm{Det}\Delta_\beta^{-1/2}$. Altogether the Jacobians give

$$\mathrm{Det}\Delta_0^{1/2} \, \mathrm{Det}\Delta_\beta^{-1/2}. \tag{5.99}$$

Multiplying all these factors one remains with

$$Z = \frac{\sqrt{\mathrm{Det}\Delta_1}}{\sqrt{\mathrm{Det}\Delta_2}} \,. \tag{5.100}$$

The one-loop effective action at vanishing fluctuation field is therefore

$$\Gamma(\bar{g}) = S(\bar{g}) + \frac{1}{2}\mathrm{Tr}\log\left(\frac{\Delta_2}{\mu^2}\right) - \frac{1}{2}\mathrm{Tr}\log\left(\frac{\Delta_1}{\mu^2}\right) \,. \tag{5.101}$$

This result had been first obtained by Christensen and Duff [77] in the Feynman-de Donder gauge $\alpha = \beta = 1$, where the operators Δ_2, Δ_1 and Δ_β all differ from the corresponding Lichnerowicz Laplacian by -2Λ. The derivation given here proves explicitly the independence of the effective action from the gauge parameters. Note that this works somewhat differently for α and β: the gauge parameter α appears only in the prefactor of the kinetic operators and therefore automatically drops out of the effective action. The gauge parameter β appears only in Δ_β and there is a cancellation of these determinants between the contributions from the gauge fixing action, the ghost action and the Jacobians.

While the final result is very simple and elegant, it comes from the cancellation of several determinants. This suggests that the description of the theory that we are using is vastly redundant. The situation is even worse if one works with the "unhatted" variables ξ_μ and χ. The reader may check as an exercise that this leads to the same final result, but with more cancellations. A geometrically-motivated and somewhat simpler alternative to the standard Faddeev-Popov gauge-fixing procedure, has been discussed by Bern, Blau and Mottola [179,180]. In the next section we describe a choice of gauge that leads to an equally economical formulation.

The evaluation of the divergences of (5.101) requires knowledge of the heat kernel of the Laplacians acting on the differentially-constrained fields $h_{\mu\nu}^{TT}$, ξ_μ etc. These will be discussed in section 5.5.

5.4.5 *Physical gauge*

We now start from the quadratic action written in terms of the "hatted" York variables $h_{\mu\nu}^{TT}$, $\hat{\xi}_\mu$, $\hat{\sigma}$ and h, Eq. (5.74). In these variables there is no Jacobian to be taken into account. Instead of adding to this action a gauge fixing term, we shall choose a "physical gauge" where we simply put to zero some gauge degrees of freedom [181–183].

Let us return to the transformation properties of the fields under transverse and longitudinal infinitesimal diffeomorphisms, given in Eqs. (5.11), (5.24). We see that $\hat{\xi}_\mu$ gets shifted under transverse infinitesimal diffeomorphisms while $\hat{\sigma}$ and h get shifted under longitudinal infinitesimal diffeomorphisms. This means that one can simply set these variables to zero as a gauge choice. We will choose

$$\hat{\xi}_\mu = 0 \; ; \qquad h = 0 \; . \tag{5.102}$$

There is a subtlety concerning the second of these choices: on compact manifolds without boundary it has to be weakened to $h =$ constant. The reason for this is that the total volume is diffeomorphism invariant and hence an observable. An infinitesimal deformation of the metric that changes the total volume cannot be a gauge deformation. Hence all modes of h with the property that $\int d^d x \sqrt{g} h = 0$ are gauge modes, but the constant mode is not. This implies that there cannot exist any infinitesimal diffeomorphism that annihilates h. Indeed, under an infinitesimal diffeomorphism ϵ^μ, $\delta\sqrt{g} = \sqrt{g}\,\nabla_\mu \epsilon^\mu$, then integrating we find $0 = \delta V = \int d^d x \sqrt{g}\,\nabla_\mu \epsilon^\mu$, so on a compact manifold the integral of the divergence of a vectorfield is zero.[3] In the following we shall ignore this topological subtlety.

In the physical gauge, one can simply remove from the quadratic action (5.74) the terms involving the fields $\hat{\xi}_\mu$ and h. The former is already absent on-shell, so we just cancel the last line. Rewritten in terms of the variable $\hat{\sigma}$ defined in (5.23), the quadratic action becomes

$$S^{(2)}(h;\bar{g}) = \frac{1}{4\kappa^2} \int d^d x \sqrt{\bar{g}} \left[\frac{1}{2} h_{\mu\nu}^{TT} (\Delta_{L2} - 2\Lambda) h^{TT\mu\nu} - \frac{3}{16} \hat{\sigma}\Delta_{L0}\hat{\sigma} \right] \; . \tag{5.103}$$

---

[3]If this is not obvious, one may think of the simple example of the circle, with coordinate $0 < \varphi < 2\pi$. The unique component of a vectorfield on S^1 must be periodic and this implies that the integral of the divergence is zero: $0 = \epsilon^1(2\pi) - \epsilon^1(0) = \int_0^{2\pi} dx (d\epsilon^1/d\varphi)$.

Note that this agrees with (5.76), when we take into account that

$$s = \frac{\sqrt{\Delta_{L0}}}{\sqrt{\Delta_{L0} - \frac{\bar{R}}{d-1}}} \hat{\sigma}$$

when $h = 0$.

When a field transforming by a shift is set to a constant, there is no associated ghost. In the present case, because of the way we defined the gauge parameters in (5.10), the field transforming by a shift is not h but $\frac{1}{\sqrt{\Delta_{L0}}}h$ and so there is a Jacobian factor associated to this transformation: $\mathrm{Det}(\sqrt{\Delta_{L0}}) = (\mathrm{Det}\Delta_{L0})^{1/2}$. One can also see this as the ghost determinant associated to the gauge fixing condition $F = h$, via the standard rule $\Delta_{FP} = \frac{\delta F}{\delta \psi}$. Similarly, given that the field transforming by a shift is ξ_μ, the gauge condition $\hat{\xi}_\mu = 0$ gives a determinant $\mathrm{Det}\left(\sqrt{\Delta_{L1} - \frac{2\bar{R}}{d}}\right) = \mathrm{Det}\left(\Delta_{L1} - \frac{2\bar{R}}{d}\right)^{1/2}$. Altogether in the physical gauge one has the following ghost determinants

$$\mathrm{Det}\Delta_{L0}^{1/2}\mathrm{Det}\left(\Delta_{L1} - \frac{2\bar{R}}{d}\right)^{1/2}. \tag{5.104}$$

We can now collect the various contributions to the partition function. The fields $h_{\mu\nu}^{TT}$ and $\hat{\sigma}$ give

$$\mathrm{Det}(\Delta_{L2} - 2\Lambda)^{-1/2}\,\mathrm{Det}\Delta_{L0}^{-1/2} \tag{5.105}$$

and the only other contribution are the ghosts (5.104). The factors of $\mathrm{Det}\Delta_{L0}$ cancel out and the rest agrees with (5.100). This is clearly a much more economical way of arriving at the result.

5.4.6 *Exponential parametrization*

In this section we digress briefly to discuss an alternative to the standard linear splitting $g_{\mu\nu} = \bar{g}_{\mu\nu} + h_{\mu\nu}$. We can parametrize the metric as

$$g_{\mu\nu} = \bar{g}_{\mu\rho}(e^h)^\rho{}_\nu \tag{5.106}$$

where h in the exponent is a mixed tensor $h^\rho{}_\nu$ such that $h_{\mu\nu} = \bar{g}_{\mu\rho}h^\rho{}_\nu$ is symmetric. The formal advantage of this parametrization is that the exponential of a matrix always has positive definite eigenvalues, so the signature of g is guaranteed to be the same as the signature of $\bar{g}$. This is not so significant in a perturbative one-loop evaluation, where the fluctuation field is assumed to be small, but may be important in a non-perturbative setting. We have

$$g_{\mu\nu} = \bar{g}_{\mu\nu} + h_{\mu\nu} + \frac{1}{2}h_{\mu\lambda}h^\lambda{}_\nu + \dots \tag{5.107}$$

$$g^{\mu\nu} = \bar{g}^{\mu\nu} - h^{\mu\nu} + \frac{1}{2}h^{\mu\lambda}h_\lambda{}^\nu + \dots \tag{5.108}$$

In contrast to the usual linear split, here also the covariant metric is non-polynomial in the quantum field $h^\mu{}_\nu$.

For a one-loop calculation we need the expansion of the action to second order in the fluctuation. It is not necessary to repeat the whole calculation of the second variation. One can start from the expansion based on the linear decomposition $g_{\mu\nu} = \bar{g}_{\mu\nu} + \delta g_{\mu\nu}$ and replace

$$\delta g_{\mu\nu} \to h_{\mu\nu} + \frac{1}{2}h_{\mu\rho}h^\rho{}_\nu \ . \tag{5.109}$$

We have

$$S(g) = S(\bar{g}) + \int dx\sqrt{\bar{g}}\, G^{\mu\nu}\delta g_{\mu\nu} + \frac{1}{2}\int dx\sqrt{\bar{g}}\,\delta g_{\mu\nu}H^{\mu\nu\rho\sigma}\delta g_{\rho\sigma} + \dots$$

$$= S(\bar{g}) + \int dx\sqrt{\bar{g}}\, G^{\mu\nu}h_{\mu\nu} + \frac{1}{2}\int dx\sqrt{\bar{g}}\, h_{\mu\nu}H'^{\mu\nu\rho\sigma}h_{\rho\sigma} + \dots$$

where $H'^{\mu\nu\rho\sigma} = H^{\mu\nu\rho\sigma} + G^{\mu\rho}\bar{g}^{\nu\sigma}$. In the expansion of the Hilbert action, the Hessian in exponential parametrization differs from the one in linear parametrization by:

$$\frac{1}{2\kappa^2}\int d^dx\sqrt{\bar{g}}\frac{1}{2}\left[\frac{1}{2}\bar{g}^{\mu\nu}(2\Lambda - \bar{R}) + \bar{R}^{\mu\nu}\right]h_{\mu\rho}h^\rho{}_\nu \ . \tag{5.110}$$

Proceeding as in section 5.4.1, with the assumption of an almost on-shell background, one arrives at the following York-decomposed Hessian:

$$S^{(2)}(h;\bar{g}) = \frac{1}{4\kappa^2}\int dx\sqrt{\bar{g}}\left[\frac{1}{2}h_{\mu\nu}^{TT}\left(\Delta_{L2} - \frac{2\bar{R}}{d}\right)h^{TT\mu\nu}\right.$$

$$\left. - \frac{(d-1)(d-2)}{2d^2}s\left(\Delta_{L0} - \frac{\bar{R}}{d-1}\right)s - \frac{d-2}{4d}Eh^2\right] . \tag{5.111}$$

This formula has been derived under the same assumptions as (5.75). We note the following differences. The vector ξ_μ is completely absent from the quadratic action. In (5.75) its contribution was proportional to the equation of motion, but in the exponential parametrization this has been cancelled by an opposite contribution coming from the term linear in $\delta g_{\mu\nu}$. Likewise, there are no contributions proportional to E in the $h_{\mu\nu}^{TT}$ and s sectors. The kinetic operators of the fields $h_{\mu\nu}^{TT}$ and s are different but agree on-shell. The only term that is not written in terms of the gauge-invariant fields $h_{\mu\nu}^{TT}$ and s is now in the trace sector h and is again proportional to the equation of motion.

The structure of (5.111) strongly suggest the partial gauge choice $h = 0$. This fixes one of the three gauge degrees of freedom of GR, leaving a residual gauge freedom of volume-preserving diffeomorphisms, parametrized by a transverse vector ϵ_μ^T as in (5.11,5.24). This can be gauge-fixed as follows. Define the longitudinal and transverse projectors $\Pi_\mu{}^\nu = \bar{\nabla}_\mu\frac{1}{\bar{\nabla}^2}\bar{\nabla}^\nu$ and $\delta_\mu^\nu - \Pi_\mu{}^\nu$. We define the gauge condition

$$F_\mu = (\delta_\mu^\nu - \Pi_\mu{}^\nu)\bar{\nabla}_\rho h^\rho{}_\nu = \bar{\nabla}_\rho h^\rho{}_\mu - \bar{\nabla}_\mu\frac{1}{\bar{\nabla}^2}\bar{\nabla}_\rho\bar{\nabla}_\sigma h^{\rho\sigma}$$

$$= \left(\bar{\nabla}^2 + \frac{\bar{R}}{d}\right)\xi_\mu = -\left(\Delta_{L1} - \frac{2\bar{R}}{d}\right)\xi_\mu \ , \tag{5.112}$$

where passing to the second line we used (5.72) and (5.73). The corresponding Faddeev-Popov operator is again of the same form.

Specializing to $d = 4$, the full gauge-fixed action reads

$$\frac{1}{4\kappa^2} \int d^4x \sqrt{\bar{g}} \left[\frac{1}{2} h^{TT}_{\mu\nu} \left(\Delta_{L2} - \frac{2\bar{R}}{d} \right) h^{TT\mu\nu} - \frac{3}{16} \hat{\sigma} \Delta_{L0} \hat{\sigma} + \frac{1}{\alpha} \hat{\xi}_\mu \left(\Delta_{L1} - \frac{\bar{R}}{2} \right) \hat{\xi}^\mu \right]$$

$$+ \int dx \sqrt{\bar{g}} \left[\bar{C}^{T\mu} \left(\Delta_{L1} - \frac{\bar{R}}{2} \right) C^T_\mu + \tau \Delta_{L0} \tau \right]. \tag{5.113}$$

In the second term the gauge condition $h = 0$ has been used to replace $s \to \Delta_{L0} \sigma$ and then σ has been replaced by $\hat{\sigma}$. The third term is the gauge-fixing term for volume-preserving diffeomorphisms. (We note that in flat space it is equivalent to writing $h_{\mu\nu} P^{(1)\mu\nu\rho\sigma} h_{\rho\sigma}$, where $P^{(1)}$ is the spin projector (2.48).) The second line contains the transverse ghost and antighost fields $\bar{C}^{T\mu}$ and C^T_ν, and a real anti-commuting ghost τ that exponentiates the ghost for the condition $h = 0$, discussed in the previous section. Since we use the variables $h^{TT}_{\mu\nu}$, $\hat{\sigma}$ and $\hat{\xi}_\mu$, there are no Jacobians to worry about.

The remarkable virtue of the exponential parametrization, together with the partial gauge condition $h = 0$, is that the separation between physical and unphysical degrees of freedom works also off-shell. We see that the determinants produced by the integrations over $\hat{\sigma}$ and τ cancel, while those coming from $\hat{\xi}_\mu$, $\bar{C}^{T\mu}$ and C^T_ν partially cancel. Altogether the off-shell partition function is $\mathrm{Det}\Delta_1^{1/2}\mathrm{Det}\Delta_2^{-1/2}$, just like in (5.100), but the operators are witten in terms of $\bar{R}$ instead of Λ. The effective action, off-shell, is equal to

$$\Gamma(\bar{g}) = S(\bar{g}) + \frac{1}{2} \mathrm{Tr} \log \left(\frac{\Delta_{L2} - \frac{\bar{R}}{2}}{\mu^2} \right) - \frac{1}{2} \mathrm{Tr} \log \left(\frac{\Delta_{L1} - \frac{\bar{R}}{2}}{\mu^2} \right). \tag{5.114}$$

It is independent of the gauge parameter α and it agrees with (5.101) on-shell.

5.5 Spectral geometry of differentially constrained fields

As mentioned in section 5.1, in the evaluation of the effective action in gauge theories it is sometimes convenient to decompose the gauge field into its longitudinal and transverse parts, and likewise in gravity it is convenient to decompose the graviton field $h_{\mu\nu}$ according to the York decomposition. Then, the fields in the functional integral satisfy various algebraic and differential constraints (traceless-ness, transversality etc.). In order to apply the general formalism of chapter 3 to this situation, one needs the heat kernel expansion for Laplace-type operators acting on such differentially constrained fields.

We will discuss separately the case of the Bochner Laplacian and of the Lichnerowicz Laplacian. We start with the latter, because it has somewhat simpler properties, and we begin from the case of one-forms. In much of the following the topology of the manifold plays a subtle but important role. Things are simpler when the spectrum is continuous, as is generally the case on non-compact manifolds, and a bit trickier when the manifold is compact and without boundary, and the spectrum discrete. In this section we shall mostly ignore such subtleties, so that the results given apply in the non-compact case. They will be properly treated in section 5.6.1, when we discuss the heat kernel on the sphere.

5.5.1 *Lichnerowicz Laplacians*

According to the Hodge theorem, a one-form on a compact Riemannian manifold without boundary can be uniquely decomposed into a harmonic, a closed and a co-closed part. The harmonic forms are eigenfunctions of the Laplacian with eigenvalue zero. If there are no such zero modes, then the closed forms are also exact and the Hodge decomposition coincides with the conventional decomposition of a vector into its transverse and longitudinal parts, Eq. (5.1). Using (5.57) we see that the spectrum of the Laplacian on closed one-forms coincides with the spectrum of the Laplacian on zero-forms, with the exception of the zero-forms such that $df = 0$, which do not correspond to any one-form. It is important to treat these spurious modes properly. If the manifold is non-compact and zero is the lowest end of a continuous spectrum, the single constant zero-mode is of measure zero in the spectrum and removing it does not change the trace. In this case we can write for the heat kernel of Δ_{L1}

$$\mathrm{Tr}\, e^{-t\Delta_{L1}}\,|_{A_\mu} = \mathrm{Tr}\, e^{-t\Delta_{L1}}\,|_{A_\mu^T} + \mathrm{Tr}\, e^{-t\Delta_{L0}}\,|_\Phi \ . \tag{5.115}$$

On the other hand if the manifold is compact, without boundary and the spectrum of the Laplacian is discrete, then the contribution of the spurious mode makes a nontrivial contribution:

$$\mathrm{Tr}\, e^{-t\Delta_{L1}}\,|_{A_\mu} = \mathrm{Tr}\, e^{-t\Delta_{L1}}\,|_{A_\mu^T} + \mathrm{Tr}\, e^{-t\Delta_{L0}}\,|_\Phi - 1 \ . \tag{5.116}$$

We can use these formulae, together with the known expansion of the heat kernels of Δ_{L1} and Δ_{L0} on unconstrained forms, which are given by the general formula (3.45), to gain information on the heat kernel of Δ_{L1} acting on transverse vectors, which we will denote Δ_{L1}^T. From (5.115) one finds

$$b_n(\Delta_{L1}^T) = b_n(\Delta_{L1}) - b_n(\Delta_{L0}) \ . \tag{5.117}$$

This formula holds without further corrections in the non-compact case. In the compact case the spurious mode, being independent of t, gives an additional contribution to the coefficient b_d.

The heat kernel coefficients for the Lichnerowicz Laplacians acting on scalars and vectors can be computed from the general formula (3.45). In view of later applications we will assume that the metric satisfies the Einstein condition (5.68).

Then $R_{\mu\nu}R^{\mu\nu} = R^2/d$, and one gets

$$b_0(\Delta_{L0}) = 1$$

$$b_2(\Delta_{L0}) = \frac{1}{6}R \tag{5.118}$$

$$b_4(\Delta_{L0}) = \frac{1}{180}R_{\mu\nu\rho\sigma}R^{\mu\nu\rho\sigma} + \frac{5d-2}{360d}R^2$$

$$b_0(\Delta_{L1}) = d$$

$$b_2(\Delta_{L1}) = \frac{d-6}{6}R \tag{5.119}$$

$$b_4(\Delta_{L1}) = \frac{d-15}{180}R_{\mu\nu\rho\sigma}R^{\mu\nu\rho\sigma} + \frac{5d^2-62d+180}{360d}R^2 .$$

Note that these formulae hold both for compact and non-compact manifolds. Inserting in (5.117) and ignoring the spurious modes, we find

$$b_0(\Delta_{L1}^T) = d-1 ,$$

$$b_2(\Delta_{L1}^T) = \frac{d-7}{6}R , \tag{5.120}$$

$$b_4(\Delta_{L1}^T) = \frac{d-16}{180}R_{\mu\nu\rho\sigma}R^{\mu\nu\rho\sigma} + \frac{5d^2-67d+182}{360d}R^2 .$$

Let us now come to the case of symmetric tensors. Using (5.70), we see that the spectrum of Δ_{L2} on symmetric tensors is the union of the spectrum of Δ_{L2} on transverse traceless symmetric tensors, the spectrum of Δ_{L1} on transverse vectors and two copies of the spectrum of Δ_{L0} on scalars. Also in this case there can be spurious modes, namely modes of ξ_μ, σ or h that contribute zero to $h_{\mu\nu}$.

Taking the covariant divergence of the Killing equation one sees that a Killing vector always satisfies $\nabla^2 K_\mu + R_\mu{}^\nu K_\nu = 0$, so on an Einstein manifold Killing vectors are eigenvectors of Δ_{L1} with eigenvalue $2R/d$. If ξ_μ is a Killing vector it contributes zero to $h_{\mu\nu}$, so such modes are not part of the spectrum of Δ_{L2}.

Now consider the equation

$$\nabla_\mu\nabla_\nu\sigma - \frac{1}{d}g_{\mu\nu}\nabla^2\sigma = 0 . \tag{5.121}$$

One obvious solution that is always present and is an eigenfunction of Δ_{L0} with eigenvalue zero, is a constant. There may be further solutions. Consider a conformal Killing vector that is not a Killing vector. It satisfies the equation

$$\nabla_\mu K_\nu + \nabla_\nu K_\mu = \frac{2\lambda}{d}g_{\mu\nu}\nabla_\rho K^\rho ,$$

with $\nabla^\rho K_\rho \neq 0$, so it is not transverse. Assume that $K_\nu = \nabla_\nu\sigma$, for some function σ. Then the conformal Killing equation implies (5.121), so if σ is an eigenvector of Δ_{L0}, it is a spurious mode.

The existence of such spurious modes depends on the metric and therefore has to be discussed on a case-by-case basis. If the spectra are continuous, a finite number of modes is of measure zero and such effects can be neglected. In such a case,

$$\left.\mathrm{Tr}\,e^{-s\Delta_{L2}}\right|_{h_{\mu\nu}} = \left.\mathrm{Tr}\,e^{-s\Delta_{L2}}\right|_{h_{\mu\nu}^T} + \left.\mathrm{Tr}\,e^{-s\Delta_{L1}}\right|_{\xi} + \left.\mathrm{Tr}\,e^{-s\Delta_{L0}}\right|_{h} + \left.\mathrm{Tr}\,e^{-s\Delta_{L0}}\right|_{\sigma} . \tag{5.122}$$

Denoting $\Delta_{L2}{}^{TT}$ the Lichnerowicz Laplacian acting on transverse traceless tensors, we arrive at the following formula for its heat kernel coefficients:

$$b_n(\Delta_{L2}{}^{TT}) = b_n(\Delta_{L2}) - b_n(\Delta_{L1}{}^T) - 2b_n(\Delta_{L0}) . \tag{5.123}$$

Let us derive explicit formulae for the first heat kernel coefficients. Using (3.45) we have on symmetric tensors

$$b_0(\Delta_{L2}) = \frac{d(d+1)}{2} ,$$

$$b_2(\Delta_{L2}) = \frac{d^2 - 11d - 24}{12} R , \tag{5.124}$$

$$b_4(\Delta_{L2}) = \frac{d^2 - 29d + 480}{360} R_{\mu\nu\rho\sigma} R^{\mu\nu\rho\sigma} + \frac{5d^3 - 117d^2 + 478d + 2160}{720d} R^2 .$$

Using this, and (5.118,5.120) one gets

$$b_0(\Delta_{L2}{}^{TT}) = \frac{(d+1)(d-2)}{2} ,$$

$$b_2(\Delta_{L2}{}^{TT}) = \frac{d^2 - 13d - 14}{12} R , \tag{5.125}$$

$$b_4(\Delta_{L2}{}^{TT}) = \frac{d^2 - 31d + 508}{360} R_{\mu\nu\rho\sigma} R^{\mu\nu\rho\sigma} + \frac{5d^3 - 127d^2 + 592d + 1804}{720d} R^2 .$$

5.5.2 Bochner Laplacians

One can obtain similar formulae for other Laplacians. In later applications the Bochner Laplacian is often used. To distinguish it from the Lichnewrowicz Laplacian we will denote it $\Delta_B = -\nabla^2$. The Bochner Laplacian does not generally satisfy simple relations such as (5.61)–(5.65). However, commuting covariant derivatives, one has

$$\nabla^\mu \nabla^2 A_\mu = \nabla^2 \nabla^\mu A_\mu + \nabla_\rho (R^{\rho\mu} A_\nu) .$$

On an Einstein manifold the last term becomes $\frac{R}{d} \nabla^\rho A_\rho$. Thus, on an Einstein manifold, the Bochner Laplacian acting on a transverse vector gives a transverse vector. On the other hand, acting on a scalar

$$\Delta_B \nabla_\mu \Phi = \nabla_\mu \left(\Delta_B - \frac{R}{d} \right) \Phi. \tag{5.126}$$

Therefore one can write for the heat kernel of Δ_B acting on vectors, on a non-compact manifold

$$\mathrm{Tr}\, e^{-t\Delta_B} \big|_{A_\mu} = \mathrm{Tr}\, e^{-t\Delta_B} \big|_{A_\mu^T} + \mathrm{Tr}\, e^{-t\left(\Delta_B - \frac{R}{d}\right)} \big|_\Phi \tag{5.127}$$

while in the compact case

$$\mathrm{Tr}\, e^{-t\Delta_B} \big|_{A_\mu} = \mathrm{Tr}\, e^{-t\Delta_B} \big|_{A_\mu^T} + \mathrm{Tr}\, e^{-t\left(\Delta_B - \frac{R}{d}\right)} \big|_\Phi - e^{t\frac{R}{d}}. \tag{5.128}$$

Note that unlike the case of the Lichnerowicz Laplacian (5.116), the spurious constant scalar mode gives rise to a t- and R-dependent contribution.

Denoting Δ_{B1}^T the Bochner Laplacian acting on transverse vectors, we arrive at

$$b_n(\Delta_{B1}^T) = b_n(\Delta_{B1}) - b_n\left(\Delta_{B0} - \frac{R}{d}\right) . \tag{5.129}$$

In the compact case the spurious mode, being of order t^k with $k \geq 0$, makes additional contributions to all b_n for $n \geq d$.

A similar argument works for symmetric tensors, when using the York decomposition (5.8). On an Einstein space one can use the equations

$$\Delta_{B2}(\nabla_\mu \xi_\nu + \nabla_\nu \xi_\mu) = \nabla_\mu\left(\Delta_{B1} - \frac{d+1}{d(d-1)}R\right)\xi_\nu + \nabla_\nu\left(\Delta_{B1} - \frac{d+1}{d(d-1)}R\right)\xi_\mu \tag{5.130}$$

and

$$\Delta_{B2}\left(\nabla_\mu\nabla_\nu - \frac{1}{d}g_{\mu\nu}\nabla^2\right)\sigma = \left(\nabla_\mu\nabla_\nu - \frac{1}{d}g_{\mu\nu}\nabla^2\right)\left(\Delta_{B0} - \frac{2}{d-1}R\right)\sigma \tag{5.131}$$

to relate the spectrum of various operators on vectors and scalars to the spectrum of Δ_B on tensors. When spurious modes are absent or can be ignored, one finds for the heat kernel

$$\mathrm{Tr}\, e^{-t\Delta_B}\Big|_{h_{\mu\nu}} = \mathrm{Tr}\, e^{-t\Delta_B}\Big|_{h_{\mu\nu}^T} + \mathrm{Tr}\, e^{-t\left(\Delta_B - \frac{(d+1)R}{d(d-1)}\right)}\Big|_\xi$$
$$+ \mathrm{Tr}\, e^{-t\left(\Delta_B - \frac{2R}{d-1}\right)}\Big|_\sigma + \mathrm{Tr}\, e^{-t(\Delta_B)}\Big|_h , \tag{5.132}$$

whence

$$b_n(\Delta_{B2}^{TT}) = b_n(\Delta_{B2}) - b_n\left(\Delta_{B1} - \frac{(d+1)R}{d(d-1)}\right) - b_n\left(\Delta_{B0} - \frac{2R}{d-1}\right) - b_n(\Delta_{B0}) . \tag{5.133}$$

5.6 Heat kernels of maximally symmetric spaces

In this section we consider the sphere and the hyperboloid, which are the simply connected Euclidean maximally symmetric spaces with positive and negative curvature respectively. The Riemann and Ricci tensors of a maximally symmetric space are entirely determined by the curvature scalar via

$$R_{\mu\nu\rho\sigma} = \frac{R}{d(d-1)}(g_{\mu\rho}g_{\nu\sigma} - g_{\nu\rho}g_{\mu\sigma}) ; \qquad R_{\mu\nu} = \frac{R}{d}g_{\mu\nu} . \tag{5.134}$$

Explicit metrics have been given at the end of section 5.2. If the radius of the sphere is r, the curvature scalar is $R = \pm d(d-1)/r^2$. However, it is more convenient to express the heat kernels as function of the constant R. The d-sphere is compact and its volume is

$$V(S^d) = (4\pi)^{d/2}\frac{\Gamma(d/2)}{\Gamma(d)}r^d = \frac{2\pi^{\frac{d+1}{2}}}{\Gamma\left(\frac{d+1}{2}\right)}\left(\frac{d(d-1)}{R}\right)^{d/2} . \tag{5.135}$$

On maximally symmetric spaces, the quadratic curvature invariants are related by (3.177).

One can calculate the heat kernel coefficients by specializing the general formulas like (3.45), but these formulas are only available for a few of the lowest coefficients. There is an alternative method that allows one to make some useful cross checks on these low coefficients and to take the expansion much further. It is based on the fact that the spectra of Laplace-type operators on maximally symmetric spaces are known exactly.

5.6.1 *Sphere*

5.6.1.1 *Spectra*

The derivation of the spectrum of the Laplace-Beltrami operator on S^2 is familiar from quantum mechanics. Here we give the generalization to S^d and also list the spectra of the Bochner Laplacian on vectors and tensors.

Consider $d+1$-dimensional Euclidean space with coordinates $z_1, \ldots, z_{d+1}$. We introduce hyperspherical coordinates $(r, \theta_1, \ldots, \theta_{d-1}, \varphi)$ where $0 < \theta_i < \pi$ and $0 < \varphi < 2\pi$. We denote the angular coordinates collectively by ω. We embed S^d in R^{d+1} by fixing r. The line element of the Euclidean metric g_E, written in hyperspherical coordinates, reads

$$ds_E^2 = dr^2 + r^2 d\Omega^2 \tag{5.136}$$

where $d\Omega^2$ is the line element of the induced metric g in the unit sphere. In particular the volume element is $\sqrt{\det g_E} = r^d \sqrt{\det g}$. The Euclidean Laplace-Beltrami operator

$$\Delta_E = \sum_{i=1}^{d+1} \frac{\partial^2}{\partial z_i^2}$$

can be written in hyperspherical coordinates

$$\Delta_E = -\frac{1}{\sqrt{g_E}} \partial_\mu \sqrt{g_E} g_E^{\mu\nu} \partial_\nu = -\frac{1}{r^d} \partial_r r^d \partial_r + \frac{1}{r^2} \Delta_S \ , \tag{5.137}$$

where

$$\Delta_S = -\frac{1}{\sqrt{g}} \partial_\mu \sqrt{g} g^{\mu\nu} \partial_\nu \tag{5.138}$$

is the Laplace-Beltrami operator constructed with the induced metric on the unit sphere.

Let $Y_\ell(z)$ be a homogeneous polynomial in $(z_1, \ldots, z_{d+1})$ of degree ℓ. In spherical coordinates it is a function $Y_\ell(r, \omega)$ and homogeneity implies that $Y_\ell(r, \omega) = (r/r_0)^\ell Y_\ell(r_0, \omega)$. It is entirely determined by its restriction to the sphere, so we can think of it equivalently as a function on R^{d+1} or as a function on S^d. Direct calculation yields

$$\frac{1}{r^d} \partial_r r^d \partial_r Y_\ell = \frac{1}{r^2} \ell(\ell + d - 1) Y_\ell \ ,$$

thus $\Delta_E Y_\ell = 0$ if and only if Y_ℓ is an eigenfunction of Δ_S with eigenvalue $\lambda_\ell = \ell(\ell + d - 1)$:

$$\Delta_S Y_\ell = \lambda_\ell Y_\ell \ . \tag{5.139}$$

There is therefore a one-to-one correspondence between harmonic homogeneous polynomials on R^{d+1} of degree ℓ and eigenfunctions of Δ_S with eigenvalue λ_ℓ. To count the multiplicity of the eigenvalue λ_ℓ we thus have to count the harmonic homogeneous polynomials of degree ℓ.

Let M_ℓ be the space of homogeneous polynomials of degree ℓ. One can choose a basis in M_ℓ consisting of homogeneous monomials $z_1^{n_1} \ldots z_{d+1}^{n_{d+1}}$, with $n_1 + \ldots + n_{d+1} = \ell$. The dimension of M_ℓ is the number of ways of partitioning ℓ in the sum of $d + 1$ non-negative integers, and equals

$$\frac{1}{d!}(\ell + 1)(\ell + 2) \ldots (\ell + d) \ .$$

Now, using the flat coordinate representation of the Laplacian, Δ_E is a linear map from M_ℓ to $M_{\ell-2}$. The kernel of this map is the space H_ℓ of harmonic homogeneous polynomials, whose dimension we want to determine. It can be shown that Δ_E is surjective, so $M_{\ell-2} = M_\ell / H_\ell$. From here we obtain that the dimension of H_ℓ is the difference of the dimensions of M_ℓ and $M_{\ell-2}$, that can be written in the form

$$m_\ell = \frac{(2\ell + d - 1)(\ell + d - 2)!}{\ell!(d-1)!} \ . \tag{5.140}$$

One can determine in a similar way the eigenvalues and multiplicities of the Bochner Laplacian $\Delta_{Bs} = -\nabla^2$ on S^d acting on fields of transverse vectors (spin-1 fields) and transverse traceless symmetric tensors (spin-2 fields) [184]. The results are listed in the following table.

Table 5.1 Eigenvalues and multiplicities of the Laplacian on the d-sphere

Spin s	Eigenvalue $\lambda_l(d, s)$	Multiplicity $D_l(d, s)$
0	$\frac{l(l+d-1)}{d(d-1)}R;\ l = 0, 1 \ldots$	$\frac{(2l+d-1)(l+d-2)!}{l!(d-1)!}$
1	$\frac{l(l+d-1)-1}{d(d-1)}R;\ l = 1, 2 \ldots$	$\frac{l(l+d-1)(2l+d-1)(l+d-3)!}{(d-2)!(l+1)!}$
2	$\frac{l(l+d-1)-2}{d(d-1)}R;\ l = 2, 3 \ldots$	$\frac{(d+1)(d-2)(l+d)(l-1)(2l+d-1)(l+d-3)!}{2(d-1)!(l+1)!}$

5.6.1.2 *Heat kernels*

Knowing the spectra, we can calculate the heat kernel coefficients on the four-sphere. For example, the heat kernel of the Laplace-Beltrami operator Δ_0 acting on scalars is defined by the sum

$$K_{\Delta_0}(t) = \sum_{n=0}^{\infty} f(n, t) \ , \quad \text{where} \quad f(n, t) = \frac{1}{6}(n+1)(n+2)(2n+3)e^{-t\frac{R}{12}n(n+3)} \ .$$

This sum cannot be calculated in closed form but its small-t expansion can be calculated using the Euler-Maclaurin formula (see section 5.8). We have

$$K_{\Delta_0}(t) = \int_0^\infty dx f(x) + \frac{1}{2}f(0) - \frac{B_2}{2!}f'(0) - \frac{B_4}{4!}f^{(3)}(0) - \frac{B_5}{5!}f^{(4)}(0) - \frac{B_6}{6!}f^{(5)}(0) - \cdots$$

The integral gives $\frac{4(6+t\,R)}{R^2 t^2}$, which, when Laurent expanded, correctly reproduces the first two terms of the asymptotic expansion. Adding also the finite correction terms we get

$$K_{\Delta_0}(t) = \frac{V(S^4)}{(4\pi t)^2}\left(1 + \frac{R}{6}t + \frac{29R^2}{2160}t^2 + \frac{37R^3}{54432}t^3 + \frac{149R^4}{6531840}t^4 + \cdots\right). \qquad (5.141)$$

Using the relations (3.177), the first three terms are seen to agree with the general formula (3.36).

The operator $\Delta_0 + aR$ has eigenvalues $\frac{R}{12}n(n+3) + aR$, and the same multiplicities as Δ_0. Proceeding in the same way we arrive at

$$K_{\Delta_0 + aR}(t) = \frac{V(S^4)}{(4\pi t)^2}\left(1 + \frac{R(1 - 6a)}{6}t + \frac{R^2(29 - 360a + 1080a^2)}{2160}t^2\right.$$
$$\left. + \frac{R^3(185 - 3654a + 22680a^2 - 45360a^3)}{272160}t^3 + \cdots\right). \qquad (5.142)$$

For $a = 1/6$ this is the conformal operator and the first three terms can be checked against the second row in table 3.1.

Let us now do the same calculation with the Bochner Laplacian acting on a spin-1 field, which corresponds to a transverse vector. Its heat kernel is

$$K_{\Delta_{B_1^T}}(t) = \sum_{n=0}^\infty f(n,t) \ , \quad \text{where} \ \ f(n,t) = \frac{1}{2}n(n+3)(2n+3)e^{-t\frac{R}{12}(n^2 + 3n - 1)} \ .$$

The integral gives $\frac{24(3+t\,R)}{R^2 t^2}e^{-t\,R/4}$; Laurent expanding and adding the finite corrections we find

$$K_{\Delta_{B_1^T}}(t) = \frac{V(S^4)}{(4\pi t)^2}\left(3 + \frac{R}{4}t - \frac{7R^2}{1440}t^2 - \frac{541R^3}{362880}t^3 + \cdots\right). \qquad (5.143)$$

We can verify this formula using (5.128). Adding to $K_{\Delta_{B_1^T}}$ the heat kernel of $\Delta_0 - \frac{R}{4}$, which according to (5.142) is

$$K_{\Delta_{B_0} + \frac{R}{4}}(t) = \frac{V(S^4)}{(4\pi t)^2}\left(1 + \frac{5R}{12}t + \frac{373R^2}{4320}t^2 + \frac{12899R^3}{1088640}t^3 + \cdots\right),$$

and subtracting $e^{-tR/d} \approx 1 - tR/d + \cdots$, one obtains

$$K_{\Delta_{B_1}}(t) = \frac{V(S^4)}{(4\pi t)^2}\left(4 + \frac{2R}{3}t + \frac{43R^2}{1080}t^2 - \frac{R^3}{17010}t^3 + \cdots\right). \qquad (5.144)$$

The first three terms of this expansion agree with the fourth row of table 3.1, upon using the properties (3.177). We see that keeping more terms in the Euler-Maclaurin formula we can easily calculate the higher coefficients of the heat kernel expansion.

Next we consider the Bochner Laplacian acting on spin-2 fields (transverse traceless tensors) on the sphere. Using the Table 5.1, its heat kernel is

$$K_{\Delta_B{}_2^{TT}}(t) = \sum_{n=0}^{\infty} f(n,t) \,, \quad \text{where} \quad f(n,t) = \frac{5}{6}(n-1)(n+4)(2n+3)e^{-t\frac{R}{12}(n^2+3n-2)} \,.$$

The integral gives $\frac{60(2+tR)}{R^2 t^2}e^{-2tR/3}$; Laurent expanding and adding the finite corrections we find

$$K_{\Delta_B{}_2^{TT}}(t) = \frac{V(S^4)}{(4\pi t)^2}\left(5 - \frac{5R}{6}t - \frac{R^2}{432}t^2 + \frac{311R^3}{54432}t^3 + \ldots\right) \,. \tag{5.145}$$

Again, we can check the first three terms using (5.132), which on the d-sphere has to be replaced by

$$\text{Tr}\, e^{-t(-\nabla^2)}\Big|_{h_{\mu\nu}} = \text{Tr}\, e^{-t(-\nabla^2)}\Big|_{h_{\mu\nu}^T} + \text{Tr}\, e^{-t\left(-\nabla^2 - \frac{(d+1)R}{d(d-1)}\right)}\Big|_{\xi}$$
$$+ \text{Tr}\, e^{-t\left(-\nabla^2 - \frac{2R}{d-1}\right)}\Big|_{\sigma} + \text{Tr}\, e^{-t(-\nabla^2)}\Big|_{h}$$
$$- e^{t\frac{2R}{d-1}} - (d+1)e^{t\frac{R}{d-1}} - \frac{d(d+1)}{2}e^{t\frac{2R}{d(d-1)}} \,. \tag{5.146}$$

due to the spurious modes. The first term in the last line corresponds to the constant mode of σ. The second term corresponds to the $d+1$ modes of σ that are proportional to the Cartesian coordinates of the embedding $\mathbf{R}^n$; they correspond to the lowest nonzero eigenvalue of $-\nabla^2$, equal to $R/(d-1)$ (hence they are eigenfunctions of $-\nabla^2 - \frac{2}{d-1}R$ with eigenvalue $-R/(d-1)$) and also do not contribute to the spectrum of tensors. The last term corresponds to the $d(d+1)/2$ Killing vectors of the sphere, which are eigenvectors of $-\nabla^2$ on vectors with eigenvalue R/d (and hence of $-\nabla^2 - \frac{d+1}{d(d-1)}R$ with negative eigenvalue $-2/d(d-1)$).

The first term in the r.h.s. of (5.146) is given by (5.145). For the second term we use

$$K_{(\Delta_{B1} - \frac{5R}{12})^T}(t) = e^{t\frac{5R}{12}} K_{\Delta_{B1}^T}(t) \tag{5.147}$$

and equation (5.143) to obtain

$$K_{(\Delta_{B1} - \frac{5R}{12})^T}(t) = \frac{V(S^4)}{(4\pi t)^2}\left(3 + \frac{3R}{2}t + \frac{259R^2}{720}t^2 + \frac{4931R^3}{90720}t^3 + \ldots\right) \,. \tag{5.148}$$

The third term in the r.h.s. (corresponding to the scalar field σ) can be obtained from (5.142) with $a = -2/3$:

$$K_{\Delta_{B0} - \frac{2R}{3}}(t) = \frac{V(S^4)}{(4\pi t)^2}\left(1 + \frac{5R}{6}t + \frac{749R^2}{2160}t^2 + \frac{26141R^3}{272160}t^3 + \ldots\right) \,. \tag{5.149}$$

The fourth term is given by (5.141). Finally adding the expansions of the exponentials, one obtains

$$K_{\Delta_{B2}}(t) = \frac{V(S^4)}{(4\pi t)^2}\left(10 + \frac{5R}{3}t + \frac{11R^2}{216}t^2 - \frac{1343R^3}{136080}t^3 + \ldots\right) \,. \tag{5.150}$$

The first three terms agree with table 3.1. Note in particular that the contribution of the exponentials to the coefficient b_4 is just the number of the spurious modes $1 + 5 + 10 = 16$.

Table 5.2 gives the coefficients of the heat kernel expansion for $-\nabla^2$ acting on various types of fields. Some of these results are already contained in table 3.1, but on the sphere it is possible to go much beyond b_4. Here we limit ourselves to results that will be needed in section 7.7.

Table 5.2   Heat kernel coefficients on S^4

	b_0	b_2	b_4	b_6	b_8
$s = 0$	1	$\dfrac{R}{6}$	$\dfrac{29R^2}{2160}$	$\dfrac{37R^3}{54432}$	$\dfrac{149R^4}{6531840}$
$s = 1$	3	$\dfrac{R}{4}$	$-\dfrac{7R^2}{1440}$	$-\dfrac{541R^3}{362880}$	$-\dfrac{157R^4}{2488320}$
Vector	4	$\dfrac{2R}{3}$	$\dfrac{43R^2}{1080}$	$-\dfrac{R^3}{17010}$	$-\dfrac{2039R^4}{13063680}$
$s = 2$	5	$-\dfrac{5R}{6}$	$-\dfrac{R^2}{432}$	$\dfrac{311R^3}{54432}$	$\dfrac{109R^4}{1306368}$
Tensor	10	$\dfrac{5R}{3}$	$\dfrac{11R^2}{216}$	$-\dfrac{1343R^3}{136080}$	$-\dfrac{2999R^4}{3265920}$

5.6.2   *Hyperboloid*

5.6.2.1   *Spectra*

The d-dimensional hyperboloid is non-compact and the spectrum of the Bochner Laplacians is continuous. For fields of spin $s = 0, 1, 2$ it can be parametrized by a continuous dimensionless parameter σ (which replaces the discrete label n of the sphere). The eigenvalue corresponding to σ is

$$\lambda_\sigma = \frac{|R|}{d(d-1)} \left(\sigma^2 + s + \frac{(d-1)^2}{4} \right) \tag{5.151}$$

and the spectral density is [185]

$$\rho(\sigma) = A_d\, g(s) \left[\sigma^2 + \left(s + \frac{d-3}{2} \right)^2 \right] \prod_{j=0}^{(d-5)/2} (\sigma^2 + j^2) \qquad \text{for } d \text{ odd} \tag{5.152}$$

$$\rho(\sigma) = A_d\, g(s) \left[\sigma^2 + \left(s + \frac{d-3}{2} \right)^2 \right] \sigma \tanh(\pi\sigma) \prod_{j=1/2}^{(d-5)/2} (\sigma^2 + j^2) \qquad \text{for } d \text{ even} \tag{5.153}$$

where A_d is an overall normalization constant,

$$g(s) = \frac{(2s + d - 3)(s + d - 4)!}{(d - 3)!\, s!}$$

and for $d = 3, 4$ the products have to be omitted. With these data one can construct traces of functions of Laplacians, for example the zeta function is

$$\zeta(u) = \int_0^\infty d\sigma \rho(\sigma) \lambda_\sigma^{-u} .$$

5.6.2.2   *Heat kernels*

The calculations on the hyperboloid are simpler than those on the sphere, for two reasons. First, since the spectrum is continuous, the isolated modes that have to be removed are of measure zero and therefore can be completely ignored. Second, it is not necessary to use the Euler-Maclaurin formula to convert sums into integrals: the traces are already given by integrals. We work in $d = 4$. The three spins $s = 0, 1, 2$ can be treated all at once. The spin-dependent factor in the measure $g(s) = 2s + 1$ is just the number of components of a transverse (for $s = 1$) or transverse-traceless (for $s = 2$) field. The heat kernel is given by the integral:

$$K_{\Delta_{B_s}}(t) = A_4(2s + 1) \int d\sigma \left[\sigma^2 + \left(s + \frac{1}{2} \right)^2 \right] \sigma \tanh(\pi\sigma) e^{\frac{1}{12} R t (\sigma^2 + s + \frac{9}{4})} \quad (5.154)$$

Note that the exponent has the correct negative sign since $R < 0$. Were it not for the hyperbolic tangent, this integral has the same form as the ones we did on the sphere. We note that for large σ the hyperbolic tangent tends indeed to one, so the complications come from the infrared region. We can split

$$\tanh(\pi\sigma) = 1 - \frac{2}{1 + e^{2\pi\sigma}}$$

and correspondingly the integral is split in two integrals. The first can be performed in closed form without difficulty. In the second we expand $e^{\frac{1}{12} R t \sigma^2}$ in Taylor series in t. The exponential in the denominator renders all terms convergent.

The normalization A_d can be fixed in such a way that the leading term of the expansion of the heat kernel for $s = 0$ is $\frac{V(H^4)}{(4\pi t)^{d/2}}$, which is formally infinite, with coefficient one. This fixes all other coefficients. The result is

$$K_{\Delta_{B_s}}(t) = \frac{V(H^4)}{(4\pi t)^{d/2}} (1 + 2s) \left[1 + \frac{R}{12}(2 - s^2)t + \frac{R^2}{4320}(58 - 10s - 85s^2 - 30s^3)t^2 \right.$$

$$+ \frac{R^3}{1088640}(740 - 504s - 2562s^2 - 1680s^3 - 315s^4)t^3$$

$$\left. + \frac{R^4}{52254720}(1192 - 2448s - 9860s^2 - 8988s^3 - 3255s^4 - 420s^5)t^4 + \ldots \right] .$$

$$(5.155)$$

This formula yields directly the first, second and fourth rows in table 5.3. The rows for the vector field and for the symmetric tensor field can be obtained by using equations (5.127) and (5.132). Notice that these rows, as well as the first, turn out to be identical to the analogous rows in table 5.2.

Table 5.3 Heat kernel coefficients on H^4

	b_0	b_2	b_4	b_6	b_8
$s = 0$	1	$\dfrac{R}{6}$	$\dfrac{29R^2}{2160}$	$\dfrac{37R^3}{54432}$	$\dfrac{149R^4}{6531840}$
$s = 1$	3	$\dfrac{R}{4}$	$-\dfrac{67R^2}{1440}$	$-\dfrac{4321R^3}{362880}$	$-\dfrac{3397R^4}{2488320}$
Vector	4	$\dfrac{2R}{3}$	$\dfrac{43R^2}{1080}$	$-\dfrac{R^3}{17010}$	$-\dfrac{2039R^4}{13063680}$
$s = 2$	5	$-\dfrac{5R}{6}$	$-\dfrac{271R^2}{432}$	$-\dfrac{7249R^3}{54432}$	$-\dfrac{22571R^4}{1306368}$
Tensor	10	$\dfrac{5R}{3}$	$\dfrac{11R^2}{216}$	$-\dfrac{1343R^3}{136080}$	$-\dfrac{2999R^4}{3265920}$

5.7 Formula for functional traces

In chapter 3 we have encountered the trace of the heat kernel, which is the trace of the exponential of a differential operator, as well as the one-loop effective action, which is the trace of the logarithm of a differential operator. In the next chapter we will encounter the trace of a rather general function of a differential operator. We will need a general tool to deal with such expressions. In the rare cases when the spectrum of the differential operator is known, one can directly calculate the spectral sum, using the Euler-Maclaurin formula when the spectrum is discrete. When the spectrum is not known, one can still calculate the first few terms in a derivative expansion. We give here a general formula for such an expansion. Aside from application to functional renormalization, this formula is also used in the definition of the so-called spectral action [186].

As in section 3.2 we consider a covariant Laplacian $\Delta = -\nabla^2 \mathbf{1} + \mathbf{E}$, where ∇ is a gauge- and gravitational covariant derivative and $\mathbf{E}$ is a linear map acting on the spacetime and internal indices of the fields. The trace of a function W of the operator Δ can be written as

$$\mathrm{Tr}W(\Delta) = \sum_n W(\lambda_n) \, , \tag{5.156}$$

where λ_n are the eigenvalues of Δ (degenerate eigenvalues are summed separately). Introducing the Laplace anti-transform $\tilde{W}(s)$

$$W(z) = \int_0^\infty ds \, e^{-zs} \tilde{W}(s) \tag{5.157}$$

we can rewrite (5.156) as

$$\mathrm{Tr}W(\Delta) = \int_0^\infty ds \, \mathrm{Tr}K(s)\tilde{W}(s) \, , \tag{5.158}$$

where $\mathrm{Tr}K(s) = \sum_i e^{-s\lambda_i}$ is the trace of the heat kernel of Δ. We assume that there are no negative and zero eigenvalues; if present, these will have to be dealt with separately. If we are interested in the local behavior of the theory (*i.e.* the behavior at length scales much smaller than the typical curvature radius) we can use the asymptotic expansion (3.34) and then evaluate each integral separately. Then we get

$$\mathrm{Tr}W(\Delta) = \frac{1}{(4\pi)^{\frac{d}{2}}}\left[Q_{\frac{d}{2}}(W)B_0(\Delta) + Q_{\frac{d}{2}-1}(W)B_2(\Delta) + \ldots\right.$$

$$\left. + Q_0(W)B_d(\Delta) + Q_{-1}(W)B_{d+2}(\Delta) + \ldots\right], \qquad (5.159)$$

where

$$Q_n(W) = \int_0^\infty ds\, s^{-n}\tilde{W}(s) . \qquad (5.160)$$

In the case of four dimensional field theories, it is enough to consider integer values of n. However, in odd dimensions half-integer values of n are needed. Furthermore, one is also sometimes interested in the analytic continuation of results to arbitrary real dimensions. We will therefore need expressions for (5.160) that hold for any real n.

If we denote $W^{(i)}$ the i-th derivative of W, we have from (5.157)

$$W^{(i)}(z) = (-1)^i\int_0^\infty ds\, s^i e^{-zs}\tilde{W}(s) . \qquad (5.161)$$

This formula can be extended to the case when i is a real number to define a notion of "noninteger derivative". From this it follows that for any real i

$$Q_n(W^{(i)}) = (-1)^i Q_{n-i}(W) . \qquad (5.162)$$

For n a positive integer one can use the definition of the Gamma function to rewrite (5.160) as a Mellin transform:

$$Q_n(W) = \frac{1}{\Gamma(n)}\int_0^\infty dz\, z^{n-1}W(z) , \qquad (5.163)$$

while for m a positive integer or $m = 0$

$$Q_{-m}(W) = (-1)^m W^{(m)}(0) . \qquad (5.164)$$

More generally, for n a positive real number we can define $Q_n(W)$ by Eq. (5.163), while for n real and negative we can choose a positive integer k such that $n + k > 0$; then we can write the general formula

$$Q_n(W) = \frac{(-1)^k}{\Gamma(n+k)}\int_0^\infty dz\, z^{n+k-1}W^{(k)}(z) . \qquad (5.165)$$

This reduces to the two cases mentioned above when n is integer. In the case when n is a negative half integer $n = -\frac{2m+1}{2}$ we will set $k = m + 1$ so that we have

$$Q_{-\frac{2m+1}{2}}(W) = \frac{(-1)^{m+1}}{\sqrt{\pi}}\int_0^\infty dz\, z^{-1/2}f^{(m+1)}(z) . \qquad (5.166)$$

The evaluation of the Q-functionals for some specific functions W that will be needed later is left to section 6.9.

Let us make a simple consistency check on formula (5.159). Suppose that W is a function of the operator $\Delta + q\mathbf{1}$, where q is a real number. We can also think of this as a function of the operator Δ, by defining $\bar{W}(z) = W(z + q)$. We can calculate $\mathrm{Tr}(\bar{W}(\Delta))$ and $\mathrm{Tr}(W(\Delta + q\mathbf{1}))$ using (5.159). In the former case one uses the heat kernel coefficients of Δ, in the latter of $\Delta + q\mathbf{1}$. These heat kernel coefficients of Δ are related as follows:

$$\mathrm{Tr}e^{-s(\Delta+q\mathbf{1})} = e^{-sq}\mathrm{Tr}e^{-s\Delta} = \frac{1}{(4\pi)^{d/2}}\sum_{k,\ell=0}^{\infty}\frac{(-1)^{\ell}}{\ell!}\mathrm{tr}\,B_k(\Delta)q^{\ell}s^{k+\ell-2}. \qquad (5.167)$$

The expansion of $\mathrm{Tr}(\bar{W}(\Delta))$ involves $Q_n(\bar{W})$. By Taylor-expanding it in q we get

$$\begin{aligned}
Q_n(\bar{W}) &= \frac{1}{\Gamma(n)}\int_0^{\infty}dz\,z^{n-1}W(z+q)\\
&= \frac{1}{\Gamma(n)}\int_0^{\infty}dz\,z^{n-1}\left(W(z)+qW'(z)+\frac{1}{2!}q^2W''(z)+\frac{1}{3!}q^3W'''(z)\ldots\right)\\
&= Q_n(W)+qQ_n(W')+\frac{1}{2!}q^2Q_n(W'')+\frac{1}{3!}q^3Q_n(W''')+\ldots\\
&= Q_n(W)-qQ_{n-1}(W)+\frac{1}{2!}q^2Q_{n-2}(W)-\frac{1}{3!}q^3Q_{n-3}(W)\ldots \qquad (5.168)
\end{aligned}$$

where in the last step we have used Eq. (5.162). Using (5.159) for the function $\bar{W}$ we then have

$$\begin{aligned}
\mathrm{Tr}\bar{W}[\Delta] &= \frac{1}{(4\pi)^{d/2}}\left[Q_{\frac{d}{2}}(\bar{W})B_0(\Delta)+Q_{\frac{d}{2}-1}(\bar{W})B_2(\Delta)+\ldots+Q_0(\bar{W})B_{2d}(\Delta)+\ldots\right]\\
&= \frac{1}{(4\pi)^{d/2}}\left[\left(Q_{\frac{d}{2}}(W)-qQ_{\frac{d}{2}-1}(W)+\frac{1}{2!}q^2Q_{\frac{d}{2}-2}(W)-\frac{1}{3!}q^3Q_{\frac{d}{2}-3}(W)+\ldots\right)B_0(\Delta)\right.\\
&\quad +\left(Q_{\frac{d}{2}-1}(W)-qQ_{\frac{d}{2}-2}(W)+\frac{1}{2!}q^2Q_{\frac{d}{2}-3}(W)-\frac{1}{3!}q^3Q_{\frac{d}{2}-4}(W)+\ldots\right)B_2(\Delta)\\
&\quad +\ldots\\
&\quad \left.+\left(Q_0(W)-qQ_{-1}(W)+\frac{1}{2!}q^2Q_{-2}(W)-\frac{1}{3!}q^3Q_{-3}(W)+\ldots\right)B_{2d}(\Delta)+\ldots\right].(5.169)
\end{aligned}$$

We can now collect the terms that have the same Q-functionals. They correspond to the anti-diagonal lines in (5.169). Using (5.167), one recognizes that the coefficient of $Q_{\frac{d}{2}-k}$ is $B_{2k}(\Delta+q\mathbf{1})$. Therefore

$$\begin{aligned}
\mathrm{Tr}\bar{W}[\Delta] &= \frac{1}{(4\pi)^{d/2}}\left[Q_{\frac{d}{2}}(\bar{W})B_0(\Delta+q\mathbf{1})+Q_{\frac{d}{2}+1}(\bar{W})B_2(\Delta+q\mathbf{1})\right.\\
&\quad \left.+\ldots+Q_0(\bar{W})B_{2d}(\Delta+q\mathbf{1})+\ldots\right], \qquad (5.170)
\end{aligned}$$

which coincides term by term with the expansion of $\mathrm{Tr}W[\Delta+q]$.

5.8 Appendix: The Euler-Maclaurin formula

If a and b are integers with $b > a$ and f is a smooth function, the Euler-Maclaurin formula says that

$$\sum_{\ell=a}^{b} f(\ell) = \int_a^b dx f(x) + \frac{1}{2}(f(a) + f(b)) + \sum_{k=2}^{n} \frac{B_k}{k!} \left(f^{(k-1)}(x) \right) \Big|_a^b - R \, , \quad (5.171)$$

where B_n are the Bernoulli numbers and R is a remainder for which bounds can be given, and is small under typical circumstances. By neglecting the remainder, the Euler-Maclaurin formula can be used to approximate sums by integrals: the difference between the two is expressed in terms of the values of the function and its derivatives at the extremes of the interval.

The Euler-Maclaurin formula can be obtained by repeated integrations by parts. We sketch the proof here. Let us begin by defining the Bernoulli polynomials recursively by the rules

$$B_0(x) = 1$$
$$B_n'(x) = n \, B_{n-1}(x) \quad\quad (5.172)$$
$$\int_0^1 dx B_n(x) = 0 \, .$$

The first few Bernoulli polynomials are

$$B_0(x) = 1$$
$$B_1(x) = x - \frac{1}{2}$$
$$B_2(x) = x^2 - x + \frac{1}{6}$$
$$B_3(x) = x^3 - \frac{3}{2}x^2 + \frac{1}{2}x$$
$$B_4(x) = x^4 - 2x^3 + x^2 - \frac{1}{30}$$

We see that

$$B_1(1) = -B_1(0) = \frac{1}{2}$$
$$B_2(1) = B_2(0) = \frac{1}{6}$$
$$B_3(1) = B_3(0) = 0$$
$$B_4(1) = B_4(0) = -\frac{1}{30}$$

One can show in general that for $n > 1$,

$$B_n(1) = B_n(0) = B_n \, . \quad\quad (5.173)$$

In particular $B_n(1) = B_n(0) = 0$ for $n > 1$ odd.

Given two functions f and g on the unit interval $[0, 1]$, we have

$$f(x)g^{(n)}(x) = (-1)^n f^{(n)}(x)g(x) - \sum_{k=1}^n (-1)^k \left(f^{(k-1)}(x)g^{(n-k)} \right)' ,$$

thus integrating

$$\int_0^1 dx f(x)g^{(n)}(x) = (-1)^n \int_0^1 dx f^{(n)}(x)g(x) - \sum_{k=1}^n (-1)^k \left(f^{(k-1)}(x)g^{(n-k)}(x) \right) \Big|_0^1 .$$

$$(5.174)$$

Now assume that g is the Bernoulli polynomial B_n. Using the recursive rule (5.172) one finds

$$g^{(n-k)} = n(n-1)\ldots(k+1)B_k(x) . \tag{5.175}$$

Inserting in (5.174) one finds

$$\int_0^1 dx f(x) = \frac{(-1)^n}{n!} \int_0^1 dx f^{(n)}(x)B_n(x) - \sum_{k=1}^n \frac{(-1)^k}{k!} \left(f^{(k-1)}(x)B_k(x) \right) \Big|_0^1 .$$

$$(5.176)$$

Remembering that odd Bernoulli numbers are zero, we can drop the $(-1)^k$ in the sum and we remain with

$$\int_0^1 dx f(x) = \frac{1}{2}(f(1) - f(0)) - \sum_{k=2}^n \frac{B_k}{k!} \left(f^{(k-1)}(x) \right) \Big|_0^1 + R_1 , \tag{5.177}$$

where

$$R_1 = \frac{(-1)^n}{n!} \int_0^1 dx f^{(n)}(x)B_n(x)$$

is a "remainder". Adding $f(1)/2$ to both sides, this can be rewritten as

$$f(1) = \int_0^1 dx f(x) + \frac{1}{2}(f(1) - f(0)) + \sum_{k=2}^n \frac{B_k}{k!} \left(f^{(k-1)}(x) \right) \Big|_0^1 - R_1$$

$$= \int_0^1 dx f(x) + \sum_{k=1}^n \frac{B_k}{k!} \left(f^{(k-1)}(x) \right) \Big|_0^1 - R_1 . \tag{5.178}$$

Assuming that $a = 0$, this is an estimate for the first term in the sum (5.171). In order to write the other terms one can define the periodic Bernoulli functions $P_n(x)$ by $P_n(x) = B_n(x - [x])$, where $[x]$ denotes the largest integer smaller than x. Because of (5.173), all periodic Bernoulli functions are continuous, except for P_1, and for any integer m they satisfy

$$\int_m^{m+1} dx P_n(x) = 0 .$$

Using the same methods as above, one can see that for any integer ℓ,

$$f(\ell) = \int_{\ell-1}^\ell dx f(x) + \frac{1}{2}(f(\ell) - f(\ell - 1)) + \sum_{k=2}^n \frac{B_k}{k!} \left(f^{(k-1)}(x) \right) \Big|_{\ell-1}^\ell - R_\ell ,$$

where

$$R_\ell = \frac{(-1)^n}{n!} \int_{\ell-1}^{\ell} dx f^{(n)}(x) P_n(x) . \tag{5.179}$$

Summing several such terms we obtain

$$\sum_{\ell=a+1}^{b} f(\ell) = \int_{a}^{b} dx f(x) + \frac{1}{2}(f(b) - f(a)) + \sum_{k=2}^{n} \frac{B_k}{k!} \left(f^{(k-1)}(x) \right)\Big|_{a}^{b} - R , \tag{5.180}$$

where

$$R = \frac{(-1)^n}{n!} \int_{a}^{b} dx f^{(n)}(x) P_n(x) . \tag{5.181}$$

Finally adding $f(a)$ to both sides, one arrives at the form (5.171).

Chapter 6

The functional renormalization group equation

In chapter 3 we introduced the (Euclidean) Schwinger-de Witt method to calculate one-loop divergences in QFT. It depends on a genuine mass cutoff Λ,[1] but is nevertheless well suited to discuss gauge theories, because one can retain background gauge invariance. One of the main results of this type of calculation is the beta function, which, for dimensionless couplings, coincides with the coefficient of the logarithmic divergence. For example, in a scalar theory the beta function of the quartic coupling is the coefficient of $(\log \Lambda)\phi^4$ in Γ_Λ, and in Yang-Mills theory the beta function can be read from the coefficient of $(\log \Lambda)F^a_{\mu\nu}F^{a\mu\nu}$.

Given that divergences are unphysical features, one wonders whether it would be possible to obtain the same beta functions in a more direct way. The answer is positive: instead of computing Γ_Λ one could compute $\Lambda \frac{d\Gamma_\Lambda}{d\Lambda}$. This functional is finite and the beta functions are just the coefficients of ϕ^4 and $F^a_{\mu\nu}F^{a\mu\nu}$ in $\Lambda \frac{d\Gamma_\Lambda}{d\Lambda}$.

The derivative $\Lambda \frac{d\Gamma_\Lambda}{d\Lambda}$ can be called the "beta functional" of the theory and we shall see that there is a remarkably simple, exact formula for it, from which the beta function of any coupling can be read off by a suitable "projection" technique. This is in essence the idea of the functional renormalization group. The next three sections contain general definitions and a derivation of the exact renormalization group equation. As in chapter 3, before applying this formalism to gravity we shall discuss scalar and Yang-Mills theories and the effect of free matter fields in an external metric. We then write the ERGE for gravity and use it to calculate the beta functions of Newton's constant and of the cosmological constant.

6.1 The Effective Average Action (EAA)

We start from the Euclidean partition function of a real scalar field ϕ in flat space:

$$Z(j) = \int (d\phi) \exp\left(-S(\phi) + \int dx\, j\phi\right). \tag{6.1}$$

Unlike the discussion in section 3.1, here S could contain also arbitrary interaction terms. We add to the bare action S an infrared "cutoff" or "regulator" term $\Delta S_k(\phi)$

---

[1] Not to be confused with the cosmological constant.

of the form:

$$\Delta S_k(\phi) = \frac{1}{2} \int d^d x \, \phi \, R_k(\Delta) \, \phi \tag{6.2}$$

We call $Z_k(j)$ the partition function computed with the action $S + \Delta S_k$. In (6.2) the kernel $R_k(\Delta)$ must be chosen so as to suppress the contribution to the functional integral of the field modes a_n corresponding to eigenvalues λ_n smaller than the cutoff scale k^2. We will call k the cutoff, $\Delta S_k(\phi)$ the cutoff action and Δ the cutoff operator. We will also often write z for the argument of R_k when we think of it as an ordinary function rather than as a kernel. One can think of z as an eigenvalue of Δ; in flat space, it is just the momentum squared. The functional form of $R_k(z)$ is quite arbitrary except for a few basic requirements:

- for fixed k it is a monotonically decreasing function of z;
- for fixed z it is a monotonically increasing function of k;
- $\lim_{k \to 0} R_k(z) = 0$ for all z;
- for $z > k^2$, R_k goes to zero sufficiently fast, typically as an exponential;
- $R_k(0) = k^2$.

The first two conditions are obvious properties of a cutoff. The third guarantees that $Z_k \to Z$ for $k \to 0$. The fourth condition ensures that high momentum modes are integrated out unsuppressed. The fifth and last condition provides a sort of normalization. For certain limited purposes, one may sometimes forgo the last two conditions and consider cutoffs that either do not decrease very fast for large momenta or diverge when $z \to 0$.

Both z and the function $R_k(z)$ have dimension of mass squared, so we can write

$$R_k(z) = k^2 r(y) \; ; \qquad y = z/k^2 \; , \tag{6.3}$$

where r is a dimensionless "cutoff profile". The following are typical choices:

$$r(y) = \frac{y}{e^y - 1} \tag{6.4}$$

$$r(y) = \frac{y^2}{e^{y^2} - 1} \tag{6.5}$$

$$r(y) = (1 - y)\theta(1 - y) \, . \tag{6.6}$$

These functions are plotted in Figure 6.1. The third choice has been argued to provide "optimized" results, in a suitable sense [187]. For certain purposes its non-differentiability is an issue, but it has the great advantage of allowing an analytic evaluation of momentum integrals.

We emphasize that the addition of the cutoff function has not modified the vertices. Its only effect is to replace the original inverse propagator z by the cutoff inverse propagator

$$P_k(z) = z + R_k(z) \tag{6.7}$$

The original "bare" propagator $G(z) = 1/z$ is therefore replaced by

$$G_k(z) = \frac{1}{P_k(z)} \ . \tag{6.8}$$

This function is also shown in Figure 6.1, for the three different cutoff shapes in (6.4), (6.5), (6.6). For modes with eigenvalues $z \gg k^2$ the propagation is unaffected, while below the cutoff scale their propagation is increasingly suppressed as if they were massive particles of mass $\sim k$.

Note that k plays the role of an *infrared* cutoff: its effect is to give a mass of order k to the modes with $\sqrt{|z|} < k$, and no mass to the modes with $\sqrt{|z|} > k$. However, curing IR divergences is not its primary purpose: rather, it is a way of introducing explicitly a k dependence in the functional integral.

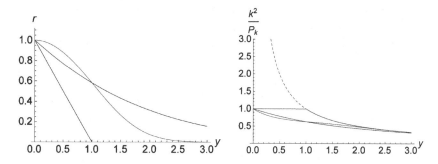

Fig. 6.1 Left: the cutoff profile functions (6.5), (6.4) and (6.6), from top to bottom near $y = 0$. Right: The massless propagator in units of k^2, $1/y$, (dashed) together with the modified propagators obtained using the cutoffs on the left.

We now have a scale-dependent partition function generalizing (6.1):

$$Z_k(j) = \int (d\phi) \exp\left(-S(\phi) - \Delta S_k(\phi) + \int dx\, j\phi\right) \ . \tag{6.9}$$

Next we define the scale-dependent generating functional of connected correlation functions as:

$$W_k(j) = \log Z_k(j) \ . \tag{6.10}$$

We then define the Legendre transform of W_k by

$$\tilde{\Gamma}_k(\varphi) = -W_k(j_\varphi) + \int dx\, j_\varphi \varphi \ , \tag{6.11}$$

where j_φ is obtained inverting the function φ_j, defined from W_k as in (3.15). Finally we define the Effective Average Action (EAA) Γ_k by subtracting from $\tilde{\Gamma}_k$ the cutoff that was introduced in the beginning [188]:

$$\Gamma_k(\varphi) = \tilde{\Gamma}_k(\varphi) - \Delta S_k(\varphi) \ . \tag{6.12}$$

We notice that when $k \to 0$, $\Delta S_k \to 0$ and therefore Γ_k reduces to the ordinary effective action Γ, where all fluctuations have been integrated out unsuppressed.

The virtue of this definition is that the EAA satisfies a very simple functional equation, as we shall see next.

6.2 The Wetterich equation

The rationale behind the definition (6.12) becomes clearer if we evaluate the EAA at one loop. Recall that the ordinary one-loop effective action is given by

$$\Gamma^{(1)} = S + \frac{1}{2}\mathrm{Tr}\log\frac{\delta^2 S}{\delta\phi\delta\phi} \ . \tag{6.13}$$

Since the only difference between Γ and Γ_k is the addition of ΔS_k to the action and the subtraction of ΔS_k from the Legendre transform, the one-loop EAA is

$$\Gamma_k^{(1)} = S + \Delta S_k + \frac{1}{2}\mathrm{Tr}\log\frac{\delta^2(S + \Delta S_k)}{\delta\phi\delta\phi} - \Delta S_k = S + \frac{1}{2}\mathrm{Tr}\log\left(\frac{\delta^2 S}{\delta\phi\delta\phi} + R_k\right) \ . \tag{6.14}$$

Note that the term ΔS_k has canceled out in the r.h.s, so that the only modification has been the replacement of the bare inverse propagator by the cutoff inverse propagator. This motivates the definition (6.12).

If we now take the derivative with respect to $t = \log k$ we obtain the following equation:

$$\frac{d\Gamma_k^{(1)}}{dt} = \frac{1}{2}\mathrm{Tr}\left(\frac{\delta^2 S}{\delta\phi\delta\phi} + R_k\right)^{-1}\frac{dR_k}{dt} \ . \tag{6.15}$$

This r.h.s. of this equation may be called the one-loop "beta functional" of the theory.

In the r.h.s. of (6.15) there appears the bare inverse propagator $\frac{\delta^2 S}{\delta\phi\delta\phi}$. One may guess that the "renormalization group improvement" of this equation, namely the equation obtained by replacing S by Γ_k in the r.h.s., gives a more accurate description of physics. We shall now show that this improved equation is actually exact.

Let us begin by deriving the functional W_k:

$$\frac{dW_k}{dt} = -\frac{d}{dt}\langle\Delta S_k\rangle = -\frac{1}{2}\mathrm{Tr}\langle\phi\phi\rangle\frac{dR_k}{dt} \ ,$$

where the trace is an integration over coordinate and momentum space. Then using (6.12) and (3.16) we have

$$\frac{d\Gamma_k[\varphi]}{dt} = -\frac{dW_k}{dt} - \frac{d\Delta S_k[\varphi]}{dt}$$

$$= \frac{1}{2}\mathrm{Tr}(\langle\phi\phi\rangle - \langle\phi\rangle\langle\phi\rangle)\frac{dR_k}{dt}$$

$$= \frac{1}{2}\mathrm{Tr}\frac{\delta^2 W_k}{\delta j\delta j}\frac{dR_k}{dt}$$

From (6.11), the first functional derivative of $\tilde{\Gamma}_k$ gives

$$\frac{\delta\tilde{\Gamma}_k}{\delta\varphi} = j \ ,$$

so comparing

$$\frac{\delta^2\tilde{\Gamma}_k}{\delta\varphi\delta\varphi} = \frac{\delta j}{\delta\varphi} \ ; \quad \frac{\delta^2 W_k}{\delta j\delta j} = \frac{\delta\varphi}{\delta j}$$

one finds that

$$\frac{\delta^2 W_k}{\delta j \delta j} = \left(\frac{\delta^2 \tilde{\Gamma}_k}{\delta \varphi \delta \varphi} \right)^{-1} .$$

Using this identity and reexpressing $\tilde{\Gamma}_k$ in terms of Γ_k by means of (6.12), we arrive at the equation [189–191]:

$$\frac{d\Gamma_k}{dt} = \frac{1}{2} \mathrm{Tr} \left(\frac{\delta^2 \Gamma_k}{\delta \varphi \delta \varphi} + R_k \right)^{-1} \frac{dR_k}{dt} . \tag{6.16}$$

This is variously referred to as the Wetterich equation, or the Exact Renormalization Group Equation (ERGE), or the Functional Renormalization Group Equation (FRGE), or the $1PI$ flow equation.

At this point, several comments are in order. First, as mentioned above, the ERGE is identical to the one-loop equation except for the replacement of the bare action by Γ_k in the r.h.s.. That the Wetterich equation has the structure of a one-loop equation can be made manifest by the graphical representation in Figure 6.2. This should not be understood in the usual sense of Feynman diagrams, though,

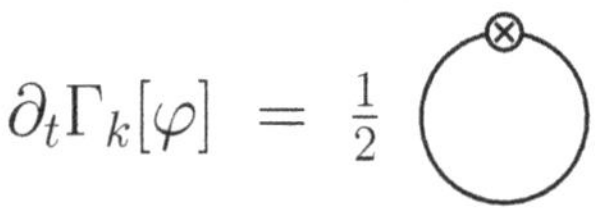

$$\partial_t \Gamma_k[\varphi] = \frac{1}{2} \; \bigotimes$$

Fig. 6.2 Graphical representation of the ERGE (6.16). The continuous line represents the complete propagator while the crossed circle represents the insertion of $\partial_t R_k$.

because the continuous line represents the exact propagator of the theory

$$\left(\frac{\delta^2 \Gamma_k}{\delta \phi \delta \phi} + R_k \right)^{-1} ,$$

and not the perturbative one.

Second, note that the equation does not contain any reference to a bare action anymore. This means that the equation contains no reference to UV physics: the derivative of Γ_k at the the scale k only depends on the EAA at the scale k. One could say that the equation is "local in momentum space" (with the degree of locality depending on how sharply the cutoff function R_k decreases).

Third, and most important, even though the cutoff that has been introduced in the definition of the functional integral was an IR cutoff, the trace in the r.h.s. of (6.16) is UV-finite. This is because the beta functional can be viewed as the difference of the EAA's with infinitesimally different IR cutoffs. In the difference, the UV divergences cancel exactly and we remain with a finite rest that depends only on the behavior of the degrees of freedom with momenta close to k. At a more technical level, due to the requirement that the cutoff decreases fast for $z > k^2$, the expression $k\frac{dR_k}{dk}$ itself decreases fast for $z > k^2$, and acts as a UV cutoff in the trace on the r.h.s. of (6.16).

Finally, we mention that the ERGE is closely related to Wilson's RG [209]. The basic object in Wilson's discussion is the partition function

$$Z = \int (d\phi)e^{-S_\Lambda(\phi)} \ ,$$

where Λ is a UV cutoff and S_Λ an action holding at the scale Λ. Wilson's RG describes the way in which the action S_Λ must change when Λ is lowered infinitesimally, in such a way that Z remains the same. This amounts to performing the functional integral "one momentum shell at the time", rather than all at once. At each step in this process, one uses in the integral the Wilsonian action at the scale Λ, rather than the bare action (which corresponds to the Wilsonian action at the initial scale $\Lambda = \Lambda_{UV}$). The definition of the EAA is an implementation of the same idea for the $1PI$-generating functional.

In Wilson's approach, a renormalization group transformation consists of one such infinitesimal functional integration, followed by two further transformations. The first is a rescaling of physical momenta by a factor $1 - \delta\Lambda/\Lambda$ and physical lengths by a factor $1 + \delta\Lambda/\Lambda$. This transformation brings the cutoff back to the original value. Second, one assumes that in the initial action S the kinetic term is canonically normalized. Integration over a momentum shell will generally change the coefficient of the kinetic term. At each step one rescales the field by a suitable factor so that the kinetic term remains canonically normalized. This changes the dimension of the field by an amount known as "anomalous dimension". These additional transformations will be discussed in the next section.

6.3   Beta functions

The functional W_k, and consequently also the EAA Γ_k, are ill-defined. In general, to define them one would need to introduce a UV cutoff. The standard perturbative approach to QFT is to make sense of the functional integral by means of regularization and renormalization. This can be done in a meaningful way for renormalizable theories and more generally for effective field theories. The ERGE provides an alternative way of studying a QFT.

By means of a series of formal manipulations we have shown that the derivative of Γ_k with respect to the cutoff must satisfy a simple first order differential equation, whose r.h.s. is a well-defined functional trace, free of UV and IR divergences. We can therefore forget the formal derivation and take this equation as the primary tool to calculate the effective action and thereby to define the QFT. We start at some initial scale Λ_{UV} with a specific form of the EAA. We integrate the flow equation with this initial condition from $k = \Lambda_{UV}$ down to $k = 0$. The EAA at $k = 0$ is the standard EA, out of which all the information on the theory can be extracted. We can thus view the ERGE as a way of calculating the effective action by solving a well-defined first-order differential equation, rather than performing an ill-defined functional integral.

Of course the divergences of the theory are still present and can be found by investigating the limit $\Lambda_{UV} \to \infty$. If we want to know what initial condition would yield the same effective action starting from a scale $\Lambda'_{UV} > \Lambda_{UV}$, we have to integrate the flow equation from Λ_{UV} to Λ'_{UV}. So, the UV behavior of the theory can be explored by solving the ERGE in the direction of increasing k. Several behaviors are possible in principle and will be discussed in section 7.1. In the rest of this chapter we will develop the necessary techniques by revisiting the examples given earlier in chapter 3.

To this end we begin by defining the "theory space" to be the space of all functionals of φ. We will not attempt to give a precise definition of this functional space. For many practical purposes the following is adequate. We call "local" a functional containing at most finitely many derivatives and "semi-local" a functional containing infinitely many positive powers of derivatives. For example $\varphi \frac{1}{\Box + m^2} \varphi$ is semi-local, since the fraction can be expanded as a geometric series. An example of a "non-local" functional is $\varphi \frac{1}{\Box} \varphi$. A typical way of approximating the Wetterich equation is the expansion in derivatives. In this case the appropriate functional space would be the space of semi-local functionals of the form

$$\Gamma_k(\varphi) = \sum_i g_i(k) \mathcal{O}_i(\varphi) \tag{6.17}$$

where $\mathcal{O}_i[\varphi]$ are integrals of monomials in the field and its derivatives, and $g_i(k)$ are running couplings. Taking the derivative of (6.17) we have

$$\frac{d\Gamma_k}{dt} = \sum_i \beta_i \mathcal{O}_i \ , \tag{6.18}$$

where

$$\frac{dg_i}{dt} = \beta_i(g_j, k) \tag{6.19}$$

are the beta functions. Note that in general they can be functions of all the couplings as well as the cutoff k.

One can compare Eqs. (6.18) and (6.16). If we are able to expand the trace in the r.h.s. of (6.16) on the basis of operators $\mathcal{O}_i$, then the coefficient of the operator $\mathcal{O}_i$ is the beta function of g_i. In this way, with some work, one can extract explicit beta functions from the abstract form of the Wetterich equation. We shall see several examples of this procedure in the sequel.

Now we have to address the issues of the rescalings of momenta and fields, which was mentioned in the end of the preceding section. The first of these, the rescaling of physical lengths and momenta (and more generally of all dimensionful quantities, each with a power of $\delta k/k$ equal to its mass dimension) is solved very simply by working with dimensionless variables. If the coupling g_i has mass dimension d_i, we define the dimensionless parameter

$$\tilde{g}_i = k^{-d_i} g_i \ . \tag{6.20}$$

It is just the coupling g_i measured in units of the cutoff k.[2] From now on we are going to take the $\tilde{g}_i$ as coordinates of theory space. These variables avoid the need to perform explicitly a rescaling of physical lengths and momenta, because they would not be affected by such a rescaling. However, their beta functions are different from the ones of the dimensionful couplings g_i. From dimensional analysis we can write $\beta_i(g_j, k) = k^{d_i} \alpha_i(\tilde{g}_j)$ where $\alpha_i(\tilde{g}_j) = \beta_i(g_j k^{-d_j}, 1)$. The beta functions of the dimensionless variables are given by

$$\tilde{\beta}_i(\tilde{g}_j) = \frac{d\tilde{g}_i}{dt} = -d_i \tilde{g}_i + \alpha_i(\tilde{g}_j) \ , \tag{6.21}$$

where the first term contains the classical scaling and the second contains the contribution of quantum fluctuations. The RG flow equation in theory space is then

$$\frac{d\tilde{g}_i}{dt} = \tilde{\beta}_i(\tilde{g}_j(t)) \ . \tag{6.22}$$

Note that α_i can only depend on k (or t) implicitly via its arguments, so that (6.22) is an autonomous system of equations. This is an additional advantage of working with dimensionless variables.

There is then the issue of the normalization of the field. Usually a QFT contains a kinetic term of the form $\frac{1}{2} Z_\varphi (\partial \varphi)^2$ and by a simple rescaling of the field (and appropriate redefinitions of all the other couplings) one can set $Z_\varphi = 1$. In the original Wilsonian definition of the RG, one rescales the fields at each step in order to maintain this normalization condition. This makes the fields k-dependent. An alternative is to assume the fields as k-independent and treat $Z_\varphi(k)$ like any one of the couplings g_i. From the ERGE one can then read the beta function of Z_φ, which is usually reexpressed as an anomalous dimension

$$\eta_\varphi = -\frac{d \log Z_\varphi}{d \log k} \ . \tag{6.23}$$

One could introduce powers of $\sqrt{Z_\varphi}$ in the definition of the $\tilde{g}_i$ in such a way that the canonical normalization is automatically maintained. This would produce additional terms proportional to η_φ in (6.21). The two procedures are equivalent. We will work with scale-independent fields.

The normalization of the kinetic term is just one aspect of a much more general issue. One says that a parameter in the action is "redundant" or "inessential" if it can be eliminated from the action by a field redefinition. The wave function renormalization constant is just the simplest example. As we shall see in section 7.1, it is important to keep track of the redundant couplings, because their behavior at fixed points is not restricted as the other couplings.

---

[2]The need to work with dimensionless variables becomes very clear as soon as one tries to solve a QFT problem numerically with a computer. Our use of the cutoff as the unit of mass is parallel to the use of the lattice spacing as the universal unit of length in lattice QFT.

6.4 Scalar potential interactions

Our main task in this section is to rederive the beta function (3.51) from the ERGE. This may sound like using a sledgehammer to break a nut, but we shall see that (i) the calculation is no harder than the one in section 3.2.2 and it has the advantage that one always deals only with mathematically well-defined quantities; (ii) with very little additional effort one can obtain an infinite amount of additional information, namely the beta function for the whole potential.

The basic trick is to "truncate" the general functional (6.17) to a manageable subset of couplings. This may mean a single coupling, finitely many couplings or sometimes infinite classes of couplings. Here we are in the latter situation, and we can write a flow equation for all the couplings of the potential. Thus, we truncate the EAA to the form

$$\Gamma_k(\varphi) = \int d^d x \left[\frac{1}{2} (\partial_\mu \varphi)^2 + V_k(\varphi^2) \right] . \tag{6.24}$$

Note that this is a truncation in the sense that the beta functions of the couplings that we are neglecting are certainly not zero. It can be better justified when it is viewed as the first term in an expansion in derivatives. It is then called the Local Potential Approximation (LPA) (at the next order of the derivative expansion one would also consider the flow of $Z_k(\varphi)(\partial\varphi)^2$).

We will now insert the ansatz (6.24) in the ERGE and extract the beta functions of the potential. Since we are in flat spacetime, the calculations are made easier by the availability of the Fourier transform. In momentum space the cutoff action (6.2) can be written

$$\Delta S_k(\varphi) = \frac{1}{2} \int \frac{d^d q}{(2\pi)^d} \varphi(-q) R_k(q^2) \varphi(q) . \tag{6.25}$$

and the inverse propagator is

$$\frac{\delta^2 \Gamma_k}{\delta\varphi \delta\varphi} = q^2 + 2V_k' + 4\varphi^2 V_k'' , \tag{6.26}$$

where a prime denotes the derivative with respect to φ^2. The modified inverse propagator is thus

$$\frac{\delta^2 \Gamma_k}{\delta\varphi \delta\varphi} + R_k = P_k(q^2) + 2V_k' + 4\varphi^2 V_k'' , \tag{6.27}$$

where we recall that $P_k(z) = z + R_k(z)$ (for any argument). With these definitions the ERGE can be written

$$\partial_t \Gamma_k = \frac{1}{2} \text{Tr} \left(\frac{\partial_t P_k}{P_k + 2V_k' + 4\varphi^2 V_k''} \right) . \tag{6.28}$$

(We have used $\partial_t R_k = \partial_t P_k$.) The trace involves an integration over spacetime and over momenta. For any function W we have

$$\text{Tr}(W(-\partial^2)) = \int d^d x \int \frac{d^d q}{(2\pi)^d} W(q^2) = \frac{1}{(4\pi)^{d/2}} \int d^d x \, Q_{\frac{d}{2}}(W) . \tag{6.29}$$

In the last step we used (5.135) to perform the angular integration, which gives

$$\frac{1}{(2\pi)^d}\text{Vol}(S^{d-1}) = \frac{1}{(4\pi)^{d/2}}\frac{2}{\Gamma(d/2)} ,$$

and we defined the Q-functionals

$$Q_n[W] = \frac{1}{\Gamma(n)}\int_0^\infty dz\, z^{n-1}W(z) , \tag{6.30}$$

where W is viewed as a function of $z = q^2$. The Q-functionals contain the integration over the modulus of the momentum. If we now restrict ourselves to constant scalar fields, we can remove a volume factor from both sides of the ERGE and we obtain the k-dependence of the potential as:

$$\partial_t V_k = \frac{1}{2}\frac{1}{(4\pi)^{d/2}}Q_{\frac{d}{2}}\left(\frac{\partial_t P_k}{P_k + 2V_k' + 4\varphi^2 V_k''}\right) . \tag{6.31}$$

This is a flow equation for the potential. It is still a bit implicit, because we have not yet performed the integration that is contained in the definition of the Q-functional. Nevertheless, in order to make a connection with familiar formulae let us consider what happens when the potential is Taylor expanded

$$V(\varphi^2) = \sum_{n=0}^N \frac{\lambda_{2n}}{(2n)!}\varphi^{2n} . \tag{6.32}$$

The coupling constants can be extracted from the potential by

$$\lambda_{2n} = \frac{(2n)!}{n!}\frac{\partial^n V}{\partial(\varphi^2)^n}\bigg|_{\varphi=0} , \tag{6.33}$$

and the beta functions can be extracted from the "beta functional" (6.31) by

$$\beta_{2n} = \partial_t\lambda_{2n} = \frac{(2n)!}{n!}\frac{\partial^n \partial_t V_k}{\partial(\varphi^2)^n}\bigg|_{\varphi=0} . \tag{6.34}$$

Explicitly, the first few beta functions are given by

$$\beta_0 = \frac{1}{2(4\pi)^{d/2}}Q_{\frac{d}{2}}\left(\frac{\partial_t P_k}{P_k + \lambda_2}\right) , \tag{6.35}$$

$$\beta_2 = \frac{1}{2(4\pi)^{d/2}}\left[-\lambda_4 Q_{\frac{d}{2}}\left(\frac{\partial_t P_k}{(P_k + \lambda_2)^2}\right)\right] , \tag{6.36}$$

$$\beta_4 = \frac{1}{2(4\pi)^{d/2}}\left[-\lambda_6 Q_{\frac{d}{2}}\left(\frac{\partial_t P_k}{(P_k + \lambda_2)^2}\right) + 6\lambda_4^2 Q_{\frac{d}{2}}\left(\frac{\partial_t P_k}{(P_k + \lambda_2)^3}\right)\right] , \tag{6.37}$$

$$\beta_6 = \frac{1}{2(4\pi)^{d/2}}\left[-\lambda_8 Q_{\frac{d}{2}}\left(\frac{\partial_t P_k}{(P_k + \lambda_2)^2}\right) + 30\lambda_4\lambda_6 Q_{\frac{d}{2}}\left(\frac{\partial_t P_k}{(P_k + \lambda_2)^3}\right)\right.$$
$$\left. -90\lambda_4^3 Q_{\frac{d}{2}}\left(\frac{\partial_t P_k}{(P_k + \lambda_2)^4}\right)\right] . \tag{6.38}$$

Some comments are in order at this point. First we observe that each term in the r.h.s. can be represented as a diagram with $2n$ external legs, one loop and an insertion of the cutoff on one of the internal lines. This is in line with the one-loop structure of the Wetterich equation. However, here the internal lines are the full propagators of the theory and the vertices are the full vertices.

Second, we note that strictly speaking it is not consistent to truncate the potential to a finite polynomial, except in the quadratic case. Indeed, if all the $\lambda_{2n} = 0$ for $n > 1$, the theory is free and all the beta functions vanish identically. However, as soon as one of the couplings is nonzero, then all the other ones will be turned on by the flow. Thus, in truncating the potential to a polynomial one ignores terms in the action that have no *a priori* reason to be small. Nevertheless, it is sometimes useful to consider truncations of this type and one can see *a posteriori* that they yield valuable information.

Third, the value of the Q-functionals depends in general on the shape of the cutoff function R_k. There are however some Q-functionals which are independent of it and hence carry special significance. We shall now discuss these cases.

As explained in the preceding section, the flow in theory space has to be described in terms of dimensionless coordinates. Thus, we define the dimensionless couplings

$$\tilde{\lambda}_{2n} = k^{(d-2)n-d} \lambda_{2n} . \tag{6.39}$$

Their beta functions are

$$\partial_t \tilde{\lambda}_{2n} = ((d-2)n - d)\tilde{\lambda}_{2n} + k^{(d-2)n-d}\beta_{2n} . \tag{6.40}$$

At the same time, the cutoff can be written as in (6.3). In terms of the dimensionless squared momentum $y = z/k^2$, the Q-functionals can be written

$$Q_n\left(\frac{\partial_t R_k}{(P_k + A)^\ell}\right) = \frac{2}{\Gamma(n)} k^{2(n-\ell+1)} \int_0^\infty dy \, y^{n-1} \frac{r(y) - yr'(y)}{(y + r(y) + \tilde{A})^\ell} \tag{6.41}$$

where $\tilde{A} = A/k^2$. Now we observe that if $\tilde{A} = 0$, and $\ell = n+1$, the integrand is a total derivative:

$$y^{n-1}\frac{r(y) - yr'(y)}{(y + r(y))^{n+1}} = \frac{1}{n}\frac{d}{dy}\left(\frac{y}{y+r(y)}\right)^n . \tag{6.42}$$

The general conditions on the cutoff R_k discussed in section 6.1 imply that $\lim_{y\to\infty} r(y) = 0$ and $r(0) = 1$. Therefore

$$Q_n\left(\frac{\partial_t P}{P^{n+1}}\right) = \frac{2}{\Gamma(n+1)} , \tag{6.43}$$

independent of the shape of the cutoff. These are the only "universal" Q-functionals. Notice that they are also the only dimensionless ones. All the dimensionful ones depend on the shape of the cutoff.

We can evaluate them explicitly if we choose the so-called optimized cutoff

$$R_k(z) = (k^2 - z)\theta(k^2 - z) ; \qquad r(y) = (1 - y)\theta(1 - y) . \tag{6.44}$$

It corresponds to the blue curve in figure 6.1. With this cutoff $\partial_t R_k = 2k^2\theta(k^2 - z)$. Since the integrals are all cut off at $z = k^2$ by the theta function in the numerator, we can simply use $P_k(z) = k^2$ in the denominators. We find

$$Q_n\left(\frac{\partial_t P}{(P+A)^\ell}\right) = \frac{2}{n!}\frac{1}{(1+\tilde{A})^\ell}k^{2(n-\ell+1)} \, . \tag{6.45}$$

Then the beta functions of the dimensionless couplings $\tilde{\lambda}_{2n}$ are

$$\tilde{\beta}_0 = -d\tilde{\lambda}_0 + c_d\frac{1}{1+\tilde{\lambda}_2} \, ,$$

$$\tilde{\beta}_2 = -2\tilde{\lambda}_2 - c_d\frac{\tilde{\lambda}_4}{(1+\tilde{\lambda}_2)^2} \, ,$$

$$\tilde{\beta}_4 = (d-4)\tilde{\lambda}_4 + c_d\left[-\frac{\tilde{\lambda}_6}{(1+\tilde{\lambda}_2)^2} + 6\frac{\tilde{\lambda}_4^2}{(1+\tilde{\lambda}_2)^3}\right] \, , \tag{6.46}$$

$$\tilde{\beta}_6 = (2d-6)\tilde{\lambda}_6 + c_d\left[-\frac{\tilde{\lambda}_8}{(1+\tilde{\lambda}_2)^2} + 30\frac{\tilde{\lambda}_4\tilde{\lambda}_6}{(1+\tilde{\lambda}_2)^3} - 90\frac{\tilde{\lambda}_4^3}{(1+\tilde{\lambda}_2)^4}\right] \, ,$$

where

$$c_d = \frac{1}{(4\pi)^{d/2}}\frac{1}{\Gamma\left(\frac{d}{2}+1\right)} \, . \tag{6.47}$$

We note once again that if $\tilde{\lambda}_{2n} = 0$ for $n > 1$, all the beta functions vanish except for $\tilde{\beta}_2 = -2\tilde{\lambda}_2$, which implies the classical scaling of the squared mass $\tilde{\lambda}_2 = \lambda_2/k^2$ with constant λ_2. If one of the $\tilde{\lambda}_{2n}$ with $n > 2$ is nonzero, then all the others will be turned on as soon as one starts integrating the flow equations.

We note the ubiquitous appearance of the denominators $1 + \tilde{\lambda}_2$. These mark the occurrence of a threshold at $k^2 = \lambda_2$. In fact, for $k^2 \gg \lambda_2$ the denominators become equal to one, whereas for $k^2 \ll \lambda_2$ the denominators become large and the beta functions go to zero. This is a manifestation of the phenomenon of decoupling: at energies lower than the mass of a certain particle species, the loops containing that particle are suppressed.

Finally, we observe that the only universal beta function is the one of φ^4 theory in $d = 4$, in the limit where $\tilde{\lambda}_2 \to 0$:

$$\beta_4 = \frac{6\lambda_4^2}{32\pi^2}Q_2\left(\frac{\partial_t P_k}{P_k^3}\right) = \frac{3\lambda_4^2}{16\pi^2} \, . \tag{6.48}$$

This is the familiar one-loop result.

The true power of functional renormalization consists in its ability to deal with infinitely many couplings at once. In the scalar case, this can be brought to good use in the treatment of the Wilson-Fisher fixed point, which is a fixed point in three dimensions [192]. A discussion of this point would take us too far afield. The interested reader is referred to one of the reviews of the subject [193–196]. In the next section, we shall turn instead to the formulation of the ERGE for a gauge theory.

6.5 Flow equation for Yang-Mills theory

The main aim of this section is to rederive the one-loop beta function of the gauge coupling from a flow equation. We will first define the EAA and its exact functional RG equation and then evaluate it in a suitable approximation. In the course of this section we will encounter new issues that considerably complicate the evaluation of the flow. We generally follow the approach of [197–199]. A somewhat different approach to the RG flow of gauge theories has been described in [200].

6.5.1 *The background ERGE*

We use the background field method, introduced in section 3.2.3. At this stage we do not specify the form of the action, nor of the gauge fixing and ghost terms, but we assume that the action is a functional of the full field $S(\bar{A} + a)$ and that the gauge fixing and ghost terms $S_{GF}(a; \bar{A})$ and $S_{gh}(a, \bar{c}, c; \bar{A})$ break the quantum gauge invariance (3.55) but are invariant under the background gauge transformations (3.56).

To define the EAA for YM theory we have to add to the action a cutoff term. In order to maintain background gauge invariance, the cutoff cannot be simply a function of q^2 but must be defined by means of a covariant differential operator. As we have seen in the derivation of the ERGE, it is important to choose a cutoff action that is purely quadratic in the quantum field. In order to achieve this in a gauge theory this covariant differential operator has to be constructed with the background covariant derivative $\bar{D}_\mu$. There is more than one natural choice. For simplicity let us start by assuming that the cutoffs are functions of the simple covariant Laplacian $-\bar{D}^2$. The cutoff action will have the form

$$\Delta S_k(a, \bar{c}, c; \bar{A}) = \frac{1}{2} \int d^d x \, a_\mu g^{\mu\nu} \mathcal{R}_k(-\bar{D}^2) a_\nu + \int d^d x \, \bar{c} \, \mathcal{R}_k^{(gh)}(-\bar{D}^2) c \ . \tag{6.49}$$

Gauge indices are suppressed for notational simplicity, but to keep track of this, a cutoff kernel that is a matrix in some space (in this case the Lie algebra of the gauge group) is designated $\mathcal{R}_k$. The symbol R_k is reserved for real functions, such as occurred in the case of the scalar field. The cutoff partition function is

$$Z_k(j_a, \bar{J}, J; \bar{A}) = e^{W_k(j_a, \bar{J}, J; \bar{A})}$$

$$= \int (da \, d\bar{c} \, dc) \, e^{-S(a; \bar{A}) - S_{GF}(a; \bar{A}) - \Delta S_k(a, \bar{c}, c; \bar{A}) + \int d^d x (j_a a + \bar{J}\bar{c} + Jc)} \ . \tag{6.50}$$

By performing the Legendre transformation and subtracting the cutoff one arrives at the Yang-Mills EAA $\Gamma_k(a, \bar{c}, c; \bar{A})$, where for notational simplicity we have denoted the classical fields in the argument of the EAA by the same symbol as the corresponding quantum fields.

By following the same steps as in the scalar case, one shows that this functional obeys the ERGE

$$\frac{d\Gamma_k(\varphi; \bar{A})}{dt} = \frac{1}{2} \text{Tr} \left(\frac{\delta^2(\Gamma_k + \Delta S_k)}{\delta\varphi\delta\varphi} \right)^{-1} \frac{d}{dt} \frac{\delta^2 \Delta S_k}{\delta\varphi\delta\varphi} \ , \tag{6.51}$$

where we have denoted collectively the quantum fields $\varphi = (a_\mu, \bar{c}, c)$. Since the cutoff term does not contain mixed a-c and a-$\bar{c}$ terms, the trace can be written as the sum of two terms

$$\frac{d\Gamma_k(\varphi; \bar{A})}{dt} = \frac{1}{2} \mathrm{Tr} \left(\frac{\delta^2(\Gamma_k + \Delta S_k)}{\delta\varphi\delta\varphi} \right)^{-1}_{aa} \frac{d\mathcal{R}_k}{dt} - \mathrm{Tr} \left(\frac{\delta^2(\Gamma_k + \Delta S_k)}{\delta\varphi\delta\varphi} \right)^{-1}_{\bar{c}c} \frac{d\mathcal{R}_k^{(gh)}}{dt} .$$

(6.52)

Note that the ghost action is bilinear in the ghost fields but also contains the covariant derivative with respect to the full gauge field A_μ. When expanded, this contains interaction terms between the gauge field a_μ and the ghosts. Therefore, the Hessian contains mixed a-c and a-$\bar{c}$ terms, which enter into the traces in (6.52).

It is important to stress that it is not possible to write a flow equation for a functional of a single field $A_\mu = \bar{A}_\mu + a_\mu$. The gauge fixing and cutoff actions introduce separate dependences on the background and quantum fluctuation, so that the EAA must necessarily be a separate functional of these two arguments. The source of this is in the cutoff and gauge fixing terms. In the limit $k \to 0$ only the gauge fixing remains and although the EA is still formally a function of two arguments, this has no effect on physical observables.

Let us define $\bar{\Gamma}_k(\bar{A}) = \Gamma_k(0, 0, 0; \bar{A})$ the gauge invariant functional obtained by putting the expectation values of the quantum fields to zero. We can then split the EAA into a gauge invariant part depending only on the full field A_μ and a remainder $\hat{\Gamma}_k$:

$$\Gamma_k(a, \bar{c}, c; \bar{A}) = \bar{\Gamma}_k(\bar{A} + a) + \hat{\Gamma}_k(a, \bar{c}, c; \bar{A}) ,$$

(6.53)

which implies in particular that

$$\hat{\Gamma}_k(0, 0, 0; \bar{A}) = 0 .$$

(6.54)

We can then write a flow equation for the functional $\bar{\Gamma}_k$ by putting all the fluctuation fields φ to zero. In this case $D_\mu \to \bar{D}_\mu$ in the ghost operator, and the mixed a-c and a-$\bar{c}$ terms go to zero. Then (6.52) simplifies to

$$\frac{d\bar{\Gamma}_k(\bar{A})}{dt} = \frac{1}{2} \mathrm{Tr} \left(\frac{\delta^2\Gamma_k}{\delta a \delta a} + \mathcal{R}_k \right)^{-1} \frac{d\mathcal{R}_k}{dt} - \mathrm{Tr} \left(\frac{\delta^2\Gamma_k}{\delta \bar{c} \delta c} + \mathcal{R}_k^{(gh)} \right)^{-1} \frac{d\mathcal{R}_k^{(gh)}}{dt} ,$$

(6.55)

where the first term in the r.h.s. is due entirely to the gauge field fluctuations and the second to the ghosts. Note that on the right we still have the Hessian of the full Γ_k, so (6.55) is not a closed equation for $\bar{\Gamma}_k$. In order to calculate the flow of $\bar{\Gamma}_k$ one needs some information about the Hessian of $\hat{\Gamma}_k$.

Short of solving the full ERGE, one can get some insight from the structure of the one-loop EAA at vanishing fluctuation:

$$\bar{\Gamma}_k^{(1)}(\bar{A}) = S(\bar{A}) + \frac{1}{2} \mathrm{Tr} \log \left(\frac{\delta^2(S + S_{GF} + \Delta S_k)}{\delta a \delta a} \right) \Bigg|_{a=0} - \mathrm{Tr} \log \frac{\delta^2(S_{gh} + \Delta S_k)}{\delta \bar{c} \delta c} ,$$

(6.56)

which is a generalization of (3.62). Note that since S is a function of the full field $A = \bar{A} + a$, its Hessians with respect to a and $\bar{A}$ are the same. We can thus write

$$\bar{\Gamma}_k^{(1)}(\bar{A}) = S(\bar{A}) + \frac{1}{2}\mathrm{Tr}\log\left(\frac{\delta^2 S(\bar{A})}{\delta\bar{A}\delta\bar{A}} + \frac{\delta^2 S_{GF}}{\delta a \delta a}\bigg|_{a=0} + \mathcal{R}_k\right) - \mathrm{Tr}\log\left(\frac{\delta^2 S_{gh}}{\delta\bar{c}\delta c} + \mathcal{R}_k^{(gh)}\right).$$

$$(6.57)$$

If the gauge fixing condition is linear in a, the gauge fixing action is quadratic in a and therefore we can remove the suffix $a = 0$. If we now perform a "renormalization group improvement" similar to the one discussed for the scalar case, by replacing $S(\bar{A})$ by $\bar{\Gamma}_k(\bar{A})$ in the r.h.s., we obtain a closed equation for $\bar{\Gamma}_k$ with fixed gauge fixing and ghost actions

$$\frac{d\bar{\Gamma}_k(\bar{A})}{dt} = \frac{1}{2}\mathrm{Tr}\left(\frac{\delta^2\bar{\Gamma}_k}{\delta\bar{A}\delta\bar{A}} + \frac{\delta^2 S_{GF}}{\delta a \delta a} + \mathcal{R}_k\right)^{-1}\frac{d\mathcal{R}_k}{dt} - \mathrm{Tr}\left(\frac{\delta^2 S_{gh}}{\delta\bar{c}\delta c} + \mathcal{R}_k^{(gh)}\right)^{-1}\frac{d\mathcal{R}_k^{(gh)}}{dt}.$$

$$(6.58)$$

This suggests taking

$$\hat{\Gamma}_k(a, \bar{c}, c; \bar{A}) = S_{GF}(a; \bar{A}) + S_{gh}(\bar{c}, c; \bar{A}) ,$$

$$(6.59)$$

which is also consistent with (6.54), and ignoring the running of $\hat{\Gamma}_k$. Unlike the scalar case, where the RG improvement led directly to the full ERGE, this is still an approximate equation, but it is more manageable than the exact one. In this section we are only interested in reproducing the one-loop beta function, and (6.58) will be sufficient for this purpose.

6.5.2 *One-loop beta function*

In order to read off the beta function of the gauge coupling g we make the truncation

$$\bar{\Gamma}_k(\bar{A}) = \frac{1}{4g(k)^2}\int d^d x\, \bar{F}_{\mu\nu}^a \bar{F}^{\mu\nu a} .$$

$$(6.60)$$

Since the Yang-Mills Hessian (3.59) has an overall prefactor $1/g^2$, it is convenient to have such a prefactor also in the cutoff action. Thus we define

$$\mathcal{R}_k^{ab}(z) = \frac{1}{g^2}\delta^{ab} R_k(z) ,$$

$$(6.61)$$

where R_k is a scalar function with the properties listed in section 6.1. Let us return for a moment to the issue of the choice of a covariant differential operator to be taken as the argument of the cutoff. In the truncation considered here, the natural choice to be used in the ghost term is $-\bar{D}^2$ but for the a_μ sector one faces a choice:

- a "type I" cutoff defined using $z = -\bar{D}^2$, or
- a "type II" cutoff defined using the kinetic operator Δ given in (3.60) .

Let us consider first a type II cutoff, which leads to somewhat simpler algebra. The cutoff action is

$$\Delta S_k(a, \bar{c}, c; \bar{A}) = \frac{1}{2}\int d^d x\, a_\mu g^{\mu\nu} R_k(\Delta) a_\nu + \int d^d x\, \bar{c}\, \mathcal{R}_k^{(gh)}(-\bar{D}^2) c .$$

$$(6.62)$$

When we introduce this in (6.58) we have

$$\frac{d\mathcal{R}_k}{dt} = \frac{1}{g^2}\left(\frac{dR_k}{dt} - \eta R_k\right) ,$$

where $\eta = \frac{2}{g}\frac{dg}{dt}$. Thus we find

$$\frac{d\bar{\Gamma}_k(\bar{A})}{dt} = \frac{1}{2}\mathrm{Tr}\frac{\partial_t R_k - \eta R_k}{P_k(\Delta)} - \mathrm{Tr}\frac{\partial_t R_k}{P_k(-\bar{D}^2)} . \tag{6.63}$$

where Δ is given by (3.60) and the function P_k was defined in (6.7). The overall factors of g^2 have canceled between numerator and denominator.

The trace of a function of a differential operator can be evaluated using formula (5.159). This gives, for the r.h.s. of (6.63):

$$\frac{d\bar{\Gamma}_k(\bar{A})}{dt} = \frac{1}{2}\frac{1}{(4\pi)^{d/2}}\left[Q_{\frac{d}{2}}(W_a)B_0(\Delta) + Q_{\frac{d}{2}-1}(W_a)B_2(\Delta) + Q_{\frac{d}{2}-2}(W_a)B_4(\Delta) + \ldots\right.$$
$$\left.-2\left(Q_{\frac{d}{2}}(W_{gh})B_0(-\bar{D}^2) + Q_{\frac{d}{2}-1}(W_{gh})B_2(-\bar{D}^2) + Q_{\frac{d}{2}-2}(W_{gh})B_4(-\bar{D}^2) + \ldots\right)\right], \tag{6.64}$$

where $W_a = \frac{\partial_t R_k - \eta R_k}{P_k}$ and $W_{gh} = \frac{\partial_t R_k}{P_k}$. We look for the terms that are quadratic in F, and these are contained in the b_4's. The necessary heat kernel coefficients have been computed for the case $d = 4$ in (3.63,3.64). In general dimension we have

$$\mathrm{tr}\mathbf{E}^2 = 4C_2 F^a_{\mu\nu}F^{a\mu\nu} , \qquad \mathrm{tr}\Omega_{\mu\nu}\Omega^{\mu\nu} = -dC_2 F^a_{\mu\nu}F^{a\mu\nu}$$

and therefore from (3.45)

$$b_4(\Delta) = \frac{24-d}{12}C_2 F^a_{\mu\nu}F^{a\mu\nu}$$

and for the ghost operator

$$b_4(-\bar{D}^2) = -\frac{1}{12}C_2 F^a_{\mu\nu}F^{a\mu\nu} .$$

Comparing the terms proportional to the Yang-Mills action on both sides of the equation one gets

$$-\frac{1}{2g^3}\beta_g\int d^d x\, F^a_{\mu\nu}F^{a\mu\nu} = \frac{1}{(4\pi)^2}\frac{26-d}{12}C_2 Q_{d/2-2}\left(\frac{\partial_t R_k}{P_k}\right)\int d^d x\, F^a_{\mu\nu}F^{a\mu\nu}$$

whence one gets the beta function

$$\beta_g = -\frac{1}{(4\pi)^{d/2}}\frac{26-d}{12}C_2 Q_{d/2-2}\left(\frac{\partial_t R_k}{P_k}\right)g^3 . \tag{6.65}$$

The Q-functional generally depends on the shape of the profile function R_k. For the optimized cutoff it is evaluated in (6.150), leading to

$$\beta_g = -\frac{1}{(4\pi)^{d/2}}\frac{(26-d)(d-2)}{12\Gamma(d/2)}C_2 k^{d-4}g^3 . \tag{6.66}$$

In particular when $d = 4$ it is shown in (6.148) that $Q_0\left(\frac{\partial_t R_k}{P_k}\right) = 2$, independently of the shape of the cutoff,[3] so that the beta function agrees with the standard one-loop result (3.68).

---

[3]This can also be seen as an analytic continuation of the result (6.43), that was derived for $n \geq 1$.

It is instructive to see also how the calculation proceeds with a type I cutoff

$$\Delta S_k(a, \bar{c}, c; \bar{A}) = \frac{1}{2g^2} \int d^d x \, a_\mu g^{\mu\nu} \mathcal{R}_k(-\bar{D}^2) a_\nu + \int d^d x \, \bar{c} \mathcal{R}_k^{(gh)}(-\bar{D}^2) c \ . \quad (6.67)$$

Instead of (6.63) we arrive at

$$\frac{d\bar{\Gamma}_k(\bar{A})}{dt} = \frac{1}{2} \text{Tr} \frac{\partial_t R_k - \eta R_k}{P_k(-\bar{D}^2) + \mathbf{E}} - \text{Tr} \frac{\partial_t R_k}{P_k(-\bar{D}^2)} \ . \quad (6.68)$$

The ghost contribution is the same as before, but the gauge field contribution is non-polynomial in F, so to extract the term proportional to F^2 we have to expand the denominator. The relevant terms are

$$\frac{1}{2} \left[\text{Tr} \left(\frac{\partial_t R_k}{P_k} \right) + \text{Tr} \left(\frac{\partial_t R_k}{P_k^3} \mathbf{E}^2 \right) \right] \ .$$

While the first term can be treated by the methods developed so far, the second is of a different type, since it involves a trace of a function of $-\bar{D}^2$ with an insertion of $\mathbf{E}^2$. For this, knowledge of the coefficients of the expansion of the trace of the heat kernel is not sufficient: one needs also information of the non-diagonal terms of the expansion. In this case, however, the necessary information is very simple. Since the insertion is already quadratic in F, we only need the information on the non-diagonal part of the heat kernel at zeroth order in F. This is contained in the coefficient $\mathbf{b}_0 = 1$. (We denote $\mathbf{b}_n$ the untraced coefficients of the expansion, so that $b_n = \text{tr}\mathbf{b}_n$.) Therefore the relevant terms are

$$\frac{1}{2} \frac{1}{(4\pi)^{d/2}} \left[Q_{d/2-2} \left(\frac{\partial_t R_k}{P_k} \right) \text{tr} \mathbf{B}_4(-\bar{D}^2) + Q_{d/2} \left(\frac{\partial_t R_k}{P_k^3} \right) \text{tr}(\mathbf{B}_0(-\bar{D}^2)\mathbf{E}^2) \right] \ ,$$

where

$$\text{tr}(\mathbf{b}_0(-\bar{D}^2)\mathbf{E}^2) = \text{tr}\mathbf{E}^2 = 4C_2 F_{\mu\nu}^a F^{a\mu\nu} \ , \qquad \text{tr}\mathbf{b}_4(-\bar{D}^2) = -\frac{d}{12} C_2 F_{\mu\nu}^a F^{a\mu\nu} \ .$$

This, together with the explicit formula for the Q-functionals (6.150), leads to

$$\beta_g = -\frac{1}{(4\pi)^{d/2}} \frac{192 - d(d-2)^2}{12d\Gamma(d/2)} C_2 k^{d-4} g^3 \ . \quad (6.69)$$

The result is seen to be quite different except for $d = 4$, where the one-loop beta function is universal.

Let us conclude with some observations. First, taking into account the η-term in (6.63),(6.68) allows one to go beyond the one-loop approximation. An evaluation of the corresponding trace leads, in $d = 4$ and with type II cutoff, to [201]

$$\beta_g = \frac{ag^3}{1 - bg^2} \ ; \qquad \text{where } a = -\frac{11}{3} \frac{C_2}{(4\pi)^2} \quad b = \frac{10}{3} \frac{C_2}{(4\pi)^2} \ . \quad (6.70)$$

A completely different reasoning based on higher loop results and on the exact beta function of super-Yang-Mills theory [202] leads to a beta function of the same form, with the same universal a and $b = \frac{34}{11} \frac{C_2}{(4\pi)^2}$, which is numerically quite close.

The second observation is that in $d > 4$ the gauge coupling has dimension $4-d < 0$, making the theory power-counting non-renormalizable. In the parametrization of the Yang-Mills theory space one has to define the dimensionless coupling $\tilde{g} = gk^{d-4}$. The beta function of this coupling is

$$\tilde{\beta}_g = (4-d)\tilde{g} + ag^3 \, ,$$

where a is the constant appearing in the beta function. We have seen that a is not universal, but asymptotic freedom means that $a < 0$ as long as d is not too different from four. Then, the beta function has a nontrivial zero at $g > 0$. This nontrivial fixed point becomes negative when the dimension is sufficiently large. The dimension where this happens depends on the scheme. With the type II scheme it happens in $d = 26$. This is related to an old observation of Nepomechie [203].

6.6 Gaussian matter fields coupled to an external metric

We shall now follow the same logic as in chapter 3: before discussing the ERGE for gravity, we shall consider the technically simpler situation of massless Gaussian matter fields (*i.e.* matter fields with quadratic actions, without self-interactions) minimally coupled to an external metric. This provides a gentle introduction to the technicalities that are needed also in quantum gravity.

As we saw in section 3.2, matter loops induce divergences that depend on the external metric. Hence, when the metric is made dynamical, they will also contribute to the beta functions of the couplings appearing in the gravitational action. We will compute here the matter contributions to the beta functions of the gravitational couplings.

The matter fields consist of N_S scalar fields ϕ^i, N_D Dirac fields ψ^i and N_V abelian gauge fields A^i_μ, together with their ghosts and antighosts c and $\bar{c}$, all coupled to an external metric $g_{\mu\nu}$. The action is given by

$$S_{\text{matter}} = S_S(\phi; g) + S_D(\bar{\psi}, \psi; g) + S_V(A, \bar{c}, c; g)$$

where

$$S_S(\phi; g) = \frac{1}{2} \int d^dx \sqrt{g}\, g^{\mu\nu} \sum_{i=1}^{N_S} \partial_\mu \phi^i \partial_\nu \phi^i$$

$$S_D(\bar{\psi}, \psi; g) = i \int d^dx \sqrt{g} \sum_{i=1}^{N_D} \bar{\psi}^i \slashed{D} \psi^i,$$

$$S_V(A, \bar{c}, c; g) = \frac{1}{4} \int d^dx \sqrt{g} \sum_{i=1}^{N_V} g^{\mu\nu} g^{\kappa\lambda} F^i_{\mu\kappa} F^i_{\nu\lambda} + \frac{1}{2\xi} \int d^dx \sqrt{g} \sum_{i=1}^{N_V} \left(g^{\mu\nu} \nabla_\mu A^i_\nu\right)^2$$

$$+ \int d^dx \sqrt{g} \sum_{i=1}^{N_V} \bar{c}_i(-\nabla^2) c_i \, . \tag{6.71}$$

In the Dirac action, $\not{D} = \gamma^a e^\mu_a \nabla_\mu$, where e^μ_a is an orthonormal frame.

The kinetic operators (inverse propagators) are differential operators of the form

$$\Delta = -\nabla^2 + \mathbf{E} \tag{6.72}$$

where ∇ is the Levi-Civita connection of g and $\mathbf{E}$ is a linear map acting on the quantum field. The endomorphisms $\mathbf{E}$ for our fields have already been given in section 3.2.4. For the scalar field $\mathbf{E} = 0$. For the Dirac fields, squaring the Dirac operator one obtains $\mathbf{E} = \frac{R}{4}\mathbf{1}$. For the Maxwell fields, choosing the Feynman gauge $\xi = 1$, the endomorphism $\mathbf{E}$ is given by the Ricci tensor acting on vectors. For the scalar ghosts, $\mathbf{E} = 0$. The spacetime dimension d is left arbitrary at this stage.

In order to write the ERGE, we have to define the cutoff. As in the Yang-Mills case, for the operator to be used in the definition of (6.2), two natural choices suggest themselves:[4]

- a "type I" cutoff defined using the Bochner Laplacian $-\nabla^2$
- a "type II" cutoff defined using for each type of field its kinetic operator Δ

A priori both choices seem equally legitimate. The type II cutoff is technically simpler and furthermore the type I cutoff gives rise to a difficulty with Dirac fields. For this reason we discuss the type II cutoff first and postpone the discussion of the type I cutoff to the following subsection.

6.6.1 *Type II cutoff*

In this case we choose a real function R_k with the properties listed in section 6.1 and for each type of field we define a modified inverse propagator

$$P_k(\Delta) = \Delta + R_k(\Delta) . \tag{6.73}$$

Using (5.159), the trace in the r.h.s. of the ERGE reduces simply to:

$$\mathrm{Tr}\frac{\partial_t R_k(\Delta)}{P_k(\Delta)} = \frac{1}{(4\pi)^{d/2}} \sum_{i=0}^{\infty} Q_{\frac{d}{2}-i}\left(\frac{\partial_t R_k}{P_k}\right) B_{2i}(\Delta) \tag{6.74}$$

where $B_{2i}(\Delta)$ are the heat kernel coefficients of the operator Δ and the Q-functionals, defined in (5.163), (5.164) are the analogs of momentum integrals in this curved spacetime setting. Note that (6.74) gives a formula for the beta function of *all* the gravitational couplings, in any dimension, albeit in a somewhat implicit form. Since the Q-functionals can always be evaluated, at least numerically, the ability to turn this formula into explicit beta functions is only limited by our knowledge of the heat kernel coefficients.

---

[4]In (6.2) it is assumed for simplicity that the operator appearing in the argument of the cutoff function is also the operator whose eigenfunctions are used as a basis in the evaluation of the functional trace. It is worth stressing that this is not necessary.

In four dimensions, using Table 3.1, the contribution of Gaussian matter to the gravitational ERGE is:

$$
\frac{d\Gamma_k}{dt} = \frac{N_S}{2}\mathrm{Tr}_{(S)}\left(\frac{\partial_t R_k(\Delta^{(S)})}{P_k(\Delta^{(S)})}\right) - \frac{N_D}{2}\mathrm{Tr}_{(D)}\left(\frac{\partial_t R_k(\Delta^{(D)})}{P_k(\Delta^{(D)})}\right)
$$

$$
+ \frac{N_V}{2}\mathrm{Tr}_{(M)}\left(\frac{\partial_t R_k(\Delta^{(M)})}{P_k(\Delta^{(M)})}\right) - N_V\mathrm{Tr}_{(gh)}\left(\frac{\partial_t R_k(\Delta^{(gh)})}{P_k(\Delta^{(gh)})}\right)
$$

$$
= \frac{1}{2}\frac{1}{(4\pi)^2}\int d^4x\,\sqrt{g}\Bigg[(N_S - 4N_D + 2N_V)\,Q_2\left(\frac{\partial_t R_k}{P_k}\right)
$$

$$
+ \frac{1}{6}R\,(N_S + 2N_D - 4N_V)\,Q_1\left(\frac{\partial_t R_k}{P_k}\right) + \frac{1}{180}\Big((3N_S + 18N_D + 36N_V)\,C^2
$$

$$
- (N_S + 11N_D + 62N_V)\,E + 5N_S R^2\Big) + \dots\Bigg], \tag{6.75}
$$

where we use the Weyl basis (2.83) for the terms quadratic in curvature and we have already used that $Q_0\left(\frac{\partial_t R_k}{P_k}\right) = 2$ for any cutoff shape (see (6.148)).

In order to have more explicit formulae, and in numerical work, one needs to calculate also the scheme-dependent Q-functionals. This requires fixing the profile R_k. We will mostly use the so-called optimized cutoff (6.149) in which the integrals are readily evaluated, see Eqs. (6.150), (6.151), (6.152). We obtain for the type II cutoff

$$
\frac{d\Gamma_k}{dt} = \frac{1}{2}\frac{1}{(4\pi)^2}\int d^4x\,\sqrt{g}\Bigg[(N_S - 4N_D + 2N_V)k^4 + \frac{1}{3}Rk^2(N_S + 2N_D - 4N_V)
$$

$$
+ \frac{1}{180}\Big((3N_S + 18N_D + 36N_V)C^2 - (N_S + 11N_D + 62N_V)E + 5N_S R^2\Big)\Bigg]. \tag{6.76}
$$

Comparing to the l.h.s. of the ERGE, written for the action (3.125) plus (2.83) with the r.h.s. given in (6.76), we can read off the beta functions of the gravitational couplings. For the cosmological constant and Newton's constant we have

$$
\frac{d}{dt}\left(\frac{2\Lambda}{16\pi G}\right) = \frac{k^d}{16\pi}A_1
$$

$$
-\frac{d}{dt}\left(\frac{1}{16\pi G}\right) = \frac{k^{d-2}}{16\pi}B_1\,, \tag{6.77}
$$

where

$$
A_1 = \frac{1}{2\pi}(N_S - 4N_D + 2N_V)\,, \tag{6.78}
$$

$$
B_1 = \frac{1}{6\pi}(N_S + 2N_D - 4N_V)\,. \tag{6.79}
$$

It is not difficult to generalize this calculation to arbitrary dimension, in which case one finds

$$
A_1 = \frac{32\pi(N_S - 2^{[d/2]}N_D + (d-2)N_V)}{(4\pi)^{d/2}d\Gamma[d/2]}\,, \tag{6.80}
$$

$$
B_1 = \frac{16\pi(N_S + 2^{[d/2]-1}N_D + (d-8)N_V)}{(4\pi)^{d/2}6\Gamma[d/2]}\,. \tag{6.81}
$$

For the quadratic terms we find instead

$$\frac{d}{dt}\left(\frac{1}{2\lambda}\right) = \frac{1}{2880\pi^2}\left(\frac{3}{2}N_S + 9N_D + 18N_M\right),$$

$$\frac{d}{dt}\left(\frac{1}{\xi}\right) = \frac{1}{2880\pi^2}\frac{5}{2}N_S, \tag{6.82}$$

$$\frac{d}{dt}\left(-\frac{1}{\rho}\right) = \frac{1}{2880\pi^2}\left(-\frac{1}{2}N_S - \frac{11}{2}N_D - 31N_M\right).$$

The following comments are in order.

First, we observe that the calculation has been done without specifying the external metric. The result for the beta functions is therefore "background-independent".

Second, the optimized cutoff has the property that $Q_{-n}\left(\frac{\partial_t R_k}{P_k}\right) = 0$ for $n \geq 1$. Thus, the sum over heat kernel coefficients on the r.h.s. of (6.74) terminates. In particular, in four dimensions, there are no terms beyond those that are explicitly written in (6.76). For more general cutoffs a calculation of beta functions for curvature scalars of cubic and higher order would require the knowledge of higher heat kernel coefficients. The coefficients B_6 and B_8 for operators of the form (6.72) are known [70, 72].

6.6.2 Type I cutoff

With a type I cutoff we use the same profile function R_k but now with $-\nabla^2$ as its argument. This implies the replacement of the inverse propagator Δ by

$$\Delta + R_k(-\nabla^2) = P_k(-\nabla^2) + \mathbf{E}. \tag{6.83}$$

Therefore the r.h.s. of the ERGE will now contain the trace $\mathrm{Tr}\frac{\partial_t R_k(-\nabla^2)}{P_k(-\nabla^2)+\mathbf{E}}$. Since $\mathbf{E}$ is linear in curvature, in the limit when the components of the curvature tensor are uniformly much smaller than k^2, we can expand

$$\frac{\partial_t R_k}{P_k + \mathbf{E}} = \sum_{\ell=0}^{\infty}(-1)^\ell \mathbf{E}^\ell \frac{\partial_t R_k}{P_k^{\ell+1}}.$$

Each one of the terms on the r.h.s. can then be evaluated in a way analogous to (5.159), so in this case we get a double series:

$$\mathrm{Tr}\frac{\partial_t R_k(-\nabla^2)}{P_k(-\nabla^2) + \mathbf{E}} = \frac{1}{(4\pi)^{d/2}}\sum_{i=0}^{\infty}\sum_{\ell=0}^{\infty}(-1)^\ell Q_{\frac{d}{2}-i}\left(\frac{\partial_t R_k}{P_k^{\ell+1}}\right)\int d^d x\,\sqrt{g}\,\mathrm{tr}\mathbf{E}^\ell\mathbf{b}_{2i}(-\nabla^2). \tag{6.84}$$

Before entering into details, let us make the following observation. As we have already seen in (6.43), the integrals $Q_n\left(\frac{\partial_t R_k}{P_k^{n+1}}\right)$ are independent of the shape of R_k. Thus, in even-dimensional spacetimes with a cutoff of type II, and using (6.148), the coefficient of the term in the sum (6.74) with $i = d/2$ is $Q_0\left(\frac{\partial_t R_k}{P_k}\right)B_d(\Delta) = 2B_d(\Delta)$.

On the other hand with a type I cutoff, using (6.147), (6.148) and (5.167) the terms with $i + \ell = \frac{d}{2}$ add up to

$$\sum_{\ell=0}^{d/2}(-1)^\ell Q_\ell \left(\frac{\partial_t R_k}{P_k^{\ell+1}}\right) \int d^d x \sqrt{g}\, \mathrm{tr} \mathbf{E}^\ell \mathbf{b}_{2i}(-\nabla^2)$$

$$= 2 \int d^d x \sqrt{g}\, \mathrm{tr} \left[\mathbf{b}_d(-\nabla^2) - \mathbf{E}\mathbf{b}_{d-2}(-\nabla^2) + \ldots + \frac{(-1)^{d/2}}{(d/2)!}\mathbf{E}^{d/2}\mathbf{b}_0(-\nabla^2)\right]$$

$$= 2B_d(-\nabla^2 + \mathbf{E})\ .$$

Therefore, in addition to being independent of the shape of the cutoff function, these terms are also the same using type I or type II cutoffs. This is a further check of the universality of the beta functions of the dimensionless couplings.

Proceding with the calculation with type I cutoffs, the ERGE reads:

$$\frac{d\Gamma_k}{dt} = \frac{N_S}{2}\mathrm{Tr}_{(S)}\left(\frac{\partial_t P_k(-\nabla^2)}{P_k(-\nabla^2)}\right) - \frac{N_D}{2}\mathrm{Tr}_{(D)}\left(\frac{\partial_t R_k(-\nabla^2)}{P_k(-\nabla^2) + \frac{R}{4}}\right)$$

$$+ \frac{N_V}{2}\mathrm{Tr}_{(M)}\left(\frac{\partial_t R_k(-\nabla^2)}{P_k(-\nabla^2) + \mathrm{Ricci}}\right) - N_V \mathrm{Tr}_{(gh)}\left(\frac{\partial_t R_k(-\nabla^2)}{P_k(-\nabla^2)}\right). \quad (6.85)$$

Expanding each trace as in (5.159), collecting terms with the same number of derivatives of the metric, and keeping terms up to four derivatives we get

$$\frac{d\Gamma_k}{dt} = \frac{1}{2}\frac{1}{(4\pi)^2}\int d^4 x \sqrt{g}\Bigg[(N_S - 4N_D + 2N_V)\, Q_2\left(\frac{\partial_t R_k}{P_k}\right)$$

$$+ \left[\frac{1}{6}Q_1\left(\frac{\partial_t R_k}{P_k}\right)N_S - \left(\frac{2}{3}Q_1\left(\frac{\partial_t R_k}{P_k}\right) - Q_2\left(\frac{\partial_t R_k}{P_k^2}\right)\right)N_D\right.$$

$$+ \left.\left(\frac{1}{3}Q_1\left(\frac{\partial_t R_k}{P_k}\right) - Q_2\left(\frac{\partial_t R_k}{P_k^2}\right)\right)N_V\right]R$$

$$+ \frac{1}{180}\Bigg((3N_S + 18N_D + 36N_V)\, C^2$$

$$- (N_S + 11N_D + 62N_V)\, E + 5N_S R^2\Bigg) + \ldots\Bigg]. \quad (6.86)$$

We see that the terms linear in curvature, which contribute to the beta function of Newton's constant, have changed. However, the terms quadratic in curvature have the same coefficients as before, confirming that the beta functions of the dimensionless couplings are scheme-independent.

Evaluating the Q-functionals with the optimized cutoff one finds

$$\frac{d\Gamma_k}{dt} = \frac{1}{2}\frac{1}{(4\pi)^2}\int d^4 x \sqrt{g}\Bigg[(N_S - 4N_D + 2N_V)k^4 + \frac{1}{3}Rk^2(N_S - N_D - N_M)$$

$$+ \frac{1}{180}\left((3N_S + 18N_D + 36N_V)C^2 - (N_S + 11N_D + 62N_V)E + 5N_S R^2\right) + \ldots\Bigg].$$

The beta functions have the same form as with the type II cutoff, except for the coefficients

$$A_1 = \frac{1}{2\pi}(N_S - 4N_D + 2N_V) , \tag{6.87}$$

$$B_1 = \frac{1}{6\pi}(N_S - N_D - N_V) . \tag{6.88}$$

while in dimension d

$$A_1 = \frac{32\pi(N_S - 2^{[d/2]}N_D + (d-2)N_V)}{(4\pi)^{d/2}d\Gamma[d/2]} , \tag{6.89}$$

$$B_1 = \frac{16\pi(N_S - 2^{[d/2]}\frac{d-3}{d}N_D + \frac{d^2-2d-12}{d}N_V)}{(4\pi)^{d/2}6\Gamma[d/2]} . \tag{6.90}$$

We can now compare the results of the type I and type II cutoffs. The A_1 coefficient is the same, because A_1 only depends on the B_0 coefficients, that simply count the number of degrees of freedom. The scalar contribution is also the same in both cases, because **E** is zero in this case. The results would differ if we considered the conformally coupled operator. In the case of the Maxwell field, the contributions to B_1 are the same in $d = 2$, they are negative above $d = 2$ and change sign, from negative to positive, when d exceeds a certain value, which is 8 for the type II cutoff and between 4 and 5 for the type I cutoff. In the case of the Dirac field, the contributions to B_1 are the same in $d = 2$; it remains positive for all d with the type II cutoff but changes sign from positive to negative at $d = 3$ with the type I cutoff. To some extent such differences are not too worrying, since they reflect inevitable ambiguities. The different signs of the Dirac field contributions in $d = 4$ and above, however, is particularly nagging. We will now argue that the correct sign is the one provided by the type II cutoff.

6.6.3 *Spectral sum for the Dirac operator*

In order to decide which one of the preceding calculations gives the correct sign for the fermionic contribution to the running of Newton's constant in $d = 4$, we shall evaluate the r.h.s. of the ERGE by an independent method.

The EAA can be defined directly in terms of the Dirac operator as

$$\Gamma_k(g) = -\mathrm{tr}\log\left(|\slashed{D}| + R_k^D(|\slashed{D}|)\right) , \tag{6.91}$$

where the cutoff R_k^D has to be a function of the modulus of the Dirac operator, since we want to suppress the modes depending on the wavelength of the corresponding eigenfunctions. This is also needed for reasons of convergence. The function $R_k^D(z)$ has to satisfy conditions similar to those spelled out in section 6.1, except that k^2 has to be replaced by k, since the operator is first order. For the explicit evaluation, we will use the optimized profile

$$R_k^D(z) = (k - z)\theta(k - z) , \quad (z > 0) . \tag{6.92}$$

Then we have

$$\mathrm{Tr}\left[\frac{\partial_t R_k^D(|\not{D}|)}{P_k^D(|\not{D}|)}\right] = \sum_n m_n \frac{\partial_t R_k^D(|\lambda_n|)}{P_k^D(|\lambda_n|)} = \sum_{\pm}\sum_n m_n \theta(k - |\lambda_n|) , \qquad (6.93)$$

where λ_n are the eigenvalues of the Dirac operator and m_n their multiplicities. In general, one does not know the spectrum of the Dirac operator and it is not possible to evaluate the trace directly from this formula. However, the spectrum of the Dirac operator is known in the case of the d-sphere: the eigenvalues and multiplicities are [204]

$$\lambda_n^{\pm} = \pm\sqrt{\frac{R}{d(d-1)}}\left(\frac{d}{2}+n\right) , \qquad m_n = 2^{\left[\frac{d}{2}\right]}\binom{n+d-1}{n} , \qquad n = 0,1,\dots . \tag{6.94}$$

With this information one can compute the trace of any function of the Dirac operator as $\mathrm{Tr}\, f(\not{D}) = \sum_{n=0}^{\infty} m_n f(\lambda_n)$ on the sphere. This is enough to unambiguously extract the first two terms in the derivative expansion on the r.h.s. of the ERGE.

The sum can be computed using the Euler-Maclaurin formula. Note that only the integral depends on R, and therefore, in dimensions $d > 2$, for the terms that we are interested in, it is enough to compute the integral. After factoring the volume, the only terms we need to compute are the 0th and 1st power of R. Using (5.135), we only have to isolate the terms in the integral proportional to $R^{-d/2}$ and $R^{1-d/2}$. The integral is

$$2^{[d/2]+1}\int_0^{k\sqrt{\frac{d(d-1)}{R}}-\frac{d}{2}} dn \binom{n+d-1}{n} \tag{6.95}$$

and changing variables $n \to n' - d/2$ it can be written as

$$\frac{2^{[d/2]+1}}{(d-1)!}\int_{\frac{d}{2}}^{k\sqrt{\frac{d(d-1)}{R}}} dn' \left(n'+\frac{d}{2}-1\right)\cdots\left(n'-\left(\frac{d}{2}-1\right)\right) \tag{6.96}$$

The terms we are interested in come from the integral of the two highest powers of n':

$$\left(n'+\frac{d}{2}-1\right)\cdots\left(n'-\left(\frac{d}{2}-1\right)\right) = n'^{d-1} - n'^{d-3}\sum_{k=1}^{\left[\frac{d-1}{2}\right]}\left(\frac{d}{2}-k\right)^2 + \cdots \tag{6.97}$$

we can rewrite the sum $\sum_{k=1}^{\left[\frac{d-1}{2}\right]}\left(\frac{d}{2}-k\right)^2 = \frac{1}{24}d(d-1)(d-2)$, and perform the integral

$$\mathrm{Tr}\left[\frac{\partial_t R_k^D}{P_k^D}\right] = \frac{2^{[d/2]+1}}{(d-1)!}\frac{1}{d}\left(k\sqrt{\frac{d(d-1)}{R}}\right)^d$$

$$-\frac{2^{[d/2]+1}}{(d-1)!}\frac{1}{d-2}\left(k\sqrt{\frac{d(d-1)}{R}}\right)^{d-2}\frac{1}{24}d(d-1)(d-2)+\cdots \tag{6.98}$$

Factoring the volume, the result is

$$\frac{d\Gamma_k}{dt} = -\mathrm{Tr}\left[\frac{\partial_t R_k^D}{P_k^D}\right] = -\frac{1}{\Gamma\left(\frac{d}{2}+1\right)}\frac{2^{\left[\frac{d}{2}\right]}}{(4\pi)^{\frac{d}{2}}}V(d)\left(k^d - \frac{d}{24}k^{d-2}R + O\left(R^2\right)\right),$$

(6.99)

where $V(d)$ is the volume of the d-sphere. The corresponding contributions to the coefficients A_1 and B_1 agree exactly with (6.80), which was obtained with the type II cutoff.

Note that computing the r.h.s. of the ERGE with a spectral sum is a much more direct procedure. It avoids the ambiguities that arise in the definition of the determinant of the Dirac operator as the square root of the determinant of its square [205], and also avoids having to use the Laplace transform and the heat kernel. The agreement of the spectral sum with the type II-heat kernel calculation is a useful consistency check and suggests that the latter gives the correct result whereas the type I cutoff does not."

There remains to understand why the type I cutoff should not be admissible in this case. It can be shown that the type I cutoff imposed on the square of the Dirac operator corresponds to a cutoff on the eigenvalues of the Dirac operator itself, that does not satisfy all the conditions that we require of a good cutoff. The interested reader is referred to [206] for details.

6.7 The ERGE for gravity

Finally we discuss the application of the functional RG to dynamical gravity [207]. After the discussion of the ERGE for Yang-Mills theory and for matter fields in an external gravitational field, the formulation of an ERGE for gravity comes quite naturally. The obvious route is to use the background field method, decomposing the dynamical metric into a fixed background and a quantum fluctuation

$$g_{\mu\nu} = \bar{g}_{\mu\nu} + h_{\mu\nu}.$$

We recall that this allows one to define an effective action that is still invariant under background diffeomorphisms. As in the Yang-Mills case, the trick is to define a cutoff action that is quadratic in h but still background gauge invariant. This is achieved by using the background field method.

It is important to stress that while in the Yang-Mills case this can be seen as a very convenient but not strictly necessary procedure, in the case of gravity it is hard to imagine an alternative. The problem is conceptual and has to do with the definition of coarse-graining. In applications of the renormalization group to flat space QFT's, it is always clear what is meant by a momentum scale, but in the presence of dynamical gravity there is no preferred definition of distance between points, nor of the norm of momentum vectors. A background reference metric gets around this difficulty. With this metric we can construct a Laplace-type operator acting in the appropriate field space. This Laplacian has a spectrum, which for the sake of simplicity we can assume here to be discrete. When we give a cutoff scale k, there is a corresponding well-defined notion of coarse-graining: the field modes corresponding to eigenvalues $\lambda_n > k$ have already been integrated over, while those corresponding to eigenvalues $\lambda_n < k$ still remain to be integrated.

The following discussion is an almost word-by-word repetition of the one in section 6.5. In order to define the gravitational EAA we start from the Euclidean partition function (3.111) and we add to the action the cutoff term

$$\Delta S_k(h, \bar{C}, C; \bar{g}) = \frac{1}{2} \int d^d x \sqrt{\bar{g}}\, h_{\mu\nu} \mathcal{R}_k^{\mu\nu\rho\sigma}(\bar{g}) h_{\rho\sigma} + \int d^d x \sqrt{\bar{g}}\, \bar{C}_\mu \bar{g}^{\mu\nu} \mathcal{R}_k^{(gh)}(\bar{g}) C_\nu \ . \tag{6.100}$$

We have indicated that the cutoff is constructed with the background metric but at this stage we do not commit ourselves to a function of a specific operator. We then define the EAA as the Legendre transform of W_k, minus the cutoff term:

$$\Gamma_k(h, \bar{C}, C; \bar{g}) = -W_k(j, \bar{J}, J; \bar{g}) + \int d^d x \sqrt{\bar{g}}(j^{\mu\nu} h_{\mu\nu} + J^\mu \bar{C}_\mu + \bar{J}^\mu C_\mu)$$
$$-\Delta S_k(h, \bar{C}, C; \bar{g}) \ . \tag{6.101}$$

By a slight abuse of language, the fields in the argument of the EAA are the expectation values of the quantum fields by the same name. It should always be clear from the context whether one or the other is meant.

By following the same steps as in the scalar case one shows that this functional obeys the ERGE

$$\frac{d\Gamma_k(\varphi; \bar{A})}{dt} = \frac{1}{2}\mathrm{Tr}\left(\frac{1}{\sqrt{\bar{g}}}\frac{\delta^2(\Gamma_k + \Delta S_k)}{\delta\varphi\delta\varphi}\right)^{-1}\frac{d}{dt}\frac{1}{\sqrt{\bar{g}}}\frac{\delta^2\Delta S_k}{\delta\varphi\delta\varphi} \ , \tag{6.102}$$

where we have denoted collectively the quantum fields $\varphi = (h_{\mu\nu}, \bar{C}_\mu, C_\mu)$. Note that factors $\sqrt{\bar{g}}$ that are needed to have tensorial objects cancel out.

The ghost action is bilinear in the ghost fields but also contains the Levi-Civita covariant derivative ∇ of the full metric field $g_{\mu\nu}$. When expanded, this contains interaction terms between $h_{\mu\nu}$ and the ghosts. Therefore, the Hessian has the general structure

$$\begin{bmatrix} \frac{\delta^2\Gamma_k}{\delta h \delta h} & \frac{\delta^2\Gamma_k}{\delta h \delta \bar{C}} & \frac{\delta^2\Gamma_k}{\delta h \delta C} \\ \frac{\delta^2\Gamma_k}{\delta \bar{C} \delta h} & 0 & \frac{\delta^2\Gamma_k}{\delta \bar{C} \delta C} \\ \frac{\delta^2\Gamma_k}{\delta C \delta h} & \frac{\delta^2\Gamma_k}{\delta C \delta \bar{C}} & 0 \end{bmatrix} \ .$$

Adding the cutoff term modifies the h-h and the $\bar{C}$-C terms but not the mixed ones. The inverse of the cutoff Hessian will be denoted G and has the structure

$$\begin{bmatrix} G_{hh} & G_{h\bar{C}} & G_{hC} \\ G_{\bar{C}h} & 0 & G_{\bar{C}C} \\ G_{Ch} & G_{C\bar{C}} & 0 \end{bmatrix} \ .$$

On the other hand the cutoff term does not contain mixed h-C and h-$\bar{C}$ terms, hence the trace can be written as the sum of two terms

$$\frac{d\Gamma_k(\varphi;\bar{g})}{dt} = \frac{1}{2}\text{Tr}\left(\frac{1}{\sqrt{\bar{g}}}\frac{\delta^2(\Gamma_k+\Delta S_k)}{\delta\varphi\delta\varphi}\right)_{hh}^{-1}\frac{d\mathcal{R}_k}{dt} - \text{Tr}\left(\frac{1}{\sqrt{\bar{g}}}\frac{\delta^2(\Gamma_k+\Delta S_k)}{\delta\varphi\delta\varphi}\right)_{\bar{C}C}^{-1}\frac{d\mathcal{R}_k^{(gh)}}{dt} .$$

$$(6.103)$$

Note that the mixed h-C and h-$\bar{C}$ terms enter into the traces in (6.103).

It is important to stress that it is not possible to write a flow equation for a functional of the full metric $g_{\mu\nu}$ alone. The gauge fixing and cutoff actions introduce separate dependences on the background and quantum fluctuation, so that the EAA must necessarily be a separate functional of these two arguments. This is due to the form of the cutoff and gauge fixing terms. In the limit $k \to 0$ the cutoff term vanishes but the gauge fixing remains. The EA is still formally a function of two arguments, but this is merely a gauge-dependent artifact that has no effect on physical observables.

Let $\bar{\Gamma}_k(\bar{g}) = \Gamma_k(0,0,0;\bar{g})$ be the gauge invariant functional obtained by putting the expectation values of the quantum fields to zero. We then split

$$\Gamma_k(h,\bar{C},C;\bar{g}) = \bar{\Gamma}_k(\bar{g}+h) + \hat{\Gamma}_k(h,\bar{C},C;\bar{g}) \qquad (6.104)$$

where the first term depends only on the full field $g_{\mu\nu}$. This implies in particular that

$$\hat{\Gamma}_k(0,0,0;\bar{g}) = 0 . \qquad (6.105)$$

We can then try to write a flow equation for the functional $\bar{\Gamma}_k$ by putting all the fluctuation fields φ to zero. In this case $\nabla_\mu \to \bar{\nabla}_\mu$ in the ghost operator, and the mixed h-C and h-$\bar{C}$ terms go to zero:

$$\begin{bmatrix} \frac{\delta^2\Gamma_k}{\delta a\delta a} & 0 & 0 \\ 0 & 0 & \frac{\delta^2\Gamma_k}{\delta c\delta c} \\ 0 & \frac{\delta^2\Gamma_k}{\delta c\delta\bar{c}} & 0 \end{bmatrix} .$$

Then (6.103) simplifies to

$$\frac{d\bar{\Gamma}_k(\bar{g})}{dt} = \frac{1}{2}\text{Tr}\left(\frac{1}{\sqrt{\bar{g}}}\frac{\delta^2\Gamma_k}{\delta h\delta h}+\mathcal{R}_k\right)^{-1}\frac{d\mathcal{R}_k}{dt} - \text{Tr}\left(\frac{1}{\sqrt{\bar{g}}}\frac{\delta^2\Gamma_k}{\delta\bar{C}\delta C}+\mathcal{R}_k^{(gh)}\right)^{-1}\frac{d\mathcal{R}_k^{(gh)}}{dt} ,$$

$$(6.106)$$

where the first term in the r.h.s. is due entirely to the metric fluctuations and the second to the ghosts. Note that on the right we still have the Hessian of the full Γ_k, so (6.106) is not a closed equation for $\bar{\Gamma}_k$.

In order to calculate the flow of $\bar{\Gamma}_k$, we consider the structure of the one-loop EAA at vanishing fluctuation:

$$\bar{\Gamma}_k^{(1)}(\bar{g}) = S(\bar{g}) + \frac{1}{2}\text{Tr}\,\log\left(\frac{1}{\sqrt{\bar{g}}}\frac{\delta^2(S+S_{GF}+\Delta S_k^{(h)})}{\delta h\delta h}\right)\bigg|_{h=0}$$

$$-\text{Tr}\,\log\left(\frac{1}{\sqrt{\bar{g}}}\frac{\delta^2(S_{gh}+\Delta S_k^{(gh)})}{\delta\bar{C}\delta C}\right) . \qquad (6.107)$$

Since S is a functional of the full field $g = \bar{g} + h$, we can write

$$\bar{\Gamma}_k^{(1)}(\bar{g}) = S(\bar{g}) + \frac{1}{2} \operatorname{Tr} \log \left(\frac{1}{\sqrt{g}} \frac{\delta^2 S(\bar{g})}{\delta \bar{g} \delta \bar{g}} + \frac{1}{\sqrt{g}} \frac{\delta^2 S_{GF}}{\delta h \delta h} \bigg|_{h=0} + \mathcal{R}_k \right)$$
$$- \operatorname{Tr} \log \left(\frac{1}{\sqrt{g}} \frac{\delta^2 S_{gh}}{\delta \bar{C} \delta C} + \mathcal{R}_k^{(gh)} \right) . \tag{6.108}$$

If the gauge fixing condition is linear in h, the gauge fixing action is quadratic in h and we can remove the suffix $h = 0$. If we now perform a "renormalization group improvement", replacing $S(\bar{g})$ by $\bar{\Gamma}_k(\bar{g})$ in the r.h.s., we obtain a closed equation for $\bar{\Gamma}_k$ with fixed gauge fixing and ghost actions:

$$\frac{d\bar{\Gamma}_k(\bar{g})}{dt} = \frac{1}{2} \operatorname{Tr} \left(\frac{1}{\sqrt{g}} \frac{\delta^2 \bar{\Gamma}_k}{\delta \bar{g} \delta \bar{g}} + \frac{1}{\sqrt{g}} \frac{\delta^2 S_{GF}}{\delta h \delta h} + \mathcal{R}_k \right)^{-1} \frac{d\mathcal{R}_k}{dt}$$
$$- \operatorname{Tr} \left(\frac{1}{\sqrt{g}} \frac{\delta^2 S_{gh}}{\delta \bar{C} \delta C} + \mathcal{R}_k^{(gh)} \right)^{-1} \frac{d\mathcal{R}_k^{(gh)}}{dt} . \tag{6.109}$$

This suggests taking

$$\hat{\Gamma}_k(h, \bar{c}, c; \bar{g}) = S_{GF}(h; \bar{g}) + S_{gh}(\bar{C}, C; \bar{g}) , \tag{6.110}$$

which is also consistent with (6.105), and ignoring the running of $\hat{\Gamma}_k$.

We will call (6.109) the "single metric flow equation". It reduces to the one-loop flow equation when the running of the couplings contained in the Hessian and/or the cutoff is neglected. It is important to keep in mind that unlike the scalar case, where the RG improvement led directly to the full ERGE, this is still an approximate equation. Its main virtue is that it is more manageable than the exact one. For the rest of this chapter we will only consider this simplified equation. As we have seen in the Yang-Mills case, it is perfectly adequate to obtain the one-loop beta functions, and it is mostly these that we shall discuss. A more systematic discussion of how to approximate the exact equation (6.103) will be given in section 7.2. It is only in sections 7.6-7.8 that we shall go a little beyond the single-metric approximation.

6.8   The Einstein-Hilbert truncation

The Einstein-Hilbert truncation consists in assuming that $\bar{\Gamma}_k$ has the form of the Hilbert action (3.125), with k-dependent cosmological constant and Newton's constant. For its simplicity, this truncation has been the testbed of every technique and approximation that has been devised to study the gravitational RG equation, and consequently it has been extensively discussed in the literature, starting from [207, 208]. In this section the simplicity of the truncation will allow us to compare the results of different cutoff schemes, a luxury that is progressively reduced going to more complicated truncations.

We will now discuss various cutoff types in turn. Besides the distinction between type I and type II, that we have already encountered in the preceding examples, one also has a choice between working with the field $h_{\mu\nu}$ (case "a") and performing the York decomposition (case "b"). In the literature all these cases, and others, have been used. We present them all for the sake of completeness, but the reader that is not especially interested in these technicalities is advised to look only at the case IIa, which is in several ways the simplest.

6.8.1 *Cutoff of type IIa*

The starting point of this calculation are the formulae for the Hessian of the Hilbert action, including the gauge fixing term in the Feynman gauge, which had been given by (3.134), (3.141), (3.144). We make a change in the notation by writing

$$\frac{1}{2\kappa^2} = Z_N . \tag{6.111}$$

The rationale for this is that in the Hessian this constant plays the role of the wave function renormalization of the graviton. Let us define the following operators acting on gravitons and on ghosts:

$$\begin{aligned}
\Delta_{2\rho\sigma}{}^{\mu\nu} &= -\bar{\nabla}^2 1_{\rho\sigma}^{\mu\nu} + W_{\rho\sigma}{}^{\mu\nu} \\
\Delta_{(gh)\mu}{}^{\nu} &= -\nabla^2 \delta_\mu^\nu - \bar{R}_\mu^\nu .
\end{aligned} \tag{6.112}$$

The operator Δ_2 agrees with the kinetic operator of the graviton, $\Delta_{(h)}$, except that the cosmological constant term has been removed.[5] It thus depends on the background field, but not on any coupling.

The type II cutoff is defined by the choice

$$\Delta S_k(h; \bar{g}) = \frac{1}{2} \int d^d x \sqrt{\bar{g}}\, h_{\mu\nu} \mathcal{R}_k^{\mu\nu\rho\sigma}\, h_{\rho\sigma} - \int d^d x \sqrt{\bar{g}}\, \bar{C}_\mu \mathcal{R}_k^{(gh)\mu}{}_\nu C^\nu , \tag{6.113}$$

where $\mathcal{R}_k$ is a function of Δ_2 and $\mathcal{R}_k^{(gh)}$ is a function of Δ_{gh}:

$$\begin{aligned}
\mathcal{R}_k^{\mu\nu\rho\sigma} &= Z_N K^{\mu\nu\rho\sigma} R_k(\Delta_2) \\
\mathcal{R}_k^{(gh)\mu}{}_\nu &= \delta_\nu^\mu R_k(\Delta_{gh}),
\end{aligned} \tag{6.114}$$

for gravitons and ghosts respectively. Here R_k is a scalar function with the properties listed in section 6.1.

For the modified inverse propagators, appearing in the r.h.s. of the ERGE, we have

$$\begin{aligned}
\frac{1}{\sqrt{\bar{g}}} \frac{\delta^2 \Gamma_k}{\delta h_{\mu\nu} \delta h_{\rho\sigma}} + \mathcal{R}_k^{\mu\nu\rho\sigma} &= Z_N K^{\mu\nu\rho\sigma} \left(P_k(\Delta_2) - 2\Lambda \right) \\
\frac{1}{\sqrt{\bar{g}}} \frac{\delta^2 \Gamma_k}{\delta \bar{C}_\mu \delta C^\nu} + \mathcal{R}_k^{(gh)\mu}{}_\nu &= P_k(\Delta_{(gh)}) \delta_\nu^\mu
\end{aligned} \tag{6.115}$$

Defining

$$\eta_N = -\frac{1}{Z_N} \frac{dZ_N}{dt} , \tag{6.116}$$

---

[5] The case when the cosmological constant is not removed will be discussed in section 6.8.5.

we then have

$$\frac{d\mathcal{R}_k^{\mu\nu\rho\sigma}}{dt} = Z_N K^{\mu\nu\rho\sigma} \left[\partial_t R_k(\Delta_2) - \eta_N R_k(\Delta_2)\right] , \tag{6.117}$$

$$\frac{d\mathcal{R}_k^{(gh)\,\mu}{}_\nu}{dt} = \partial_t R_k(\Delta_{gh})\delta^\mu{}_\nu . \tag{6.118}$$

The second heat kernel coefficients for the operators Δ_2 and Δ_{gh} are

$$b_2(\Delta_2) = \mathrm{tr}\left(\frac{\bar{R}}{6}\mathbf{1} - \mathbf{W}\right) = \frac{d(7 - 5d)}{12}\bar{R}$$

$$b_2(\Delta_{gh}) = \mathrm{tr}\left(\frac{\bar{R}}{6}\mathbf{1} + \mathrm{Ricci}\right) = \frac{d + 6}{6}\bar{R} .$$

Using the general formula (5.159) we can expand the r.h.s. of the ERGE. Collecting all terms and evaluating the traces leads to

$$\begin{aligned}
\frac{d\Gamma_k}{dt} &= \frac{1}{2}\mathrm{Tr}\frac{\partial_t R_k(\Delta_2) - \eta_N R_k(\Delta_2)}{P_k(\Delta_2) - 2\Lambda} - \mathrm{Tr}\frac{\partial_t R_k(\Delta_{(gh)})}{P_k(\Delta_{(gh)})} \\
&= \frac{1}{(4\pi)^{d/2}}\int dx\,\sqrt{g}\left\{\frac{d(d+1)}{4}Q_{\frac{d}{2}}\left(\frac{\partial_t R_k - \eta_N R_k}{P_k - 2\Lambda}\right) - dQ_{\frac{d}{2}}\left(\frac{\partial_t R_k}{P_k}\right)\right. \\
&\left.+ \left[\frac{d(7 - 5d)}{24}Q_{\frac{d}{2}-1}\left(\frac{\partial_t R_k - \eta_N R_k}{P_k - 2\Lambda}\right) - \frac{d + 6}{6}Q_{\frac{d}{2}-1}\left(\frac{\partial_t R_k}{P_k}\right)\right]\bar{R} + O(\bar{R}^2)\right\}.
\end{aligned} \tag{6.119}$$

We are now ready to extract the beta functions. The first line on the r.h.s. of (6.119) gives the beta function of $2Z_N\Lambda$, while the second gives the beta function of $-Z_N$. The beta functions can be written in the form[6]

$$\begin{aligned}
\frac{d}{dt}\left(\frac{2\Lambda}{16\pi G}\right) &= \frac{k^d}{16\pi}(A_1 + A_2\eta_N) \\
-\frac{d}{dt}\left(\frac{1}{16\pi G}\right) &= \frac{k^{d-2}}{16\pi}(B_1 + B_2\eta_N) ,
\end{aligned} \tag{6.120}$$

where

$$A_1 = \frac{16\pi(d - 3 + 8\tilde{\Lambda})}{(4\pi)^{\frac{d}{2}}\Gamma(\frac{d}{2})(1 - 2\tilde{\Lambda})}$$

$$A_2 = -\frac{16\pi(d + 1)}{(4\pi)^{\frac{d}{2}}(d + 2)\Gamma(\frac{d}{2})(1 - 2\tilde{\Lambda})}$$

$$B_1 = -\frac{4\pi(5d^2 - 3d + 24 - 8(d + 6)\tilde{\Lambda})}{3(4\pi)^{\frac{d}{2}}\Gamma(\frac{d}{2})(1 - 2\tilde{\Lambda})} \tag{6.121}$$

$$B_2 = \frac{4\pi(5d - 7)}{3(4\pi)^{\frac{d}{2}}\Gamma(\frac{d}{2})(1 - 2\tilde{\Lambda})}$$

---

[6]The definition of B_1 and B_2 given here agrees with the definition given originally in [207]. In [246], η_N was defined with the opposite sign, so the coefficients A_2 and B_2 have opposite sign to those given here.

6.8.2   *Cutoff of type Ia*

This is the scheme that was used originally in [207]. The cutoff action has again the form (6.113), but now the cutoff kernels are functions of the Bochner Laplacians, acting on symmetric tensors and vectors respectively:

$$\mathcal{R}_k^{\mu\nu\rho\sigma} = Z_N K^{\mu\nu\rho\sigma} R_k(-\bar{\nabla}^2) \,,$$
$$\mathcal{R}_k^{(gh)\mu}{}_{\nu} = \delta^\mu_\nu R_k(-\bar{\nabla}^2) \,. \tag{6.122}$$

As in the Yang-Mills case, this type of cutoff requires a bit more work. If we tried to keep the background field completely general, we would find that the kinetic operators, modified by the addition of the cutoff term,

$$\frac{1}{\sqrt{\bar{g}}} \frac{\delta^2 \Gamma_k}{\delta h_{\mu\nu} \delta h_{\rho\sigma}} + \mathcal{R}_k^{\mu\nu\rho\sigma} = Z_N K^{\mu\nu\alpha\beta} \left((P_k(-\nabla^2) - 2\Lambda) \mathbf{1}_{\alpha\beta}{}^{\rho\sigma} + W_{\alpha\beta}{}^{\rho\sigma} \right) \tag{6.123}$$

$$\frac{1}{\sqrt{\bar{g}}} \frac{\delta^2 \Gamma_k}{\delta \bar{C}_\mu \delta C^\nu} + \mathcal{R}_k^{(gh)\,\mu}{}_{\nu} = P_k(-\bar{\nabla}^2) \delta^\mu_\nu - \bar{R}^\mu_\nu \tag{6.124}$$

cannot be easily inverted. Since we are only interested in the expansion of the ERGE to first order in curvature, and remembering that W is linear in curvature, we could try to expand the propagators (*i.e.* the inverses of the above expressions) to first order in W and Ricci, respectively. This, however, leads to formulas for the trace that require information about the non-diagonal part of the heat kernel, and therefore cannot be calculated with the tools we have introduced so far.

There is an easy way out, which consists in specializing the background, for example to be a sphere. Let us go back to the Hessian written in the form (3.135). We separate the terms coming from the trace and tracefree parts of $h_{\mu\nu}$ by means of the trace projector

$$P_{\mu\nu}{}^{\rho\sigma} = \frac{1}{d} \bar{g}_{\mu\nu} \bar{g}^{\rho\sigma} \,. \tag{6.125}$$

It will be convenient to suppress indices and denote operators acting in the space of symmetric tensors by boldface letters, e.g. the trace projector is denoted $\mathbf{P}$. The tensor (3.136) can be written as

$$\mathbf{K} = \frac{1}{2} \left((\mathbf{1} - \mathbf{P}) - \frac{d-2}{2} \mathbf{P} \right)$$

and if $d \neq 2$, we can rewrite Eq. (3.135) in the form:

$$\mathcal{H} = \frac{1}{2}(\mathbf{1} - \mathbf{P}) \left(-\bar{\nabla}^2 - 2\Lambda\mathbf{1} + 2\mathbf{U} \right) - \frac{d-2}{4} \mathbf{P} \left(-\bar{\nabla}^2 - 2\Lambda\mathbf{1} - \frac{4}{d-2} \mathbf{U} \right) \,.$$

Using (3.176) the endomorphism $\mathbf{U}$ can be seen to act as a multiple of the identity in the spaces of tracefree and trace perturbations:

$$\mathbf{U} = \frac{1}{2} \left[(\mathbf{1} - \mathbf{P}) \frac{d^2 - 3d + 4}{d(d-1)} \bar{R} - \mathbf{P} \frac{(d-2)(d-4)}{2d} \bar{R} \right] \,.$$

Then we have

$$\mathcal{H} = \frac{1}{2}(1 - \mathbf{P})\left(-\nabla^2 - 2\Lambda + \frac{d^2 - 3d + 4}{d(d-1)}\bar{R}\right) - \frac{d-2}{4}\mathbf{P}\left(-\nabla^2 - 2\Lambda + \frac{d-4}{d}\bar{R}\right)$$

and finally adding the cutoff terms,

$$\frac{1}{\sqrt{g}}\frac{\delta^2\Gamma_k}{\delta h \delta h} + \mathcal{R}_k = \frac{Z_N}{2}\left[(1 - \mathbf{P})\left(P_k(-\nabla^2) - 2\Lambda + \frac{d^2 - 3d + 4}{d(d-1)}\bar{R}\right)\right.$$

$$\left. - \frac{d-2}{2}\mathbf{P}\left(P_k(-\nabla^2) - 2\Lambda + \frac{d-4}{d}\bar{R}\right)\right]. \tag{6.126}$$

Notice the way in which the sign issue of the trace part has been automatically addressed: the kinetic term of the trace has the opposite sign of the tracefree part, and the cutoff of the trace part also has the opposite sign, so that in both cases they combine in the function P_k. In this form, the kinetic operator can be easily inverted. The other piece in the tensor trace is

$$\frac{d\mathcal{R}_k}{dt} = \frac{Z_N}{2}\left((1 - \mathbf{P}) - \frac{d-2}{2}\mathbf{P}\right)[\partial_t R_k - \eta_N R_k] . \tag{6.127}$$

Putting together these pieces the approximated ERGE reads

$$\frac{d\Gamma_k}{dt} = \frac{1}{2}\text{Tr}\left((1 - \mathbf{P})\frac{\partial_t R_k - \eta_N R_k}{P_k(-\nabla^2) - 2\Lambda + \frac{d^2 - 3d + 4}{d(d-1)}\bar{R}}\right)$$

$$+ \frac{1}{2}\text{Tr}\left(\mathbf{P}\frac{\partial_t R_k - \eta_N R_k}{P_k(-\nabla^2) - 2\Lambda + \frac{d-4}{d}\bar{R}}\right) - \text{Tr}\left(\frac{\partial_t R_k}{P_k(-\nabla^2) - \frac{1}{d}\bar{R}}\right) \tag{6.128}$$

and using the traces

$$\text{tr}\,\mathbf{1} = \frac{d(d+1)}{2} \ , \quad \text{tr}\,\mathbf{P} = 1 \ , \quad \text{tr}\,(1 - \mathbf{P}) = \frac{d^2 + d - 2}{2} \ ,$$

and the heat kernel coefficients of the Bochner Laplacian, one arrives at the following expansion

$$\frac{d\Gamma_k}{dt} = \frac{1}{(4\pi)^{d/2}}\int dx\,\sqrt{g}\left\{\frac{d(d+1)}{4}Q_{\frac{d}{2}}\left(\frac{\partial_t R_k - \eta_N R_k}{P_k - 2\Lambda}\right) - dQ_{\frac{d}{2}}\left(\frac{\partial_t R_k}{P_k}\right)\right.$$

$$+ \left[\frac{d(d+1)}{24}Q_{\frac{d}{2}-1}\left(\frac{\partial_t R_k - \eta_N R_k}{P_k - 2\Lambda}\right) - \frac{d}{6}Q_{\frac{d}{2}-1}\left(\frac{\partial_t R_k}{P_k}\right)\right.$$

$$\left.\left. - \frac{d(d-1)}{4}Q_{\frac{d}{2}}\left(\frac{\partial_t R_k - \eta_N R_k}{(P_k - 2\Lambda)^2}\right) - Q_{\frac{d}{2}}\left(\frac{\partial_t R_k}{P_k^2}\right)\right]\bar{R} + O(\bar{R}^2)\right\}. \tag{6.129}$$

The beta functions are again of the form (7.25), and the coefficients A_1 and A_2 are the same as in the case of the cutoffs of type IIa. However, the coefficients B_1 and B_2 are different. With the optimized cutoff, they are

$$B_1 = \frac{-4\pi(-d^3 + 15d^2 - 12d + 48 + (2d^3 - 14d^2 - 192)\tilde{\Lambda} + (16d^2 + 192)\tilde{\Lambda}^2)}{3(4\pi)^{\frac{d}{2}}d\,\Gamma(\frac{d}{2})(1 - 2\,\tilde{\Lambda})^2}$$

$$B_2 = -\frac{4\pi\,(d^2 - 9d + 14 - 2\,(d+1)(d+2)\,\tilde{\Lambda})}{3\,(4\pi)^{\frac{d}{2}}(d+2)\,\Gamma(\frac{d}{2})(1 - 2\,\tilde{\Lambda})^2} . \tag{6.130}$$

6.8.3 Cutoff of type IIb

In this case, prior to introducing the cutoff, the fluctuation $h_{\mu\nu}$ and the ghosts are decomposed into their different spin components according to the York decomposition (section 5.1). This is advantageous because for certain backgrounds, such as maximally symmetric spaces, it leads to a partial diagonalization of the kinetic operator and it allows an exact inversion. In this section we will therefore assume that the background is a sphere; this is enough to extract exactly and unambiguously the beta functions of the cosmological constant and Newton's constant.

The gauge-fixed Hessian of the Hilbert action has already been written in York-decomposed form in (5.94). Here we shall restrict our attention to the Feynman-de Donder gauge $\alpha = \beta = 1$, which leads to a vanishing h-σ mixing term. The quadratic action can be written in the form

$$
S^{(2)} + S_{GF} = \frac{Z_N}{2} \int d^d x \sqrt{\bar{g}} \left[c_2 h_{\mu\nu}^{TT} (\Delta_2 - 2\Lambda) h^{TT\mu\nu} + c_1 \hat{\xi}_\mu (\Delta_1 - 2\Lambda) \hat{\xi}^\mu \right.
$$

$$
\left. + c_\sigma \hat{\sigma} (\Delta_\sigma - 2\Lambda) \hat{\sigma} + c_h h (\Delta_h - 2\Lambda) h \right] ,
$$

where

$$
\begin{aligned}
\Delta_2 &= -\bar{\nabla}^2 + \frac{d^2 - 3d + 4}{d(d-1)} \bar{R} , \\
\Delta_1 &= -\bar{\nabla}^2 + \frac{d-3}{d} \bar{R} , \\
\Delta_h &= -\bar{\nabla}^2 + \frac{d-4}{d} \bar{R} , \\
\Delta_\sigma &= -\bar{\nabla}^2 + \frac{d-4}{d} \bar{R}
\end{aligned}
\tag{6.131}
$$

and

$$
c_2 = \frac{1}{2} , \qquad c_1 = 2 , \qquad c_h = -\frac{d-2}{2d} , \qquad c_\sigma = \frac{d-1}{d} .
\tag{6.132}
$$

The ghost operators are

$$
\begin{aligned}
\Delta_V &= -\bar{\nabla}^2 - \frac{\bar{R}}{d} \\
\Delta_S &= -\bar{\nabla}^2 + \frac{2}{d} \bar{R}
\end{aligned}
\tag{6.133}
$$

Note that we have used the redefined variables $\hat{\xi}_\mu$, $\hat{\sigma}$, $\hat{S}$, so that there is no Jacobian to be taken into account.

The cutoff is chosen separately in each spin sector:

$$
\mathcal{R}_{k,i} = Z_N c_i R_k(\Delta_i) , \qquad i = 2, 1, h, \hat{\sigma}
$$

in such a way that

$$S^{(2)} + S_{GF} + \Delta S_k = \frac{Z_N}{2} \int d^d x \sqrt{\bar{g}} \left[c_2 h_{\mu\nu}^{TT} (P_k(\Delta_2) - 2\Lambda) h^{TT\mu\nu} \right.$$

$$+ c_1 \hat{\xi}_\mu (P_k(\Delta_1) - 2\Lambda) \hat{\xi}^\mu + c_\sigma \hat{\sigma} (P_k(\Delta_\sigma) - 2\Lambda) \hat{\sigma}$$

$$\left. + c_h h (P_k(\Delta_h) - 2\Lambda) h \right] .$$

The ERGE is then obtained straightforwardly as

$$\frac{d\Gamma_k}{dt} = \frac{1}{2} \mathrm{Tr}_{(2)} \frac{\partial_t R_k - \eta_N R_k}{P_k - 2\Lambda} + \frac{1}{2} \mathrm{Tr}'_{(1)} \frac{\partial_t R_k - \eta_N R_k}{P_k - 2\Lambda}$$

$$+ \frac{1}{2} \mathrm{Tr}_{(0)} \frac{\partial_t R_k - \eta_N R_k}{P_k - 2\Lambda} + \frac{1}{2} \mathrm{Tr}''_{(0)} \frac{\partial_t R_k - \eta_N R_k}{P_k - 2\Lambda}$$

$$- \mathrm{Tr}_{(1)} \frac{\partial_t R_k}{P_k} - \mathrm{Tr}'_{(0)} \frac{\partial_t R_k}{P_k} \qquad (6.134)$$

The first term comes from the spin-2, transverse traceless components, the second from the spin-1 transverse vector, the third and fourth from the scalars h and σ. The last two contributions come from the transverse and longitudinal components of the ghosts. A prime or a double prime indicate that the first or the first and second eigenvalues have to be omitted from the trace. The reason for this has been explained in section 5.5.

The Q-functionals for the first four and for the last two terms are the same, and summing the heat kernel coefficients of various irreducible fields just reconstructs the heat kernel coefficients of the un-decomposed fields. When explicitly evaluated, the r.h.s. gives back (6.119), so in this case the beta functions are the same as with the type IIa cutoff.

6.8.4 *Cutoff of type Ib*

This type of cutoff was introduced in [208]. As in the previous section, the quantum fields are decomposed into irreducible spin components, but then the cutoff is chosen to depend only on $-\bar{\nabla}^2$: The cutoff is chosen separately in each spin sector:

$$\mathcal{R}_{k,i} = Z_N c_i R_k(-\bar{\nabla}^2) , \qquad i = 2, 1, h, \hat{\sigma}$$

in such a way that

$$S^{(2)} + S_{GF} + \Delta S_k = \frac{Z_N}{2} \int d^d x \sqrt{\bar{g}} \left[c_2 h_{\mu\nu}^{TT} \left(P_k(\Delta_2) + \frac{d^2 - 3d + 4}{d(d-1)} \bar{R} - 2\Lambda \right) h^{TT\mu\nu} \right.$$

$$+ c_1 \hat{\xi}_\mu \left(P_k(\Delta_1) + \frac{d-3}{d} \bar{R} - 2\Lambda \right) \hat{\xi}^\mu$$

$$\left. + c_\sigma \hat{\sigma} \left(P_k(\Delta_\sigma) + \frac{d-4}{d} \bar{R} - 2\Lambda \right) \hat{\sigma} + c_h h \left(P_k(\Delta_h) + \frac{d-4}{d} \bar{R} - 2\Lambda \right) h \right] .$$

In this case the ERGE is

$$\frac{d\Gamma_k}{dt} = \frac{1}{2}\text{Tr}_{(2)}\frac{\partial_t R_k - \eta_N R_k}{P_k + \frac{d^2-3d+4}{d(d-1)}\bar{R} - 2\Lambda} + \frac{1}{2}\text{Tr}'_{(1)}\frac{\partial_t R_k - \eta_N R_k}{P_k + \frac{d-3}{d}\bar{R}} - 2\Lambda$$

$$+\frac{1}{2}\text{Tr}_{(0)}\frac{\partial_t R_k - \eta_N R_k}{P_k + \frac{d-4}{d}\bar{R} - 2\Lambda} + \frac{1}{2}\text{Tr}''_{(0)}\frac{\partial_t R_k - \eta_N R_k}{P_k + \frac{d-4}{d}\bar{R} - 2\Lambda}$$

$$-\text{Tr}_{(1)}\frac{\partial_t R_k}{P_k - \frac{\bar{R}}{d}} - \text{Tr}'_{(0)}\frac{\partial_t R_k}{P_k - \frac{2\bar{R}}{d}} \; . \tag{6.135}$$

The traces are expanded using (5.159). The necessary heat kernel coefficients were given in table (5.2). Expanding the denominators to first order in $\bar{R}$, but keeping the exact dependence on Λ as in the case of a type Ia cutoff, one obtains

$$\frac{d\Gamma_k}{dt} = \frac{1}{(4\pi)^{d/2}}\int dx\,\sqrt{g}\left\{\frac{d(d+1)}{4}Q_{\frac{d}{2}}\left(\frac{\partial_t R_k - \eta_N R_k}{P_k - 2\Lambda}\right) - dQ_{\frac{d}{2}}\left(\frac{\partial_t R_k}{P_k}\right)\right.$$

$$+\bar{R}\left[-\frac{d^4 - 2d^3 - d^2 - 4d + 2}{4d(d-1)}Q_{\frac{d}{2}}\left(\frac{\partial_t R_k - \eta_N R_k}{(P_k - 2\Lambda)^2}\right) - \frac{d+1}{d}Q_{\frac{d}{2}}\left(\frac{\partial_t R_k}{P_k^2}\right)\right.$$

$$\left.\left. + \frac{d^4 - 13d^2 - 24d + 12}{24d(d-1)}Q_{\frac{d}{2}-1}\left(\frac{\partial_t R_k - \eta_N R_k}{P_k - 2\Lambda}\right) - \frac{d^2 - 6}{6d}Q_{\frac{d}{2}-1}\left(\frac{\partial_t R_k}{P_k}\right)\right] + O(\bar{R}^2)\right\}. \tag{6.136}$$

In principle in two dimensions one has to subtract the contributions of some excluded modes. However, the contributions of these isolated modes turn out to cancel. Thus, the ERGE is continuous in the dimension also at $d = 2$.

The beta functions have again the form (7.25); the coefficients A_1 and A_2 are the same as for the type Ia cutoff but now the coefficients B_1 and B_2 are

$$B_1 = 4\pi\left(d(d-1)(d^3 - 15d^2 - 36) + 24 - 2(d^5 - 8d^4 - 5d^3 - 72d^2 - 36d + 96)\tilde{\Lambda}\right.$$

$$\left. -16(d-1)(d^3 + 6d + 12)\tilde{\Lambda}^2\right)\Big/3(4\pi)^{\frac{d}{2}}d^2(d-1)\Gamma\left(\frac{d}{2}\right)(1 - 2\tilde{\Lambda})^2 \tag{6.137}$$

$$B_2 = -4\pi\frac{d(d^4 - 10d^3 + 11d^2 - 38d + 12) - 2(d+2)(d^4 - 13\,d^2 - 24d + 12)\,\tilde{\Lambda}}{3(4\pi)^{\frac{d}{2}}(2+d)(d-1)d^2\,\Gamma(\frac{d}{2})(1 - 2\tilde{\Lambda})^2} \; .$$

6.8.5 *Spectrally adjusted cutoff*

Finally, we could define the cutoff to be a function of the whole inverse propagator $\Gamma_k^{(2)}$, only stripped of overall prefactors. In the case of the graviton, $\Gamma_k^{(2)} = Z_N\mathbf{K}(\Delta_2 - 2\Lambda)$ while for the ghosts $\Gamma_{C\bar{C}}^{(2)} = \Delta_{gh}$, where Δ_2 and Δ_{gh} were defined in (6.112). The "type III" cutoff is defined for gravitons by the choice

$$\mathcal{R}_k = Z_N\mathbf{K}R_k(\Delta_2 - 2\Lambda) \; , \tag{6.138}$$

while for ghosts it is the same as in (6.114). Unlike the previous choices, in this case the operator appearing in the cutoff kernel contains the coupling Λ and is therefore itself a function of k. Cutoffs with this property are said to be "spectrally adjusted". They pose the conceptual issue that since the operator they depend on contains running couplings, it does not provide a fixed basis of eigenfunctions in field space. The basis changes as the flow evolves, making the notion of coarse graining unclear. (This is obviously not a problem for one-loop calculations, where one keeps the couplings in the operator fixed.) Nevertheless they offer computational advantages, and are often used, so we present such a calculation here.

Since the operator in the graviton cutoff now contains the coupling Λ, the derivative of the graviton cutoff now involves an additional term:

$$\frac{d\mathbf{R}_k}{dt} = Z_N \mathbf{K}\left(\partial_t R_k(\Delta_2 - 2\Lambda) - \eta_N R_k(\Delta_2 - 2\Lambda) - 2R'_k(\Delta_2 - 2\Lambda)\partial_t\Lambda\right), \quad (6.139)$$

where R'_k denotes the partial derivative of $R_k(z)$ with respect to z. Note that the use of the chain rule in the last term is only legitimate if the t-derivative of the operator appearing as the argument of R_k commutes with the operator itself. This is the case for the operator $\Delta_2 - 2\Lambda$, since its t-derivative is proportional to the identity. The modified inverse propagator is then simply

$$\mathbf{\Gamma}_k^{(2)} + \mathbf{R}_k = Z_N \mathbf{K} P_k(\Delta_2 - 2\Lambda)$$

for gravitons, while for ghosts it is again given by Eq. (6.115). Collecting,

$$\frac{d\Gamma_k}{dt} = \frac{1}{2}\mathrm{Tr}\frac{\partial_t R_k(\Delta_2 - 2\Lambda) - \eta_N R_k(\Delta_2 - 2\Lambda) - 2R'_k(\Delta_2 - 2\Lambda)\partial_t\Lambda}{P_k(\Delta_2 - 2\Lambda)} - \mathrm{Tr}\frac{\partial_t R_k(\Delta_{(gh)})}{P_k(\Delta_{(gh)})}.$$

$$(6.140)$$

The traces over the ghosts are exactly as in the case of a cutoff of type II. As in previous cases, one should now proceed to evaluate the trace over the tensors using Eq. (5.159) and the heat kernel coefficients of the operator $\Delta_2 - 2\Lambda$. However, the situation is now more complicated because the heat kernel coefficients $B_{2k}(\Delta_2 - 2\Lambda)$ contain terms proportional to Λ^k and $\Lambda^{k-1}R$, all of which contribute to the beta functions of $2\Lambda Z_N$ and $-Z_N$. This is in contrast to the calculations with cutoffs of types I and II, where only the first two heat kernel coefficients contributed to the beta functions of $2\Lambda Z_N$ and $-Z_N$. In order to resum all these contributions, one can proceed as follows. We define the function

$$W(z) = \frac{\partial_t R_k(z) - \eta_N R_k(z) - 2R'_k(z)\partial_t\Lambda}{P_k(z)}$$

and the function $\bar{W}(z) = W(z - 2\Lambda)$. It has been shown explicitly in section 5.7 (Eq. (5.168) and following) that $\mathrm{Tr}W = \mathrm{Tr}\bar{W}$. Then, the terms without R and the terms linear in R (which give the beta functions of $2\Lambda Z_N$ and $-Z_N$ respectively) correspond to the first two lines in (5.169). In this way we obtain

$$
\frac{d\Gamma_k}{dt} = \frac{1}{(4\pi)^{d/2}} \int d^d x \sqrt{g} \left\{ \frac{d(d+1)}{4} \sum_{i=0}^{\infty} \frac{(2\Lambda)^i}{i!} Q_{\frac{d}{2}-i} \left(\frac{\partial_t R_k - \eta_N R_k - 2\partial_t \Lambda R'_k}{P_k} \right) \right.
$$

$$
- d Q_{\frac{d}{2}} \left(\frac{\partial_t R_k}{P_k} \right)
$$

$$
+ \frac{d(7-5d)}{24} R \sum_{i=0}^{\infty} \frac{(2\Lambda)^i}{i!} Q_{\frac{d}{2}-1-i} \left(\frac{\partial_t R_k - \eta_N R_k - 2\partial_t \Lambda R'_k}{P_k} \right)
$$

$$
\left. - \frac{d+6}{6} Q_{\frac{d}{2}-1} \left(\frac{\partial_t R_k}{P_k} \right) R \right\}. \tag{6.141}
$$

The remarkable property of the optimized cutoff is that in even dimensions the sums in those expressions contain only a finite number of terms; in odd dimensions the sum is infinite but can still be evaluated analytically. The necessary Q-functionals are evaluated in section 6.9. Using the results (6.150), (6.151), (6.152) and (6.153), (6.154), (6.155), (6.156) the first sum in (6.141) gives

$$
\frac{1}{(4\pi)^{d/2}} \frac{d+1}{2} \frac{(k^2 + 2\Lambda)^{d/2}}{\Gamma(d/2)} \left(2 + \frac{\eta_N}{\frac{d}{2}+1} \frac{k^2 + 2\Lambda}{k^2} + 2 \frac{\partial_t \Lambda}{k^2} \right) \int dx \sqrt{g} \tag{6.142}
$$

whereas the second sum gives

$$
\frac{1}{(4\pi)^{d/2}} \frac{d(7-5d)}{24} \frac{(k^2 + 2\Lambda)^{\frac{d-2}{2}}}{\Gamma(d/2)} \left(2 + \frac{\eta_N}{d/2} \frac{k^2 + 2\Lambda}{k^2} + 2 \frac{\partial_t \Lambda}{k^2} \right) \int dx \sqrt{g} R \tag{6.143}
$$

This resummation can actually be done also with other cutoffs.

The beta functions cannot be written in the form (6.120) anymore, because of the presence of the derivatives of Λ on the right hand side of the ERGE. Instead we have

$$
\frac{d}{dt} \left(\frac{2\Lambda}{16\pi G} \right) = \frac{k^d}{16\pi} (A_1 + A_2 \eta_N + A_3 \partial_t \tilde{\Lambda}),
$$

$$
- \frac{d}{dt} \left(\frac{1}{16\pi G} \right) = \frac{k^{d-2}}{16\pi} (B_1 + B_2 \eta_N + B_3 \partial_t \tilde{\Lambda}),
\tag{6.144}
$$

where

$$
A_1 = \frac{16\pi(-4 + (d+1)(1 + 2\tilde{\Lambda})^{\frac{d}{2}+1})}{(4\pi)^{\frac{d}{2}} \Gamma(\frac{d}{2})}
$$

$$
A_2 = -\frac{16\pi(d+1)(1 + 2\tilde{\Lambda})^{\frac{d}{2}+1}}{(4\pi)^{\frac{d}{2}}(d+2)\Gamma(\frac{d}{2})}
$$

$$
A_3 = \frac{16\pi(d+1)(1 + 2\tilde{\Lambda})^{\frac{d}{2}}}{(4\pi)^{\frac{d}{2}} \Gamma\left(\frac{d}{2}\right)}
$$

$$
B_1 = \frac{4\pi(-4(d+6) + d(7-5d)(1 + 2\tilde{\Lambda})^{\frac{d}{2}})}{3(4\pi)^{\frac{d}{2}} \Gamma(\frac{d}{2})} \tag{6.145}
$$

$$B_2 = \frac{4\pi(5d-7)(1+2\tilde{\Lambda})^{\frac{d}{2}}}{3(4\pi)^{\frac{d}{2}}\Gamma(\frac{d}{2})}$$

$$B_3 = \frac{4\pi d(7-5d)(1+2\tilde{\Lambda})^{\frac{d}{2}-1}}{3(4\pi)^{\frac{d}{2}}\Gamma\left(\frac{d}{2}\right)}$$

6.9   Appendix: Evaluation of some Q-functionals

This section is a continuation of section 5.7, where we gave general formulas for the trace of a function of a differential operator. In this chapter we have already evaluated some such traces. We collect here for convenience all the relevant formulas.

Equation (5.159) is a general formula for the derivative expansion of the trace of a function of a differential operator Δ. The r.h.s. of the ERGE consists of traces of this type, where the function to be traced is

$$W(\Delta) = \frac{\partial_t R_k(\Delta) - \eta R_k(\Delta)}{(P_k(\Delta) + q)^\ell} ,$$

where $R_k(\Delta)$ is the cutoff function and $P_k(\Delta) = \Delta + R_k(\Delta)$. We collect here the properties of the Q-functionals when W has this particular form.

As usual, it is convenient to measure everything in units of k^2. Let us define the dimensionless variable y by $z = k^2 y$ and $\tilde{q} = q/k^2$. Then $R_k(z) = k^2 r(y)$ for some dimensionless function r, $P_k(z) = k^2(y + r(y))$ and $\partial_t R_k(z) = 2k^2(r(y) - yr'(y))$. We have

$$Q_n\left(\frac{\partial_t R_k}{(P_k+q)^\ell}\right) = \frac{2}{\Gamma(n)} k^{2(n-\ell+1)} \int_0^\infty dy\, y^{n-1} \frac{r(y) - yr'(y)}{(y + r(y) + \tilde{q})^\ell} \qquad (6.146)$$

In general the integral will depend on the details of the cutoff function. However, if $q = 0$ and $\ell = n+1$ they turn out to be independent of the shape of the function. Note that they are all dimensionless. For $n > 0$, as long as $r(0) \neq 0$, we have shown in (6.43) that:

$$Q_n\left(\frac{\partial_t R_k}{P_k^{n+1}}\right) = \frac{2}{\Gamma(n)} \int_0^\infty dy\, \frac{d}{dy}\left[\frac{1}{n}\frac{y^n}{(y+r)^n}\right] = \frac{2}{n!} . \qquad (6.147)$$

For $n = 0$ we find from (5.164), as long as $r(0) \neq 0$ and $r'(0)$ is finite,

$$Q_0\left(\frac{\partial_t R_k}{P_k}\right) = 2 . \qquad (6.148)$$

Regarding the other coefficients $Q_n\left(\frac{\partial_t R_k}{(P_k+q)^\ell}\right)$, whenever explicit evaluations are necessary, we will use the so-called "optimized cutoff function" [187]

$$R_k(z) = (k^2 - z)\theta(k^2 - z) . \qquad (6.149)$$

With this cutoff $\partial_t R_k = 2k^2\theta(k^2 - z)$. Since the integrals are all cut off at $z = k^2$ by the theta function in the numerator, we can simply use $P_k(z) = k^2$ in the denominators. For $n > 0$ we have

$$Q_n\left(\frac{\partial_t R_k}{(P_k+q)^\ell}\right) = \frac{2}{n!}\frac{1}{(1+\tilde{q})^\ell} k^{2(n-\ell+1)} \qquad (6.150)$$

where $\tilde{q} = qk^{-2}$. For $n = 0$ we have

$$Q_0 \left(\frac{\partial_t R_k}{(P_k + q)^\ell} \right) = \left. \frac{\partial_t R_k}{(P_k + q)^\ell} \right|_{z=0} = \frac{2}{(1 + \tilde{q})^\ell} k^{2(-\ell+1)} . \tag{6.151}$$

Finally, owing to the fact that the function $\frac{\partial_t R_k(z)}{(P_k(z)+q)^\ell}$ is constant in an open neighborhood of $z = 0$, we have

$$Q_n \left(\frac{\partial_t R_k}{(P_k + q)^\ell} \right) = 0 \ \text{for} \ n < 0 . \tag{6.152}$$

This has the remarkable consequence that with the optimized cutoff the trace in the ERGE consists of finitely many terms.

We also need some Q-functionals of $\frac{R_k}{(P_k+q)^\ell}$. For $n > 0$ we have

$$Q_n \left(\frac{R_k}{(P_k + q)^\ell} \right) = \frac{1}{(n + 1)!} \frac{1}{(1 + \tilde{q})^\ell} k^{2(n-\ell+1)} . \tag{6.153}$$

The function $\frac{R_k(z)}{(P_k(z)+q)^\ell}$ is equal to $\frac{k^2-z}{(k^2+q)^\ell}$ in an open neighborhood of $z = 0$; therefore

$$Q_0 \left(\frac{R_k}{(P_k + q)^\ell} \right) = \left. \frac{R_k}{(P_k + q)^\ell} \right|_{z=0} = \frac{1}{(1 + \tilde{q})^\ell} k^{2(-\ell+1)} \tag{6.154}$$

$$Q_{-1} \left(\frac{R_k}{(P_k + q)^\ell} \right) = \frac{1}{(1 + \tilde{q})^\ell} k^{-2\ell} , \quad Q_n \left(\frac{R_k}{(P_k + q)^\ell} \right) = 0 \ \text{for} \ n < -1 . \tag{6.155}$$

Finally, for the type III cutoff one also needs the following

$$Q_n \left(\frac{1}{(P_k + q)^\ell} \right) = \frac{1}{n!} \frac{k^{2(n-\ell)}}{(1 + \tilde{q})^\ell} \ \text{for} \ n \geq 0 ; \quad Q_n \left(\frac{1}{(P_k + q)^\ell} \right) = 0 \ \text{for} \ n < 0 . \tag{6.156}$$

In the literature on asymptotic safety one often encounters "threshold functions" Φ_n^p and $\tilde{\Phi}_n^p$, defined by [207]

$$\Phi_n^p(\tilde{w}) = \frac{1}{\Gamma(n)} \int dy \, y^{n-1} \frac{r(y) - yr'(y)}{(y + r(y) + \tilde{w})^p} , \tag{6.157}$$

$$\tilde{\Phi}_n^p(\tilde{w}) = \frac{1}{\Gamma(n)} \int dy \, y^{n-1} \frac{r(y)}{(y + r(y) + \tilde{w})^p} , \tag{6.158}$$

where $r(y)$ is defined as in (6.3). They are instances of Q-functionals, applied to specific functions:

$$Q_n \left(\frac{\partial_t R_k}{(P_k + w)^p} \right) = 2k^{2(n+1-p)} \Phi_n^p(\tilde{w}) , \tag{6.159}$$

$$Q_n \left(\frac{R_k}{(P_k + w)^p} \right) = 2k^{2(n+1-p)} \tilde{\Phi}_n^p(\tilde{w}) , \tag{6.160}$$

with $\tilde{w} = w/k^2$.

In section 7.4 we will need to consider traces of functions of higher-order differential operators. Equation (5.159) does not depend on the order of the differential operator, but the Q-functionals do. It will be sufficient to consider the Q-functionals of the argument $W = \frac{\dot{R}_k}{P_k}$, which we denote

$$Q(p, m) = Q_m \left(\frac{\dot{R}_k}{P_k} \right), \tag{6.161}$$

when the argument of R_k is an operator of order p. We choose the cutoff profile $R_k(z) = (k^p - z)\theta(k^p - z)$, where z is a differential operator of order p. Defining $z = yk^p$, we have

$$R_k(z) = (k^p - z)\theta(k^p - z) = k^p(1 - y)\theta(1 - y), \tag{6.162}$$

$$\dot{R}_k(z) = pk^p\theta(k^p - z) = pk^p\theta(1 - y), \tag{6.163}$$

$$P_k(z) = z + R_k(z) = k^p \quad \text{for } z < k^p, \tag{6.164}$$

$$\frac{\dot{R}_k}{P_k} = p\,\theta(1 - y). \tag{6.165}$$

For $m > 0$, we find

$$Q(p, m) = \frac{1}{\Gamma(m)} \int_0^\infty dz\, z^{m-1} \frac{\dot{R}_k(z)}{P_k(z)} = \frac{k^{mp}}{\Gamma(m)} \int_0^\infty dy\, y^{m-1} p\theta(1 - y)$$

$$= \frac{pk^{mp}}{\Gamma(m)} \int_0^1 dy\, y^{m-1} = \frac{pk^{mp}}{m\Gamma(m)}. \tag{6.166}$$

Furthermore $Q(p, 0) = p$.

Chapter 7

The gravitational fixed point

In QFT the action typically contains only a small number of terms and a theory is said to be "renormalizable" if the perturbative treatment produces only divergences that are proportional to such terms. One can then absorb the divergences in renormalizations of the respective couplings. This terminology is slightly misleading, in the sense that non-renormalizable theories can also be renormalized. The problem is that one has to introduce infinitely many counterterms and accordingly infinitely many couplings have to be determined by resorting to experiment. The real issue is not the divergences, but a lack of predictivity.

As already discussed in chapter 4, this limitation is actually not so strong as to make non-renormalizable theories completely useless. For example, the perturbative QFT or gravity can be used, and is predictive, below the Planck scale. But what happens at or above the Planck scale? Do we necessarily have to give up a QFT description? Could the theory somehow heal itself and remain predictive also above the Planck scale? In the next section we will describe a non-perturbative scenario where a non-renormalizable QFT could in principle be valid and predictive up to arbitrarily high energy scales. The rest of the chapter contains several calculations supporting this scenario for gravity.

7.1 Non-perturbative renormalizability

In this section we give the definition of non-perturbative renormalizability (*a.k.a.* Asymptotic Safety).[1] It is convenient to start from the notion of "theory space" introduced in section 6.3. It is the space of functionals of the fields where the EAA takes values. It can be parametrized by the dimensionless couplings $\tilde{g}_i = k^{-d_i} g_i$, where g_i are the coefficients of monomials constructed with the fields and their derivatives, and k is the cutoff. Furthermore, independently of the renormalizability, or lack thereof, of the theory, the ERGE provides a well-defined flow equation in theory space. All points along an RG trajectory are physically equivalent, in the

---

[1] This notion is rooted in Wilson's investigations of the RG [209]. Early examples of theories of this type have been discussed in [210–212]. The term "asymptotic safety" is due to Weinberg [39].

sense that they lead to the same effective action in the limit $k \to 0$. Therefore we will generally identify "theories" as RG trajectories in theory space.

When one integrates a RG trajectory towards the UV, several types of behavior are possible. It may happen that the RG trajectory tends to infinity (*i.e.* that at least one of the coordinates $\tilde{g}_i$ diverges) at some finite scale. Examples of such a behavior are the Landau poles of QED or of ϕ^4 theory in four dimensions. In such cases the theory only makes sense for a finite range of energies and must be regarded as an effective field theory with a UV cutoff. Alternatively, the trajectory may flow towards a fixed locus, a subset of theory space that is left invariant by the flow. The simplest, zero-dimensional example of fixed locus is a fixed point (FP), namely a point where

$$\tilde{\beta}_i(\tilde{g}_{j*}) = 0 . \tag{7.1}$$

A one-dimensional fixed locus would be a limit cycle and in principle there may exist higher dimensional analogues, including the possibility of the RG trajectory ergodically filling some open subset of theory space. We will not discuss such exotic possibilities here and concentrate only on the case of fixed points.

A trajectory that reaches a FP in the UV is called a "renormalizable" or "asymptotically safe" (AS) trajectory. Such a trajectory corresponds to a UV complete theory. A physical observable such as a cross-section σ must be a function of the couplings g_i and of the external momenta. The couplings depend on a scale k and one can identify k with one of the characteristic momenta of the process. Then, from dimensional analysis, $\sigma = k^{-2}\tilde{\sigma}(\tilde{g}_i, X)$, where X denotes dimensionless kinematical variables such as angles and ratios of energies. One does not expect $\tilde{\sigma}$ to have a singularity precisely at $\tilde{g}_{i*}$, so on a renormalizable trajectory all physical observables should be well defined in the UV limit.

As already discussed in section 6.3, couplings that can be eliminated from the action by redefining the fields are called "redundant" or "inessential". They do not affect physical observables and therefore their behavior is not restricted. Only the essential couplings are required to reach a fixed point. In general, when the essential couplings approach a fixed point, the inessential ones will obey a scaling relation. For example the wave function renormalization satisfies

$$Z_\varphi(t) = Z_\varphi(0)e^{-\eta_* t} , \tag{7.2}$$

where η_* is the value of the anomalous dimension (6.23) at the fixed point.

It is also worth noting that asymptotic safety does not mean absence of UV divergences: it is only the dimensionless $\tilde{g}_i$ that must have finite limits. The couplings with positive mass dimension will have power divergences at a fixed point (for example, masses will diverge quadratically). In fact, at a fixed point every dimensionful coupling scales with the cutoff as dictated by its canonical dimension.

The fixed points determine the qualitative properties of the flow and therefore an understanding of their properties is one of the first steps in such a study. After having determined the position of all fixed points, the next step is to try and understand the nature of their basins of attraction, both in the IR and in the UV. One calls IR- (resp. UV-) critical surface of a FP the set of all points that flow to it for $t \to -\infty$ (resp. $t \to \infty$).[2] In general one cannot say much on these surfaces, but one can determine their tangent space at the FP by studying the linearized flow.

Suppose $\tilde{g}_*$ is a FP of the flow. Let $y_i = \tilde{g}_i - \tilde{g}_{i*}$ be new coordinates centered at the FP. The linearized flow equations are

$$\frac{dy_i}{dt} = M_{ij} y_j \;, \tag{7.3}$$

where

$$M_{ij} = \left. \frac{\partial \tilde{\beta}_i}{\partial \tilde{g}_j} \right|_* \tag{7.4}$$

Let S be a linear transformation that diagonalizes M:

$$S_{ik}^{-1} M_{k\ell} S_{\ell n} = \delta_{in} \lambda_n \;, \tag{7.5}$$

where λ_n are the eigenvalues of M. The linearized RG equation for the variables

$$z_i = S_{ik}^{-1} y_k \tag{7.6}$$

read

$$\frac{dz_i}{dt} = \lambda_i z_i \;, \tag{7.7}$$

so

$$z_i(t) = C_i \exp(\lambda_i t) = C_i \left(\frac{k}{k_0} \right)^{\lambda_i} \;, \tag{7.8}$$

where C_i are arbitrary constants. The scaling exponents θ_i of the fixed point are defined by $\theta_i = -\lambda_i$.

The directions z_i that correspond to positive eigenvalues (negative scaling exponent) are repelled by the FP, for growing t: they are UV repulsive and IR attractive. Such infinitesimal deformations of the FP theory are said to be *irrelevant*. The directions z_i that correspond to negative eigenvalues (positive scaling exponent) are attracted by the FP, for growing t: they are UV attractive and IR repulsive. Such infinitesimal deformations of the FP theory are said to be *relevant*. A deformation with eigenvalue zero is said to be marginal and in order to establish whether it is marginally relevant or irrelevant one has to go beyond the linearized analysis. Except for the case of the Gaussian FP, discussed below, marginal deformations are rare and in the following we will assume that they are absent.

Note that it is in general meaningless to say that a FP is an ultraviolet FP or an infrared FP. This notion only makes sense when there is a single coupling constant. In general, a fixed point will be reached in the UV along some directions and in the IR along others.

---

[2]In the literature on critical phenomena one encounters only one notion of critical surface, namely what we call here an IR-critical surface. This is because critical phenomena are related to the thermodynamic (IR) limit.

There follows from the preceding definition that the tangent space to the UV critical surface at the FP is the space spanned by the relevant directions, whereas the tangent space to the IR critical surface is the space spanned by the irrelevant directions. In particular, the dimension of the UV critical surface is equal to the number of positive scaling exponents (negative eigenvalues of the matrix M) and that of the IR critical surface is equal to the number of negative scaling exponents (positive eigenvalues of the matrix M). As we shall see below, one expects the former to be finite and the latter to be infinite. These notions are illustrated in the figure 7.1.

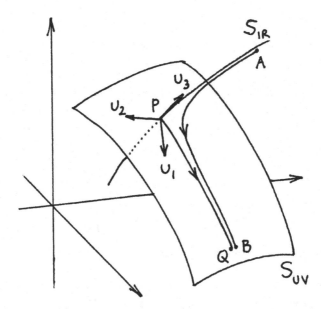

Fig. 7.1 Behavior of trajectories near a fixed point P. The arrows flow from the UV to the IR. u_1 and u_2 are relevant directions, u_3 is irrelevant. The point Q lies in the UV critical surface of P, so increasing the cutoff one is led to the fixed point P. The trajectory QP describes a UV complete theory. If one starts at the same scale from the point B, which is close to Q but does not lie in the UV critical surface, the RG trajectory will initially be close to the renormalizable trajectory QP. It will slow down near P but at some scale Λ_{max} it will be repelled in some irrelevant direction and possibly encounter a Landau pole. Thus the trajectory BA describes an EFT with cutoff at Λ_{max}. The closer the point B is to S_{UV}, the higher is the scale Λ_{max}.

In the theory of critical phenomena, the scaling exponents are directly related to important measurable quantities, the critical exponents. They should therefore be independent of immaterial details of the parametrization of the system. It is easy to see that this is indeed the case. Suppose we change the definition of the couplings $\tilde{g}_i$:

$$\tilde{g}_i' = \tilde{g}_i'(\tilde{g}_j) \ . \tag{7.9}$$

We can regard this as a coordinate transformation in theory space. The beta functions can be regarded as a vectorfield in this space, and their transformation under (7.9) is

$$\tilde{\beta}'_i = \frac{\partial \tilde{g}'_i}{\partial \tilde{g}_j} \tilde{\beta}_j \ . \tag{7.10}$$

Then, the matrix M transforms as

$$M'_{ij} = \frac{\partial \tilde{g}_k}{\partial \tilde{g}'_j} \frac{\partial^2 \tilde{g}'_i}{\partial \tilde{g}_k \partial \tilde{g}_\ell} \beta_\ell + \frac{\partial \tilde{g}'_i}{\partial \tilde{g}_\ell} M_{\ell k} \frac{\partial \tilde{g}_k}{\partial \tilde{g}'_j} \tag{7.11}$$

At a FP the first term vanishes, so the matrix M transforms as

$$M'_{ij}\Big|_* = \frac{\partial \tilde{g}'_i}{\partial \tilde{g}_\ell} M_{\ell k} \frac{\partial \tilde{g}_k}{\partial \tilde{g}'_j}\Big|_* \ . \tag{7.12}$$

This implies that the scaling exponents are the same independently of the choice of coordinates in the space of couplings. Of course, if a coordinate transformation is singular at the FP this need not be the case.

A free theory has vanishing beta functions and therefore must be a FP of the RG flow. It is called the Gaussian FP. Ordinary perturbation theory is the study of an infinitesimal neighborhood of the Gaussian FP. Let us Taylor expand the second term of (6.21) around the Gaussian FP:

$$\alpha_i(\tilde{g}_j) = \alpha_{ij}\tilde{g}_j + \alpha_{ijk}\tilde{g}_j\tilde{g}_k + \dots \ . \tag{7.13}$$

Note that there cannot be a constant term otherwise $\tilde{g}_i = 0$ would not be a FP. When used in Eq. (7.4) we find

$$M_{ij} = -d_i \delta_{ij} + \alpha_{ij} \ . \tag{7.14}$$

Consideration of the Feynman diagrams that can enter into a beta function shows that the matrix α_{ij} only has entries for $j > i$,[3] so the eigenvalues of M_{ij} are again equal to $-d_i$. There follows that at the Gaussian FP the relevant couplings are those that have positive mass dimension. So, near the origin, the UV critical surface is simply the space spanned by the (power counting) renormalizable couplings. In this way we see that asymptotic safety at a Gaussian FP is equivalent to the statement that the theory is perturbatively renormalizable and asymptotically free.

In a local QFT there are only finitely many couplings g_i with positive mass dimension and consequently finitely many relevant deformations of the Gaussian FP. This is why ordinary renormalizable theories are highly predictive. The same cannot be said in general of other fixed points, since the scaling exponents near a non-Gaussian FP may receive large quantum corrections.

We thus arrive at one of the central points of this section: the requirement that the FP has only finitely many relevant deformations. If we use the condition of UV completeness (asymptotic safety) as a selection criterion for physical theories (*i.e.*

---

[3]The reader can check this explicitly in the case of the beta functions (6.46).

RG trajectories), then the number of free parameters that are left by this condition is equal to the dimension of the UV critical surface, minus one, because points along a trajectory define the same theory. If we want this scenario to be as predictive as in the Gaussian case, we must demand that this number be finite.

This is an extremely powerful requirement, because it eliminates almost all the freedom that is available in theory space. When it holds, the condition of lying on the UV critical surface fixes the values of infinitely many couplings as functions of a finite number of couplings. In the Gaussian case, the condition is that all power-counting nonrenormalizable couplings must vanish in the UV. In the non-Gaussian case it is harder to spell it out in detail, because the relevant and irrelevant directions do not coincide with the monomials that are present in the EAA, but will rather be mixings thereof, see Eq. (7.6). Nevertheless the final effect is the same. So we see that asymptotic safety at a non-Gaussian FP leads to the same good behavior as asymptotic freedom.

The fact that perturbation theory is of limited use is a drawback in practice but not in principle. If the FP is not too far from the Gaussian one, it can be studied by perturbative methods.

Is it reasonable to expect that a non-trivial FP has finitely many relevant deformations? From the preceding discussion, it appears that the scaling exponents θ_i are equal to the canonical dimension d_i plus quantum corrections. If the FP is not too far from the Gaussian one, the quantum effects should not be too strong and one expects at most a finite number of eigenvalues to change sign.

7.2   Approximation schemes for gravity

In this chapter we shall review some of the evidence that has been gathered so far for the existence of a nontrivial gravitational fixed point. Mainly for pedagogical reasons, in much of the discussion we shall restrict ourselves to one-loop calculations. Still, we will frame the calculations in the context of the ERGE. This is in part because the ERGE provides a direct and elegant way of calculating beta functions in any theory, irrespective of its renormalizability property. Furthermore, having understood the one-loop calculations in this framework, it will be easier to generalize the discussion to more sophisticated approximations.

There are many ways of approximating the ERGE. Systematic approximations are the loop expansion, the vertex expansion (that we shall call "level expansion" below), the derivative expansion, and others. When there is no obvious small parameter, one resorts to a method called a "truncation", that we shall discuss next.

7.2.1   *Truncations*

A fixed point is a simultaneous zero of the beta functions of *all* the couplings. It could be simply defined by setting the r.h.s. of (6.109) to zero. There are no known methods to solve such an equation, and the most common procedure so far has been to resort to truncations, which consist in retaining a subset of couplings. One makes an ansatz for the functional Γ_k of the form (6.17), but now the sum is over a subset of couplings only. This is equivalent to assuming that all the remaining couplings are set to zero. Inserting the ansatz in (6.109) and comparing with (6.18), one can, with some effort, read off the beta functions of the couplings in the truncation and then look for fixed points.

One can then calculate the beta functions of the couplings that have been left out. If they are found to vanish, the truncation is said to be consistent. In this case, the subspace where the remaining couplings are zero would have the following property: starting the RG evolution from a point belonging to this subspace, one would remain forever in this subspace. Unless there are special reasons due to symmetries, this is generally not the case. The beta functions of the remaining couplings are generally nonzero and even assuming that the couplings are initially zero, they are immediately generated by the RG flow. Still, a truncation may capture some essential features of the system. Perhaps the beta functions of the remaining couplings are small, or at least the new couplings that are switched on have only a small influence on the couplings belonging to the truncation. In the case of finite-dimensional truncations, a systematic procedure would consist of the following steps:

- calculate the beta functions in a truncation containing n couplings
- find a fixed point and compute its properties, in particular scaling exponents
- add a new coupling to the truncation and repeat the analysis

The enlarged system may or may not admit a fixed point. In the latter case, the fixed point of the original truncation is likely to be a truncation artifact. In the former case, one has to compare the first n coordinates in the two truncations. If they have changed little, and if this trend persists as one enlarges the truncation, there is reason to believe that the fixed point in the truncated theory reflects the existence of a genuine fixed point in the full theory.

There are also some indirect ways of testing the quality of a truncation. For example, physical observables should be independent of the details of the cutoff procedure. This will generally not be the case in an explicit calculation based on a truncation, and the weakness of this dependence can be taken as an indicator of the quality of the truncation.

7.2.2 *The level expansion*

As explained in section 6.7, the ERGE for gravity uses the background field method and necessarily involves a functional of two fields: the background field, and the expectation value of the fluctuation field. In this and in the next three sections we shall stay within the so-called single-field truncation, which consists of studying functionals of a single metric of the form (6.104), (6.110), where only the gauge fixing term and the cutoff have separate dependence on the background and on the fluctuation fields. (More precisely, the EAA is assumed to be a functional $\bar{\Gamma}_k$ of the full metric $g_{\mu\nu}$ while the gauge-fixing, ghost and cutoff terms are quadratic in the quantum fields and non-polynomial in the background field.) The flow equation is then given by (6.109), and it is important to recall that this is *not* an exact equation. Nevertheless, we have seen that in the Yang-Mills case it gives the correct standard result when one uses a one-loop approximation, and is quantitatively close to the expected result at higher orders.

In section 7.6 we shall go a little beyond the single-field truncation, and it is useful to introduce already at this stage some notation and terminology. Aside from the ghosts, the EAA for gravity can be regarded as a functional of the background field and of $h_{\mu\nu}$. We may equally well view it as a functional of the background field and of the full metric $g_{\mu\nu} = \bar{g}_{\mu\nu} + h_{\mu\nu}$. Since these are different functionals of the respective arguments, one should denote them by different symbols. In order not to multiply the number of symbols, we shall follow M. Reuter and denote the former by $\Gamma_k(h_{\mu\nu}, \bar{C}_\mu, C^\mu; \bar{g}_{\mu\nu})$ and the latter by $\Gamma_k(g_{\mu\nu}, \bar{C}_\mu, C^\mu, \bar{g}_{\mu\nu})$. The difference is merely between a semicolon and a comma, but this should be enough.

Even though the two functionals have the same physical content, it is convenient to use one or the other notation, depending on the type of approximation that one is using. For example in [213] the "double Einstein-Hilbert" truncation has been studied, containing a Hilbert action for $g_{\mu\nu}$ and another Hilbert action for $\bar{g}_{\mu\nu}$. In this case it is more natural to use the "colon" notation. Under some circumstances (which are almost the norm in particle physics) one deals only with very few quanta at the time. This would be the case if one wanted to discuss the scattering of gravitons, for example. In this case it is obviously more appropriate to use the "semicolon" notation.

Let us now define what is meant by the "level expansion". In the semicolon formalism, Eq. (6.17) can be written more explicitly in the form

$$\Gamma_k(h_{\mu\nu}, C_\mu, \bar{C}^\mu; \bar{g}_{\mu\nu}) = \sum_{n_d=0}^{\infty} \sum_{n_h=0}^{\infty} \sum_j g_{(n_d,n_h,n_c,\bar{n}_c,j)} \mathcal{O}_{(n_d,n_h,n_c,\bar{n}_c,j)}(h_{\mu\nu}, C_\mu, \bar{C}^\mu; \bar{g}_{\mu\nu})$$

(7.15)

where n_d is the number of derivatives in $\mathcal{O}$, n_h the number of fluctuation fields, n_C and $\bar{n}_C$ the numbers of ghosts and anti-ghosts, and j and additional label that characterizes the possible ways of contracting the indices on n_d derivatives, n_h symmetric tensors, n_C ghosts and $\bar{n}_C$ antighosts. (For fixed n_d, n_h, the sum over j will be finite.) The number n_h will be referred to as the "level".

Thus, for example, consider the "Hilbert operator" $\mathcal{O}_2 = \int d^d x \sqrt{g} R$. In this case there is only one possible structure and the index j is not necessary. The functional Taylor expansion gives

$$\mathcal{O}_2 = \mathcal{O}_{(2,0,0,0)} + \mathcal{O}_{(2,1,0,0)} + \mathcal{O}_{(2,2,0,0)} + \mathcal{O}_{(2,3,0,0)} + \dots$$

(7.16)

where $\mathcal{O}_{(2,0,0,0)}$ is the Hilbert action of the background field, $\mathcal{O}_{(2,1,0,0)}$ is $\int d^d x \sqrt{\bar{g}} h_{\mu\nu} \bar{G}^{\mu\nu}$ ($\bar{G}^{\mu\nu}$ being the Einstein tensor), $\mathcal{O}_{(2,2,0,0)}$ is given by (3.126), $\mathcal{O}_{(2,3,0,0)}$ is the graviton three-valent vertex etc.

Similarly from the expansion of the cosmological term $\mathcal{O}_0 = \int d^d x \sqrt{g}$ one gets

$$\mathcal{O}_0 = \mathcal{O}_{(0,0,0,0)} + \mathcal{O}_{(0,1,0,0)} + \mathcal{O}_{(0,2,0,0)} + \mathcal{O}_{(0,3,0,0)} + \dots \tag{7.17}$$

where $\mathcal{O}_{(0,0,0,0)}$ is the cosmological term of the background field, $\mathcal{O}_{(0,1,0,0)}$ is $\int d^d x \sqrt{\bar{g}} h_{\mu\nu} \bar{g}^{\mu\nu}$, $\mathcal{O}_{(0,2,0,0)}$ is a kind of graviton mass term, also contained in (3.126), $\mathcal{O}_{(0,3,0,0)}$ is a contribution to the graviton three-valent vertex etc.

In the single-field truncation all these operators arise from the expansion of the Hilbert action and therefore have only two couplings

$$g_{(2,0,0,0)} = -Z_N = -\frac{1}{16\pi G} \quad \text{and} \quad g_{(0,0,0,0)} = \frac{2\Lambda}{16\pi G} .$$

In a loop calculation one encounters vertices with higher powers of $h_{\mu\nu}$, but their coefficients are all identified by hand with G and Λ. In a general bi-field truncation, they will all have independent couplings, and we can write

$$g_{(2,n,0,0)} = -\frac{1}{16\pi G_n} , \quad g_{(0,n,0,0)} = \frac{2\Lambda_n}{16\pi G_n}$$

where G_n will be referred to as the "level-n Newton constant" and Λ_n as the "level n cosmological constant".

A straightforward calculation of the beta functions will not give the same beta functions for all the Newton couplings. We will briefly discuss in section 8.2.3 how this proliferation of couplings must be tamed by the use of suitable Ward identities.

7.3 The single-metric Einstein-Hilbert truncation at one loop

7.3.1 *Without cosmological constant*

We begin by considering a truncation consisting of a single coupling, namely the level-zero Newton constant G (and we do not write the subscript 0 for notational simplicity). In the next subsection we will take into account the level-zero cosmological constant. Let us begin by asking whether it is reasonable to expect a fixed point for Newton's constant. The answer is positive and the basic argument runs as follows. If we are in d-dimensional spacetime, the Einstein-Hilbert action is written conventionally in the form

$$-\frac{1}{16\pi G} \mathcal{O}_2 \tag{7.18}$$

where $\mathcal{O}_2 = \int d^d x \sqrt{g} R$ has dimension $2 - d$. If we evaluate the effective action in a background metric $g_{\mu\nu}$ in the presence of an UV cutoff k, we expect a power divergence of the form $k^{d-2} \mathcal{O}_2$. This is what we found in chapter 3, but it basically just follows from dimensional analysis. In perturbation theory the beta functions are proportional to the logarithmic derivative of the effective action with respect to the cutoff. Therefore

$$k\frac{d}{dk}\left(-\frac{1}{16\pi G(k)}\right) = \frac{B_1}{16\pi}k^{d-2} \,, \tag{7.19}$$

for some numerical constant B_1. This means for Newton's constant

$$k\frac{dG}{dk} = B_1\,G^2 k^{d-2} \,. \tag{7.20}$$

Now we recall that the RG flow has to be written for the coupling measured in units of the cutoff:

$$\tilde{G} = Gk^{d-2}, \tag{7.21}$$

so we arrive at the beta function

$$k\frac{d\tilde{G}}{dk} = (d-2)\tilde{G} + B_1\tilde{G}^2 \tag{7.22}$$

which has a fixed point at $\tilde{G} = 0$ and another one at

$$\tilde{G} = -(d-2)/B_1 \,. \tag{7.23}$$

This is what is meant by a fixed point for Newton's constant. Whether this is physically viable or not depends on the sign of B_1. With this beta function, Newton's constant cannot change sign and since it is positive at low energy it should also be positive at high energy. So if $d > 2$ we have a viable fixed point provided $B_1 < 0$. We see that the existence of a nontrivial fixed point for Newton's constant is not at all unlikely, in fact it is almost a generic consequence of its dimensionality. The first task is then to determine the sign of B_1.

The coefficient B_1 has been calculated from the ERGE in a number of different schemes in section 6.8. They are given by (6.121) for a cutoff of type II (and they are the same whether the York decomposition is used or not), by (6.130) for a cutoff of type Ia (no York decomposition) and by (6.137) for a cutoff of type Ib (with York decomposition). There are many other choices that one can vary in the calculation, e.g. the gauge (here they were all calculated in the Feynman-de Donder gauge), the shape of the cutoff function R_k (here the optimized shape was used) and others. One can get a taste of the typical variability of the result by considering the plot of these coefficients B_1 as functions of the dimension, for zero cosmological constant. This is shown in Fig. (7.2). For $d = 4$ the numbers change by a factor 2, but they show a degree of universality for $d \to 2$, where the result is $B_1 = -38/3$. This is related to the fact that Newton's coupling is dimensionless in two dimensions, and we have seen in several examples that the beta functions of dimensionless couplings is independent of cutoff details.[4]

One would like to know whether B_1 is negative independently of the shape of the function R_k. A general proof is easy if we use the cutoff of type II. The coefficient

---

[4]It must be noted, that there is a difference between taking the limit $d \to 2$ of the result for general d, and calculating directly in $d = 2$, where TT fluctuations of the metric do not exist.

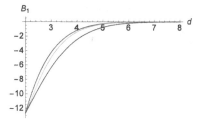

Fig. 7.2   The coefficient B_1 for $\tilde{\Lambda} = 0$ as a function of d, with cutoff of type Ia, Ib and II, from top to bottom.

B_1 is given by the terms proportional to R in Eq. (6.119), evaluated for $\eta_N = \tilde{\Lambda} = 0$. The function $\partial_t R_k(z)/P_k(z)$ is positive for all z, so $Q_{\frac{d}{2}-1}$ is positive. The coefficient of this Q-functional is

$$\frac{d(7-5d)}{24} - \frac{d+6}{6} < 0$$

so $B_1 < 0$. Somewhat less general statements can be made for the cutoffs of type I.

7.3.2   Expanding in the cosmological constant

Let us now see how things change when we consider also the cosmological constant. We define the dimensionless cosmological constant

$$\tilde{\Lambda} = \Lambda/k^2 \, , \tag{7.24}$$

then (6.120) implies

$$\frac{d\tilde{\Lambda}}{dt} = -2\tilde{\Lambda} + \frac{1}{2}(A_1 + A_2\eta_N)\tilde{G} + (B_1 + B_2\eta_N)\tilde{G}\tilde{\Lambda} \, ,$$
$$\frac{d\tilde{G}}{dt} = (d-2)\tilde{G} + (B_1 + B_2\eta_N)\tilde{G}^2 \, , \tag{7.25}$$

Here the anomalous dimension is given by Eq. (6.116) and therefore is related to the beta function of $\tilde{G}$. As we shall discuss in section 7.6, a better procedure is to evaluate the anomalous dimension with an independent calculation. Here we restrict ourselves to the one-loop approximation, which corresponds to neglecting the running of couplings in the r.h.s. of the ERGE. Thus we set $\eta_N = 0$.[5]

The coefficients A_1 and B_1 are functions of $\tilde{\Lambda}$. Since the heat kernel methods, used to derive the beta functions, amount to an expansion around zero curvature, it makes sense to expand also in the cosmological constant. This can be motivated with the equation of motion (3.128), which implies that near the mass shell Λ and

---

[5]That the anomalous dimension vanishes in the one-loop approximation is specific to the single-field (level-zero) approximation. If the wave function renormalization is treated as an independent parameter, as in section 7.6.2, one can have a nonzero anomalous dimension at one loop.

R are comparable. In the leading order of this expansion the beta functions read

$$\frac{d\tilde{\Lambda}}{dt} = -2\tilde{\Lambda} + \frac{1}{2}A_1\tilde{G} + B_1\tilde{G}\tilde{\Lambda} \,,$$

$$\frac{d\tilde{G}}{dt} = (d-2)\tilde{G} + B_1\tilde{G}^2 \,, \tag{7.26}$$

where A_1 and B_1 are evaluated at $\tilde{\Lambda} = 0$, hence are pure numbers. We call this the "simple Einstein-Hilbert flow".

These beta functions have two fixed points: the Gaussian fixed point at $\tilde{\Lambda} = 0$, $\tilde{G} = 0$, and a nontrivial fixed point at

$$\tilde{\Lambda}_* = -\frac{(d-2)A_1}{2dB_1} \,, \qquad \tilde{G}_* = -\frac{d-2}{B_1} \,, \tag{7.27}$$

The stability matrix (7.4)

$$M = \begin{pmatrix} \frac{\partial\beta_{\tilde{\Lambda}}}{\partial\tilde{\Lambda}} & \frac{\partial\beta_{\tilde{\Lambda}}}{\partial\tilde{G}} \\ \frac{\partial\beta_{\tilde{G}}}{\partial\tilde{\Lambda}} & \frac{\partial\beta_{\tilde{G}}}{\partial\tilde{G}} \end{pmatrix} \tag{7.28}$$

is equal to

$$M = \begin{pmatrix} -2 & A_1/2 \\ 0 & -(d-2) \end{pmatrix} \,, \tag{7.29}$$

at the Gaussian fixed point and

$$M = \begin{pmatrix} -d & A_1/d \\ 0 & -(d-2) \end{pmatrix} \,, \tag{7.30}$$

at the nontrivial fixed point. To understand this difference one must recall that the eigenvalues are invariant under regular coordinate transformations in the space of the couplings. If we use $2Z_N\Lambda$ and $-Z_N$ as coordinates in theory space (or more precisely their dimensionless versions) we easily see from (6.120) (with $\eta_N = 0$) that the matrix M is constant and diagonal, with eigenvalues equal to minus the canonical dimensions of $2Z_N\Lambda$ and $-Z_N$. Since the transformation between Z_N and G is invertible at the nontrivial fixed point, the scaling exponents at that point are necessarily equal to d and $d-2$. On the other hand, the transformation between Z_N and G is singular at the nontrivial fixed point. In this case the scaling exponents turn out to be equal to the canonical dimensions of Λ and G, the couplings that appear in the perturbative expansion. For both fixed points, the eigenvectors of the stability matrix are $(1,0)$ and $(A_1/2d, 1)$.

The simple Einstein-Hilbert flow can be solved exactly. The general solution is

$$\tilde{\Lambda}(t) = \frac{(\tilde{\Lambda}_0 - \frac{A_1\tilde{G}_0}{2d}(1 - e^{dt}))e^{-2t}}{1 + \frac{B_1\tilde{G}_0}{d-2}(1 - e^{(d-2)t})} \,,$$

$$\tilde{G}(t) = \frac{\tilde{G}_0 e^{(d-2)t}}{1 + \frac{B_1\tilde{G}_0}{d-2}(1 - e^{(d-2)t})} \,. \tag{7.31}$$

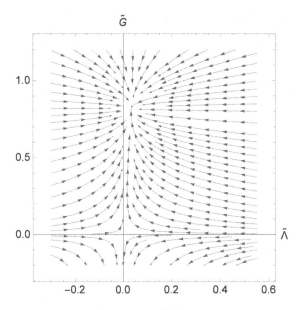

Fig. 7.3 The simple Einstein-Hilbert flow in the $\tilde\Lambda$-$\tilde G$ plane in $d = 4$ in the Feynman-de Donder gauge $\alpha = \beta = 1$ with type II cutoff.

For explicit numerical estimates one can use the coefficients A_1 and B_1 of the type II cutoff (6.121). In $d = 4$ one has $A_1 = 1/\pi$ and $B_1 = -23/3\pi$, so the flow equations are

$$\frac{d\tilde\Lambda}{dt} = -2\tilde\Lambda + \frac{1}{2\pi}\tilde G - \frac{23}{3\pi}\tilde G\tilde\Lambda \ ,$$

$$\frac{d\tilde G}{dt} = 2\tilde G - \frac{23}{3\pi}\tilde G^2 \ ,$$

(7.32)

which have a fixed point at

$$\tilde\Lambda_* = \frac{3}{92} \approx 0.0326 \ ; \qquad \tilde G_* = \frac{6\pi}{23} \approx 0.8195 \ .$$

(7.33)

This flow is shown in Fig. 7.3. Note the special trajectory joining the two fixed points. It is a straight line in the direction of the second eigenvector of the stability matrices.

When the dependence of the coefficients A_1 and B_1 on $\tilde\Lambda$ is taken into account, the flow equations become more complicated. One can still find the fixed point and study its properties analytically. In the case $d = 4$ and using again the type II cutoff, the flow equations are

$$\frac{d\tilde\Lambda}{dt} = -2\tilde\Lambda + \frac{1 + 8\tilde\Lambda}{2\pi(1 - 2\tilde\Lambda)}\tilde G - \frac{23 - 20\tilde\Lambda}{3\pi(1 - 2\tilde\Lambda)}\tilde G\tilde\Lambda \ ,$$

$$\frac{d\tilde G}{dt} = 2\tilde G - \frac{23 - 20\tilde\Lambda}{3\pi(1 - 2\tilde\Lambda)}\tilde G^2 \ ,$$

(7.34)

The non-Gaussian fixed point is now shifted to

$$\tilde\Lambda = \frac{17 - \sqrt{229}}{40} \approx 0.0467 , \qquad \tilde{G} = \frac{13\sqrt{229} - 71}{510} \approx 0.774 . \qquad (7.35)$$

and the scaling exponents are now equal to $2.31 \pm 0.382i$. The complex exponents mean that the approach to the fixed point follows spirals rather than straight lines, but since the imaginary part is quite small, the spirals are very open. This flow is shown in Fig. 7.4

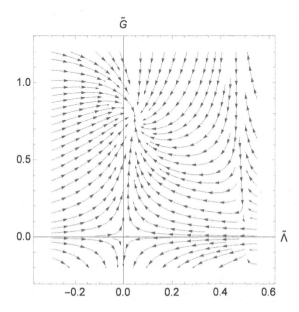

Fig. 7.4   The one-loop flow in the $\tilde\Lambda$-$\tilde{G}$ plane in $d = 4$, with type II cutoff, in the Feynman-de Donder gauge $\alpha = \beta = 1$, taking into account the full $\tilde\Lambda$-dependence of the beta functions.

The other striking feature of this flow is the singularity at $\tilde\Lambda = 1/2$, which hampers the evolution of some trajectories towards the infrared. This singularity is believed to be an artifact of the approximation made. It appears due to the $\tilde\Lambda$-dependence of the coefficients of the beta functions.

7.3.3   *Gauge dependence*

The results reported in the preceding sections were all relative to the Feynman-de Donder gauge $\beta = \frac{d}{2} - 1$, $\alpha = 1$. In this subsection we discuss their gauge dependence. For this we shall calculate the one-loop beta functions in the generic α-β gauge introduced in section 5.4.3. The starting point is the quadratic gauge-fixed action written in terms of the York variables $h_{\mu\nu}^{TT}$, $\hat\xi_\mu$, $\hat\sigma$ and h, written in Eq. (5.94).

In order to write the ERGE we have to define a cutoff term. The main nuisance is that in the generic gauge the quadratic action is not diagonal. As usual, there may be different ways to deal with this problem. The recipe that is most commonly used is to define the cutoff in such a way that the cutoff Hessian differs from the Hessian by the replacement of $-\bar{\nabla}^2$ by $P_k(-\bar{\nabla}^2)$. Formally we may write

$$\Delta S_k = (S^{(2)} + S_{GF})\Big|_{-\bar{\nabla}^2 \to P_k(-\bar{\nabla}^2)} - (S^{(2)} + S_{GF}) \,.$$

This is identical to the type Ib cutoff, introduced in section 6.8.4, on the diagonal terms.

Except for the additional complication of having to deal with a matrix in the scalar sector, the evaluation of the trace proceeds as before and one arrives at beta functions that are again of the form (7.26). When the relevant Hessian and heat kernel coefficients have been inserted in a computer, it takes relatively little effort to compute also the next term in the expansion of the beta functional:

$$\frac{d\Gamma_k}{dt} = \int d^d x \sqrt{\bar{g}} \left[\frac{A_1}{16\pi} k^d + \frac{B_1}{16\pi} k^{d-2} \bar{R} + C_1 k^{d-4} \bar{R}^2 + \ldots \right] \,. \tag{7.36}$$

While the first two terms give the coefficients of the beta functions of Λ and G, the third is a combination of the beta functions of the higher-derivative couplings. Due to the identities (3.177), on a spherical background one has a single independent invariant $\bar{R}^2$. In the Ricci-basis (2.79), its coefficient corresponds to the beta function

$$C_1 = \frac{1}{d} \beta_{a_1} + \beta_{a_2} + \frac{(d-2)(d-3)}{d(d-1)} \beta_{a_3} \,. \tag{7.37}$$

In four dimensions, integrating the flow (7.36) up to a scale Λ_{UV}, we see that

$$\Gamma_{\Lambda_{UV}} = \int d^4 x \sqrt{\bar{g}} \left[\frac{A_1}{64\pi} \Lambda_{UV}^4 + \frac{B_1}{32\pi} \Lambda_{UV}^2 \bar{R} + \frac{1}{2} C_1 \log\left(\frac{\Lambda_{UV}^2}{\mu^2}\right) \bar{R}^2 + \ldots \right] \,. \tag{7.38}$$

The coefficient C_1 is a linear combination of the coefficients of the one-loop logarithmic divergences of Einstein gravity. In this way one can obtain some information on the logarithmic divergences that holds in a more general gauge than the one used in chapter 3.

The coefficients A_1, B_1 and C_1 are very long expressions depending on Λ and on the gauge parameters α and β. We report here their complete form in $d = 4$:

$$A_1 = \frac{1}{2\pi} \left[\frac{2(3-\beta)^2 - 4\left(3 + 2\alpha - \beta^2\right) \tilde{\Lambda}}{(\beta-3)^2 - 4\left(3 + 2\alpha - \beta^2\right) \tilde{\Lambda} + 16\alpha\tilde{\Lambda}^2} + \frac{3}{1 - 2\alpha\tilde{\Lambda}} + \frac{5}{1 - 2\tilde{\Lambda}} - 8 \right] \tag{7.39}$$

$$B_1 = -\frac{m_0 + m_1\tilde{\Lambda} + m_2\tilde{\Lambda}^2 + m_3\tilde{\Lambda}^3 + m_4\tilde{\Lambda}^4 + m_5\tilde{\Lambda}^5 + m_6\tilde{\Lambda}^6 + m_7\tilde{\Lambda}^7 + m_8\tilde{\Lambda}^8}{24\pi(3-\beta)(1 - 2\tilde{\Lambda})^2(1 - 2\alpha\tilde{\Lambda})^2((3-\beta)^2 - 4(3 + 2\alpha - \beta^2)\tilde{\Lambda} + 16\alpha\tilde{\Lambda}^2)^2}$$

where

$$m_0 = 3(3 - \beta)^3(237 - 150\beta + 21\beta^2 + \alpha(62 - 36\beta + 6\beta^2))$$

$$m_1 = 4(3 - \beta)^2(-2241 + 1149\beta + 321\beta^2 - 141\beta^3$$
$$+\alpha^2(-288 + 96\beta) + \alpha(-4431 + 3335\beta - 801\beta^2 + 57\beta^3))$$

$$m_2 = 4\big(33669 - 28107\beta - 4086\beta^2 + 7218\beta^3 - 975\beta^4 - 103\beta^5$$
$$+\alpha(168354 - 181662\beta + 48660\beta^2 + 8516\beta^3 - 5622\beta^4 + 666\beta^5)$$
$$+\alpha^2(88362 - 97150\beta + 42252\beta^2 - 9572\beta^3 + 1242\beta^4 - 78\beta^5)$$
$$+\alpha^3(1512 - 936\beta + 216\beta^2 - 24\beta^3))$$

$$m_3 = 16\big(-6966 + 3870\beta + 3024\beta^2 - 1776\beta^3 - 90\beta^4 + 82\beta^5$$
$$+\alpha(-74169 + 59031\beta + 3618\beta^2 - 10814\beta^3 + 1623\beta^4 + 103\beta^5)$$
$$+\alpha^2(-101772 + 89228\beta - 20112\beta^2 - 1616\beta^3 + 1212\beta^4 - 156\beta^5)$$
$$+\alpha^3(-18576 + 13840\beta - 3840\beta^2 + 416\beta^3))$$

$$m_4 = 32(1242 - 270\beta - 900\beta^2 + 204\beta^3 + 162\beta^4 - 38\beta^5$$
$$+\alpha(32148 - 17628\beta - 10116\beta^2 + 6068\beta^3 + 144\beta^4 - 248\beta^5)$$
$$+\alpha^2(96999 - 65357\beta - 1566\beta^2 + 6778\beta^3 - 825\beta^4 - 29\beta^5)$$
$$+\alpha^3(48060 - 30220\beta + 4836\beta^2 - 20\beta^3) + \alpha^4(2448 - 752\beta)$$

$$m_5 = 256\alpha\big(-1467 + 369\beta + 840\beta^2 - 216\beta^3 - 117\beta^4 + 31\beta^5$$
$$+\alpha(-11244 + 5520\beta + 2196\beta^2 - 1156\beta^3 + 28\beta^5)$$
$$+\alpha^2(-12522 + 6394\beta + 18\beta^2 - 242\beta^3) + \alpha^3(-1872 + 560\beta))$$

$$m_6 = 512\alpha^2\big(2175 - 577\beta - 852\beta^2 + 228\beta^3 + 63\beta^4 - 17\beta^5$$
$$+\alpha(6270 - 2538\beta - 558\beta^2 + 218\beta^3) + \alpha^2(2184 - 632\beta))$$

$$m_7 = 4096\alpha^3\big(-327 + 89\beta + 66\beta^2 - 18\beta^3 + \alpha(-300 + 84\beta))$$

$$m_8 = 8192\alpha^4(69 - 19\beta)$$

The expressions simplify considerably in the case $\tilde{\Lambda} = 0$, in which case it is even possible to write them in arbitrary dimension:

$$A_1 = \frac{16\pi(d - 3)}{(4\pi)^{d/2}\Gamma[d/2]}, \tag{7.40}$$

$$B_1 = \Big[\big(d^5 - 16d^4 + 39d^3 - 96d^2 + 36d + 72\big)(d - 1)^2$$
$$-2\beta\big(d^6 - 17d^5 + 55d^4 - 123d^3 + 96d^2 + 60d - 72\big)$$
$$+\beta^2\big(d^5 - 16d^4 + 39d^3 - 60d^2 - 12d + 72\big)$$
$$\alpha\big(-6\big(4d^5 - 23d^4 + 53d^3 - 62d^2 + 36d - 8\big) + 48\beta(d - 2)(d - 1)^3$$
$$-24\beta^2(d - 2)(d - 1)^2\big)\Big]/\Big((4\pi)^{(d-1)/2}3d^2(d - 1)(d - 1 - \beta)^2\Gamma[d/2]\Big) \tag{7.41}$$

In both cases the expression for C_1 is too unwieldy to be written explicitly. It is possible to write it in the special case $d = 4$ and $\tilde{\Lambda} = 0$:

$$A_1 = \frac{1}{\pi} , \tag{7.42}$$

$$B_1 = \frac{-3(79 - 50\beta + 7\beta^2) - \alpha(62 - 36\beta + 6\beta^2)}{8\pi(3 - \beta)^2} , \tag{7.43}$$

$$
\begin{aligned}
C_1 = \frac{1}{17280\pi^2(3 - \beta)^4} \Big[& 4(17901 - 26298\beta + 14904\beta^2 - 3822\beta^3 + 431\beta^4) \\
& -180\alpha(297 - 360\beta + 176\beta^2 - 36\beta^3 + 3\beta^4) \\
& +135\alpha^2(259 - 324\beta + 162\beta^2 - 36\beta^3 + 3\beta^4) \Big] .
\end{aligned} \tag{7.44}
$$

This last formula can be compared with the calculation of the logarithmic divergences in Einstein's theory without cosmological constant, in a general two-parameter gauge, by Kallosh et al. [78]. By a flat space perturbative calculation they find

$$C_1 = \frac{1}{8\pi^2} \left(\frac{3}{2}a + \frac{1}{4}b \right) ,$$

where a and b are given in their equations (2.10) and (2.11). Taking into account that their gauge parameters a_K and b_K are related to those used here by $a_K = -1/\alpha$, $b_K = -(1 + \beta)/d$, this translates to

$$
\begin{aligned}
C_1 = \frac{1}{17280\pi^2(3 - \beta)^4} \Big[& 216 \left(252 - 381\beta + 223\beta^2 - 59\beta^3 + 7\beta^4 \right) \\
& -180\alpha(297 - 360\beta + 176\beta^2 - 36\beta^3 + 3\beta^4) \\
& +135\alpha^2(259 - 324\beta + 162\beta^2 - 36\beta^3 + 3\beta^4) \Big] .
\end{aligned} \tag{7.45}
$$

The difference with (7.44) is the constant $53/4320\pi^2$, which corresponds exactly to the last term in (3.159). We can attribute it to the different topologies in which the two results were derived.

As already mentioned at the end of section 3.5, the goal of Kallosh et al. was to find a gauge where the logarithmic divergences vanish. It is indeed easy to see that one can solve for α or β so that $C_1 = 0$. Kallosh et al. had a second independent equation and the combined system was shown to have some discrete solutions. The fact that the off-shell divergences could be eliminated by a gauge choice was seen as another proof that Einstein's theory is one-loop renormalizable.

Returning now to the flow equation and its fixed points, one can use the preceding formulae to study its gauge dependence. Let us focus on the fixed point of $\tilde{G}$, and let us restrict ourselves to the case $\Lambda = 0$. Then, the position of the fixed point is given by

$$\tilde{G}_* = -\frac{48\pi(3 - \beta)^5}{m_0} = -\frac{16\pi(3 - \beta)^2}{237 - 150\beta + 21\beta^2 + \alpha(62 - 36\beta + 6\beta^2)} . \tag{7.46}$$

Given that the calculation has been performed off-shell, the gauge dependence is not surprising. For large α the gauge condition is only weakly enforced and one expects that the results are more strongly contaminated by gauge artifacts. Therefore one places more trust on small values of α, and in particular on the Landau gauge $\alpha = 0$. There seems to be no special reason to prefer any value of β, except that the value $\beta = d - 1$ should be avoided, since the gauge condition becomes singular there, see (5.87). This is reflected by the fact that $\tilde{G}_*$ in (7.46) vanishes on the line $\beta = 3$. It also blows up when

$$\beta = \frac{75 + 18\alpha \pm 2\sqrt{6(27 - \alpha - 2\alpha^2)}}{3(7 + 2\alpha)} .$$

This singularity is shown as a dashed line in Fig. 7.5. $\tilde{G}_*$ is negative in the area delimited by it.

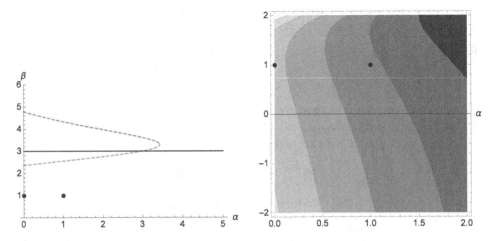

Fig. 7.5 Left: the zeroes of $\tilde{G}_*$ (black continuous line) and the singularities of $\tilde{G}_*$ in the α-β plane (dashed line). Right: level curves of $\tilde{G}_*$: $2, 1.8, 1.6, 1.4, 1.2$ (from bottom left to top right). In both cases the black dots mark the Landau-De Donder and the Feynman-De Donder gauge.

Away from these singularities, the value of $\tilde{G}_*$ is positive and relatively flat. In particular, as a function of β along the $\alpha = 0$ axis, it has two stationary points for $\beta = 1$ (the Landau-de Donder gauge), where its value is

$$\tilde{G}_* = \frac{16\pi}{35} \approx 1.436$$

and for $\beta \to \pm\infty$, where its value is

$$\tilde{G}_* = \frac{16\pi}{21} \approx 2.394 .$$

Using (5.84) it appears that taking the limit $\beta \to \pm\infty$ amounts to imposing strongly the gauge condition $h = 0$. One could invoke a "principle of minimum sensitivity" to argue that these stationary values are somehow preferred.

In any case, the position of a fixed point is never physical, and this variability should not be a cause of concern. More physical information is stored in the critical exponents, but the one-loop approximation gives only a very poor estimate. In particular the critical exponent ν, generally associated to the scaling of the two-point function near a phase transition, is given by $\nu = -1/\beta'_{\tilde{G}}(\tilde{G}_*) = 1/2$.

It is worth pointing out that there is some gauge-independent information in the beta functions. This is related to the gauge-independence of the effective action. On a spherical background the equation of motion (5.68) fixes the radius ℓ, or equivalently the scalar curvature $\bar{R} = d(d-1)/\ell^2$, as a function of Λ. Thus, on-shell the Einstein-Hilbert action is $-\frac{\Lambda V(S^d)}{8\pi G}$. Here $V(S^d)$ is the volume of the sphere. It scales as $\bar{R}^{-d/2}$, therefore, up to a purely numerical factor, the on-shell Hilbert action is proportional to the inverse of $\tau = G\Lambda^{d/2-1}$. This combination is dimensionless and therefore, as we have seen in several examples, its beta functions is expected to have some degree of universality. Using (7.26) one has

$$\partial_t(\Lambda G) = \frac{d-2}{4}G\Lambda^{d/2-2}\left(A_1 + \frac{2d}{d-2}B_1\Lambda\right).$$

(7.47)

The general expressions of A_1 and B_1 in general gauge and general dimension are too long to exhibit, but their Taylor expansion in Λ is

$$A_1 = \frac{16\pi(d-3)}{(4\pi)^{d/2}\Gamma[d/2]}$$

$$+ \frac{16\pi}{(4\pi)^{d/2}d(d-2)(d-1-\beta)^2\Gamma[d/2]}$$

$$\times \Big[2(d-1)\left((d-2)(2-2d^2+d^3) - 2(d-2)^2(d+1)\beta - (4+2d-d^2)\beta^2\right)$$

$$+\alpha(d-2)\left((-4+12d-11d^2+4d^3) - 8(d-1)^2\beta + 4(d-1)\beta^2\right)\Big]\Lambda + O(\Lambda^2)$$

$$B_1 = \frac{4\pi}{(4\pi)^{d/2}3d^2(d-1)(d-1-\beta)^2\Gamma[d/2]}$$

$$\times \Big[(d-1)^2(72+36d-96d^2+39d^3-16d^4+d^5)$$

$$-2(d-1)(72+12d-84d^2+39d^3-16d^4+d^5)\beta$$

$$+(72-12d-60d^2+39d^3-16d^4+d^5)\beta^2$$

$$+6(d-1)(d-2)(4-12d+11d^2-4d^3)\alpha$$

$$+48(d-1)^3(d-2)\alpha\beta - 24(d-1)^2(d-2)\alpha\beta^2)\Big] + O(\Lambda)$$

(7.48)

Then one finds that

$$A_1 + \frac{2d}{d-2}B_1\Lambda = \frac{16\pi(d-3)}{(4\pi)^{d/2}\Gamma[d/2]} + \frac{8\pi(24+60d-48d^2-9d^3-4d^4+d^5)}{(4\pi)^{d/2}\,3\,d(d-1)(d-2)\Gamma[d/2]}\Lambda + \dots$$

(7.49)

We see that to order Λ all dependence on the gauge parameters cancel out.

A more general analysis of the gauge- and parametrization-dependence of the one-loop divergences can be found in [214].

7.4   Higher derivative gravity at one loop

We have studied in some detail the Einstein-Hilbert truncation. As a next step we will now calculate the one-loop beta functions in gravity theories quadratic in curvature. The calculation of the logarithmic divergences in higher derivative gravity and the asymptotic freedom of the higher derivative couplings are classic results dating back to the early 1980's [115–117].[6] These results have been later checked and generalized in [215,216]. They have been rederived using the ERGE in [217–219], together with a nontrivial fixed point for $\tilde{\Lambda}$ and $\tilde{G}$. It is quite remarkable that the flow in the $\tilde{\Lambda}$-$\tilde{G}$ plane turns out to be very similar to the one that is found in the Einstein-Hilbert truncation, in spite of the very different dynamics.

The actions quadratic in curvature have been discussed in section 2.2. They do not change under Euclidean continuation. Thus, in this section we will consider Euclidean actions of the general form

$$S = \int d^d x \sqrt{g} \Big[Z_N (2\Lambda - R) + \alpha R^2 + \beta R^2_{\mu\nu} + \gamma R^2_{\mu\nu\rho\lambda} \Big], \qquad (7.50)$$

where $Z_N = 1/2\kappa^2 = 1/(16\pi G)$. For the quadratic part we will also use the Weyl basis (2.83).[7] Note that in $d = 3$, C^2 and E both vanish identically and the Weyl basis is not appropriate.

In dimensions higher than three, it is customary to define the dimensionless combinations

$$\omega \equiv -\frac{(d-1)\lambda}{\xi}, \quad \theta \equiv \frac{\lambda}{\rho}, \qquad (7.51)$$

in terms of which the action reads

$$S = \int d^d x \sqrt{g} \Big[Z_N (2\Lambda - R) + \frac{1}{2\lambda} C^2 - \frac{\omega}{(d-1)\lambda} R^2 - \frac{\theta}{\lambda} E \Big]. \qquad (7.52)$$

Note that λ is the overall factor of the curvature squared terms, while ω and θ give the relative ratio between these terms. This basis is somehow more significant physically, because, as we have seen in section 2.2, in four dimensions the Weyl squared term only contributes to the propagation of the spin-2 degree of freedom while R^2 contributes to the propagation of a scalar and E is immaterial. In practice we will first calculate the beta functions of α, β and γ and then, using (2.85), (2.86), (7.51), convert them into beta functions for λ, ω and θ.

For the calculation of the one-loop EA, or for the calculation of the right-hand-side of the ERGE, the first step is to obtain the second variation of the action.

7.4.1   *Expansion of the action*

The quadratic expansion of the Hilbert action has already been given in section 3.4. There, we have also given the variations of the curvature tensors that are needed for the quadratic expansion of the curvature squared terms.

---

[6]The first two of these references contained some errors. The correct beta functions were only established in the third one.

[7]There is a misprint in the sign of ξ in Eq. (2.5) of [219]

Before doing any integration by parts, the quadratic action $S^{(2)}$, defined as in (3.113), is equal to the sum of the integrals of the following expressions:

$$\alpha \left[\Box h \Box h - 2\Box h \bar{\nabla}_\alpha \bar{\nabla}_\beta h^{\alpha\beta} + \bar{\nabla}_\mu \bar{\nabla}_\nu h^{\mu\nu} \bar{\nabla}_\alpha \bar{\nabla}_\beta h^{\alpha\beta} - \frac{1}{2} \bar{R} \bar{\nabla}_\rho h \bar{\nabla}^\rho h \right.$$

$$-2\bar{R} \bar{\nabla}_\mu h^{\mu\nu} \bar{\nabla}_\alpha h^\alpha{}_\nu + 2\bar{R} \bar{\nabla}_\beta h \bar{\nabla}_\alpha h^{\alpha\beta} - \bar{R} \bar{\nabla}_\alpha h^{\mu\nu} \bar{\nabla}_\mu h^\alpha{}_\nu + \frac{3}{2} \bar{R} \bar{\nabla}^\rho h^{\mu\nu} \bar{\nabla}_\rho h_{\mu\nu}$$

$$+2\bar{R} h^{\mu\nu} \bar{\nabla}_\mu \bar{\nabla}_\nu h - 2\bar{R} h^{\mu\nu} \bar{\nabla}_\mu \bar{\nabla}_\alpha h^\alpha{}_\nu - 2\bar{R} h^{\mu\nu} \bar{\nabla}_\alpha \bar{\nabla}_\mu h^\alpha{}_\nu + \bar{R} h \bar{\nabla}_\alpha \bar{\nabla}_\beta h^{\alpha\beta}$$

$$+2\bar{R} h^{\mu\nu} \Box h_{\mu\nu} - \bar{R} h \Box h - 2\bar{R}_{\mu\nu} h^{\mu\nu} \bar{\nabla}_\alpha \bar{\nabla}_\beta h^{\alpha\beta} + 2\bar{R}_{\mu\nu} h^{\mu\nu} \Box h$$

$$\left. +h^{\mu\nu} \bar{R}_{\mu\alpha} \bar{R}_{\nu\beta} h^{\alpha\beta} + 2h^{\mu\nu} \bar{R}_{\mu\alpha} \bar{R} h^\alpha{}_\nu - h \bar{R} \bar{R}_{\alpha\beta} h^{\alpha\beta} + \left(\frac{1}{8} h^2 - \frac{1}{4} h_{\mu\nu} h^{\mu\nu} \right) \bar{R}^2 \right], \quad (7.53)$$

for scalar curvature squared,

$$\beta \left[\frac{1}{4} \bar{\nabla}_\rho \bar{\nabla}_\sigma h \bar{\nabla}^\rho \bar{\nabla}^\sigma h - \bar{\nabla}_\alpha \bar{\nabla}_\rho h \bar{\nabla}_\beta \bar{\nabla}^\rho h^{\alpha\beta} + \frac{1}{2} \bar{\nabla}_\alpha \bar{\nabla}_\beta h \Box h^{\alpha\beta} + \frac{1}{4} \Box h^{\mu\nu} \Box h_{\mu\nu} \right.$$

$$-\bar{\nabla}_\mu \bar{\nabla}_\alpha h^{\mu\nu} \Box h^\alpha{}_\nu + \frac{1}{2} \bar{\nabla}_\mu \bar{\nabla}_\alpha h^{\mu\nu} \bar{\nabla}_\beta \bar{\nabla}_\nu h^{\alpha\beta} + \frac{1}{2} \bar{\nabla}_\mu \bar{\nabla}_\rho h^{\mu\nu} \bar{\nabla}_\alpha \bar{\nabla}^\rho h^\alpha{}_\nu$$

$$+\frac{1}{2} \bar{R}^{\rho\sigma} \bar{\nabla}_\rho h^{\mu\nu} \bar{\nabla}_\sigma h_{\mu\nu} + \bar{R}_\beta{}^\rho \bar{\nabla}_\alpha h \bar{\nabla}_\rho h^{\alpha\beta} - \frac{1}{2} \bar{R}_{\alpha\beta} \bar{\nabla}_\rho h \bar{\nabla}^\rho h^{\alpha\beta} - 2\bar{R}_\mu{}^\rho \bar{\nabla}_\rho h^{\mu\nu} \bar{\nabla}_\alpha h^\alpha{}_\nu$$

$$+\bar{R}^{\mu\nu} \bar{\nabla}_\beta h \bar{\nabla}_\alpha h^{\alpha\beta} - \bar{R}_{\mu\alpha} \bar{\nabla}_\beta h^{\mu\nu} \bar{\nabla}_\nu h^{\alpha\beta} + \bar{R}_{\mu\alpha} \bar{\nabla}_\rho h^{\mu\nu} \bar{\nabla}^\rho h^\alpha{}_\nu$$

$$+h^{\mu\nu} \bar{R}_{\alpha\beta} \bar{\nabla}_\mu \bar{\nabla}_\nu h^{\alpha\beta} + 2h^{\mu\nu} \bar{R}_\mu{}^\rho \bar{\nabla}_\rho \bar{\nabla}_\nu h - 2h^{\mu\nu} \bar{R}_{\mu\beta} \bar{\nabla}_\alpha \bar{\nabla}_\nu h^{\alpha\beta} - 2h^{\mu\nu} \bar{R}_\mu{}^\rho \bar{\nabla}_\alpha \bar{\nabla}_\rho h^\alpha{}_\nu$$

$$+h^{\mu\nu} \bar{R}^{\rho\sigma} \bar{\nabla}_\rho \bar{\nabla}_\sigma h_{\mu\nu} + 2h^{\mu\nu} \bar{R}_{\mu\alpha} \Box h^\alpha{}_\nu - 2h^{\mu\nu} \bar{R}_\alpha{}^\rho \bar{\nabla}_\mu \bar{\nabla}_\rho h^\alpha{}_\nu$$

$$+h \bar{R}_\alpha{}^\rho \bar{\nabla}_\beta \bar{\nabla}_\rho h^{\alpha\beta} - \frac{1}{2} h \bar{R}_{\alpha\beta} \Box h^{\alpha\beta} - \frac{1}{2} h \bar{R}^{\rho\sigma} \bar{\nabla}_\rho \bar{\nabla}_\sigma h + h^{\mu\nu} \bar{R}_{\mu\alpha} \bar{R}_{\nu\beta} h^{\alpha\beta}$$

$$\left. +2h^{\mu\nu} \bar{R}_{\mu\rho} \bar{R}^\rho{}_\alpha h^\alpha{}_\nu - h \bar{R}_{\alpha\rho} \bar{R}^\rho{}_\beta h^{\alpha\beta} + \left(\frac{1}{8} h^2 - \frac{1}{4} h_{\mu\nu} h^{\mu\nu} \right) \bar{R}_{\rho\sigma} \bar{R}^{\rho\sigma} \right], \quad (7.54)$$

for Ricci curvature squared, and

$$\gamma \left[\bar{\nabla}_\alpha \bar{\nabla}_\beta h^{\mu\nu} \bar{\nabla}_\mu \bar{\nabla}_\nu h^{\alpha\beta} - \bar{\nabla}_\rho \bar{\nabla}_\alpha h^{\mu\nu} \bar{\nabla}^\rho \bar{\nabla}_\mu h^\alpha{}_\nu - \bar{\nabla}_\alpha \bar{\nabla}_\rho h^{\mu\nu} \bar{\nabla}_\mu \bar{\nabla}^\rho h^\alpha{}_\nu \right.$$

$$+\bar{\nabla}_\rho \bar{\nabla}_\sigma h^{\mu\nu} \bar{\nabla}^\rho \bar{\nabla}^\sigma h_{\mu\nu} + 4\bar{R}_{\mu\alpha\nu\rho} \bar{\nabla}_\beta h^{\mu\nu} \bar{\nabla}^\rho h^{\alpha\beta} - \bar{R}_{\mu\alpha\nu\beta} \bar{\nabla}_\rho h^{\mu\nu} \bar{\nabla}^\rho h^{\alpha\beta}$$

$$-2\bar{R}_{\mu\alpha\rho\sigma} \bar{\nabla}^\rho h^{\mu\nu} \bar{\nabla}^\sigma h^\alpha{}_\nu + 2\bar{R}_{\mu\rho\alpha\sigma} \bar{\nabla}^\sigma h^{\mu\nu} \bar{\nabla}^\rho h^\alpha{}_\nu + h^{\mu\nu} \bar{R}_{\mu\rho\alpha\sigma} (4\bar{\nabla}^\rho \bar{\nabla}^\sigma + 2\bar{\nabla}^\sigma \bar{\nabla}^\rho) h^\alpha{}_\nu$$

$$+h^{\mu\nu} \bar{R}_{\mu\alpha\rho\beta} (4\bar{\nabla}_\nu \bar{\nabla}^\rho + 2\bar{\nabla}^\rho \bar{\nabla}_\nu) h^{\alpha\beta} - 2h \bar{R}_{\alpha\rho\beta\sigma} \bar{\nabla}^\sigma \bar{\nabla}^\rho h^{\alpha\beta}$$

$$+\frac{5}{2} h^{\mu\nu} h^\alpha{}_\nu \bar{R}_{\mu\lambda\rho\sigma} \bar{R}_\alpha{}^{\lambda\rho\sigma} + \frac{1}{2} h^{\mu\nu} h^{\alpha\beta} \bar{R}_{\mu\alpha\rho\sigma} \bar{R}_{\nu\beta}{}^{\rho\sigma} - h h^{\alpha\beta} \bar{R}_{\alpha\lambda\rho\sigma} \bar{R}_\beta{}^{\lambda\rho\sigma}$$

$$\left. + \left(\frac{1}{8} h^2 - \frac{1}{4} h_{\mu\nu} h^{\mu\nu} \right) \bar{R}_{\rho\sigma\lambda\tau} \bar{R}^{\rho\sigma\lambda\tau} \right], \quad (7.55)$$

for the Riemann curvature squared.

We now integrate by parts derivatives in order to write (schematically) $S^{(2)} = \frac{1}{2} \int dx \sqrt{\bar{g}} h \mathcal{O} h$ where $\mathcal{O}$ is a fourth order operator. There is some arbitrariness in the presentation of the formula, due to the freedom of performing commutations and integrations by parts. This makes it a bit hard to compare the formulae given by different authors. For the α and β terms, the second variations given below agree with equations Eq. (3.11) or (3.15) of [220]. In order to reduce the arbitrariness, let us make some conventions. We put $h^{\mu\nu}$ on the left and $h^{\alpha\beta}$ on the right. Then, when there are several derivatives, it is convenient to always put $\bar{\nabla}_\alpha$ and $\bar{\nabla}_\beta$ on the right and $\bar{\nabla}_\mu$ and $\bar{\nabla}_\nu$ on the left, so that they form the vector combination $\bar{\nabla}_\mu h^\mu{}_\nu$ whenever possible. Also, we use the convention that in those terms where only one of the h's is traced, it stays on the left. After some manipulations, (7.53) can be rewritten in the form

$$\alpha h^{\mu\nu}\Big[\bar{\nabla}_\mu\bar{\nabla}_\nu\bar{\nabla}_\alpha\bar{\nabla}_\beta - 2\bar{g}_{\mu\nu}\Box\bar{\nabla}_\alpha\bar{\nabla}_\beta + \bar{g}_{\mu\nu}\bar{g}_{\alpha\beta}\Box^2$$
$$-\bar{g}_{\nu\beta}\bar{R}\bar{\nabla}_\mu\bar{\nabla}_\alpha - 2\bar{R}_{\mu\nu}\bar{\nabla}_\alpha\bar{\nabla}_\beta + \bar{g}_{\mu\nu}\bar{R}\bar{\nabla}_\alpha\bar{\nabla}_\beta + 2\bar{g}_{\mu\nu}\bar{R}_{\alpha\beta}\Box + \frac{1}{2}(\bar{g}_{\mu\alpha}\bar{g}_{\nu\beta} - \bar{g}_{\mu\nu}\bar{g}_{\alpha\beta})\bar{R}\Box$$
$$-\bar{g}_{\mu\nu}\bar{R}\bar{R}_{\alpha\beta} - \frac{1}{4}J_{\mu\nu\alpha\beta}\bar{R}^2 + \bar{g}_{\nu\beta}\bar{R}\bar{R}_{\mu\alpha} + \bar{R}_{\mu\nu}\bar{R}_{\alpha\beta} + \bar{R}\bar{R}_{\mu\alpha\nu\beta}$$
$$+2\bar{g}_{\mu\nu}\Box\bar{R}_{\alpha\beta} + 2\bar{g}_{\mu\nu}\bar{\nabla}_\alpha\bar{\nabla}_\beta\bar{R} - \bar{g}_{\nu\beta}\bar{\nabla}_\mu\bar{\nabla}_\alpha\bar{R} + \frac{1}{4}(3\bar{g}_{\mu\alpha}\bar{g}_{\nu\beta} + \bar{g}_{\mu\nu}\bar{g}_{\alpha\beta})\Box\bar{R}$$
$$+\bar{g}_{\nu\beta}\bar{\nabla}_\mu\bar{R}\bar{\nabla}_\alpha + 4\bar{g}_{\mu\nu}\bar{\nabla}_\rho\bar{R}_{\alpha\beta}\bar{\nabla}^\rho + 2\bar{g}_{\mu\nu}\bar{\nabla}_\alpha\bar{R}\bar{\nabla}_\beta\Big]h^{\alpha\beta}, \tag{7.56}$$

where $J_{\mu\nu\alpha\beta} = \frac{1}{2}K_{\mu\nu\alpha\beta}$ and $K^{\mu\nu,\alpha\beta}$ was defined in (3.136). The terms in the last two lines, containing one or two covariant derivatives acting on background curvatures, can be discarded because they only contribute to total derivative terms in the final results [216, 221].

Similarly one can rewrite (7.54) in the "standard" form:

$$\beta h^{\mu\nu}\Big[\frac{1}{2}\bar{\nabla}_\mu\bar{\nabla}_\nu\bar{\nabla}_\alpha\bar{\nabla}_\beta - \frac{1}{2}\bar{g}_{\mu\nu}\Box\bar{\nabla}_\alpha\bar{\nabla}_\beta - \frac{1}{2}\bar{g}_{\nu\beta}\bar{\nabla}_\mu\Box\bar{\nabla}_\alpha + \frac{1}{4}(\bar{g}_{\mu\alpha}\bar{g}_{\nu\beta} + \bar{g}_{\mu\nu}\bar{g}_{\alpha\beta})\Box^2$$
$$+\frac{1}{2}\bar{R}_{\mu\alpha}\bar{\nabla}_\nu\bar{\nabla}_\beta - 2\bar{g}_{\nu\beta}\bar{R}^\rho_\mu\bar{\nabla}_\rho\bar{\nabla}_\alpha + \bar{g}_{\mu\nu}\bar{R}^\rho_\alpha\bar{\nabla}_\rho\bar{\nabla}_\beta + \bar{R}_{\mu\alpha\nu\beta}\Box + \frac{1}{2}J_{\mu\nu\alpha\beta}\bar{R}^{\rho\lambda}\bar{\nabla}_\rho\bar{\nabla}_\lambda$$
$$+\frac{1}{2}\bar{g}_{\nu\beta}\bar{R}_{\mu\rho}\bar{R}^\rho_\alpha + \frac{1}{2}\bar{R}_{\mu\alpha}\bar{R}_{\nu\beta} + \bar{R}^\rho_\mu\bar{R}_{\rho\alpha\nu\beta} - \bar{g}_{\mu\nu}\bar{R}^{\rho\sigma}\bar{R}_{\rho\alpha\sigma\beta} + \bar{R}_{\rho\mu\sigma\nu}\bar{R}^\rho{}_\alpha{}^\sigma{}_\beta$$
$$-\frac{1}{4}J_{\mu\nu\alpha\beta}\bar{R}_{\rho\lambda}\bar{R}^{\rho\lambda}$$
$$+\frac{1}{2}\bar{g}_{\mu\nu}\Box\bar{R}_{\alpha\beta} + \frac{1}{2}\bar{g}_{\mu\nu}\bar{\nabla}_\alpha\bar{\nabla}_\beta\bar{R} + \frac{1}{8}J_{\mu\nu\alpha\beta}\Box\bar{R} + 2\bar{g}_{\nu\beta}\bar{\nabla}_\mu\bar{R}_{\alpha\rho}\bar{\nabla}^\rho$$
$$-\bar{g}_{\nu\beta}\bar{\nabla}_\rho\bar{R}_{\mu\alpha}\bar{\nabla}^\rho + \bar{\nabla}_\alpha\bar{R}_{\mu\nu}\bar{\nabla}_\beta - \frac{1}{2}\bar{\nabla}_\mu\bar{R}_{\nu\beta}\bar{\nabla}_\alpha + (\bar{\nabla}_\alpha\bar{R}_{\mu\beta} - \bar{\nabla}_\mu\bar{R}_{\alpha\beta})\bar{\nabla}_\nu$$
$$+\frac{1}{2}\bar{g}_{\mu\nu}\bar{\nabla}_\alpha\bar{R}\bar{\nabla}_\beta + \bar{g}_{\mu\nu}\bar{\nabla}_\rho\bar{R}_{\alpha\beta}\bar{\nabla}^\rho\Big]h^{\alpha\beta}. \tag{7.57}$$

Notice that if one neglects the terms of the form $\bar{\nabla}\bar{\nabla}\bar{R}$ and $\bar{\nabla}\bar{R}\bar{\nabla}$ (the last three lines), then there is some ambiguity in the form of the $\bar{R}^2$ terms, because

$$[\bar{\nabla}_\rho, \bar{\nabla}_\mu]\bar{R}^\rho{}_\nu = \bar{R}_{\mu\rho}\bar{R}^\rho{}_\nu - \bar{R}^{\rho\sigma}\bar{R}_{\mu\rho\nu\sigma}. \tag{7.58}$$

Finally we consider the terms proportional to γ. Now we encounter products of two Riemann tensors. Due to $\bar{R}_{[\mu\nu\rho]\sigma} = 0$, there are various ways of writing these products. We have

$$\bar{R}_{\mu\rho\alpha\sigma}\bar{R}_\nu{}^\sigma{}_\beta{}^\rho = \bar{R}_{\mu\rho\alpha\sigma}\bar{R}_\nu{}^\rho{}_\beta{}^\sigma - \bar{R}_{\mu\rho\alpha\sigma}\bar{R}_{\nu\beta}{}^{\rho\sigma}. \tag{7.59}$$

Furthermore, when contracted with $h^{\mu\nu}h^{\alpha\beta}$ we can replace

$$\bar{R}_{\mu\rho\nu\sigma}\bar{R}_{\alpha}{}^{\sigma}{}_{\beta}{}^{\rho} \leftrightarrow \bar{R}_{\mu\rho\nu\sigma}\bar{R}_{\alpha}{}^{\rho}{}_{\beta}{}^{\sigma} ; \qquad 2\bar{R}_{\mu\rho\alpha\sigma}\bar{R}_{\nu\beta}{}^{\rho\sigma} \leftrightarrow \bar{R}_{\mu\alpha\rho\sigma}\bar{R}_{\nu\beta}{}^{\rho\sigma} . \qquad (7.60)$$

Using these properties we choose the following basis of independent combinations:

$$\bar{R}_{\mu\rho\alpha\sigma}\bar{R}_{\nu}{}^{\rho}{}_{\beta}{}^{\sigma} ; \quad \bar{R}_{\mu\alpha\rho\sigma}\bar{R}_{\nu\beta}{}^{\rho\sigma} ; \quad \bar{R}_{\mu\rho\nu\sigma}\bar{R}_{\alpha}{}^{\rho}{}_{\beta}{}^{\sigma} ; \quad \bar{R}_{\mu\rho}\bar{R}_{\nu\alpha}{}^{\rho}{}_{\beta} . \qquad (7.61)$$

After integrations by parts and arranging in canonical order, (7.55) becomes

$$
\begin{aligned}
\gamma h^{\mu\nu} \Big[& \bar{\nabla}_{\mu}\bar{\nabla}_{\nu}\bar{\nabla}_{\alpha}\bar{\nabla}_{\beta} - 2\bar{g}_{\nu\beta}\bar{\nabla}_{\mu}\Box\bar{\nabla}_{\alpha} + \bar{g}_{\mu\alpha}\bar{g}_{\nu\beta}\Box^2 \\
& + 3\bar{R}_{\mu\alpha\nu\beta}\Box - 2\bar{g}_{\nu\beta}\bar{R}_{\mu\alpha}\Box - 2\bar{g}_{\mu\nu}\bar{R}_{\alpha\rho\beta\sigma}\bar{\nabla}^{\rho}\bar{\nabla}^{\sigma} + 2\bar{g}_{\nu\beta}\bar{R}_{\mu\rho\alpha\sigma}\bar{\nabla}^{\rho}\bar{\nabla}^{\sigma} \\
& + 4\bar{R}_{\alpha\mu\rho\nu}\bar{\nabla}^{\rho}\bar{\nabla}_{\beta} + 4\bar{R}_{\mu\alpha}\bar{\nabla}_{\nu}\bar{\nabla}_{\beta} - 4\bar{g}_{\nu\beta}\bar{R}_{\mu\rho}\bar{\nabla}^{\rho}\bar{\nabla}_{\alpha} + \bar{g}_{\mu\alpha}\bar{g}_{\nu\beta}\bar{R}_{\rho\sigma}\bar{\nabla}^{\rho}\bar{\nabla}^{\sigma} \\
& - \bar{g}_{\mu\nu}\bar{R}_{\alpha\lambda\rho\sigma}\bar{R}_{\beta}{}^{\lambda\rho\sigma} + 2\bar{R}_{\mu\alpha}\bar{R}_{\nu\beta} - 2\bar{g}_{\nu\beta}\bar{R}_{\mu\rho}\bar{R}^{\rho}{}_{\alpha} + 2\bar{g}_{\nu\beta}\bar{R}_{\mu\lambda\rho\sigma}\bar{R}_{\alpha}{}^{\lambda\rho\sigma} \\
& + 5\bar{R}_{\mu\rho\alpha\sigma}\bar{R}_{\nu}{}^{\rho}{}_{\beta}{}^{\sigma} - 4\bar{R}_{\mu\alpha\rho\sigma}\bar{R}_{\nu\beta}{}^{\rho\sigma} - 3\bar{R}_{\mu\rho\nu\sigma}\bar{R}_{\alpha}{}^{\rho}{}_{\beta}{}^{\sigma} + 3\bar{R}_{\mu\rho}\bar{R}_{\nu\alpha}{}^{\rho}{}_{\beta} \\
& - \frac{1}{4}J_{\mu\nu\alpha\beta}\bar{R}_{\rho\sigma\lambda\tau}\bar{R}^{\rho\sigma\lambda\tau} \\
& + 8(\bar{\nabla}_{\alpha}\bar{R}_{\mu\nu} - \bar{\nabla}_{\mu}\bar{R}_{\alpha\nu})\bar{\nabla}_{\beta} + 8\bar{\nabla}_{\alpha}(\bar{\nabla}_{\beta}\bar{R}_{\mu\nu} - \bar{\nabla}_{\mu}\bar{R}_{\beta\nu}) + 3\bar{\nabla}_{\rho}\bar{R}_{\mu\alpha\nu\beta}\bar{\nabla}^{\rho} \\
& + 2\bar{\nabla}_{\alpha}\bar{R}_{\beta\mu\rho\nu}\bar{\nabla}^{\rho} + 2\bar{g}_{\nu\beta}\bar{\nabla}_{\mu}\bar{R}_{\alpha\rho}\bar{\nabla}^{\rho} - 4\bar{g}_{\nu\beta}\bar{\nabla}_{\rho}\bar{R}_{\alpha\mu}\bar{\nabla}^{\rho} + \frac{1}{2}\bar{g}_{\mu\alpha}\bar{g}_{\nu\beta}\bar{\nabla}_{\rho}\bar{R}\bar{\nabla}^{\rho} \Big] h^{\alpha\beta} .
\end{aligned}
$$

$$\text{(7.62)}$$

Again one has to be careful: if one neglects terms of the type in the last two lines, which do not contribute to the final results, then the coefficients of the curvature squared terms can change, because one can add an arbitrary multiple of the combination

$$[\bar{\nabla}_{\rho}, \bar{\nabla}_{\mu}]\bar{R}_{\nu\alpha}{}^{\rho}{}_{\beta} = \bar{R}_{\mu\rho\alpha\sigma}\bar{R}_{\nu}{}^{\rho}{}_{\beta}{}^{\sigma} - \bar{R}_{\mu\alpha\rho\sigma}\bar{R}_{\nu\beta}{}^{\rho\sigma} - \bar{R}_{\mu\rho\nu\sigma}\bar{R}_{\alpha}{}^{\rho}{}_{\beta}{}^{\sigma} + \bar{R}_{\mu\rho}\bar{R}_{\nu\alpha}{}^{\rho}{}_{\beta} . \quad (7.63)$$

7.4.2 Gauge fixing

From here on the procedure to calculate the beta functions will be outlined without giving full details. The reader can find some additional intermediate steps in [219]. As usual, it is convenient to choose a linear background gauge of the form $F_{\mu} = 0$, where

$$F_{\mu} = \bar{\nabla}^{\lambda}h_{\lambda\mu} + b\bar{\nabla}_{\mu}h \qquad (7.64)$$

The gauge fixing action has the usual quadratic structure

$$S_{GF} = \frac{1}{2a}\int d^d x \sqrt{\bar{g}}\, F_{\mu}Y^{\mu\nu}F_{\nu} . \qquad (7.65)$$

The corresponding ghost operator is

$$\Delta_{gh} = \delta^\mu{}_\nu \bar{\Box} + (1 + 2b)\bar{\nabla}_\mu \bar{\nabla}^\nu + \bar{R}^\mu{}_\nu . \tag{7.66}$$

One could choose $Y^{\mu\nu} \sim \bar{g}^{\mu\nu}$ as we did in the Einstein-Hilbert truncation, but it proves more convenient to have a gauge fixing action that contains four derivatives. The general form is

$$Y_{\mu\nu} = \bar{g}_{\mu\nu}\bar{\Box} + c\bar{\nabla}_\mu \bar{\nabla}_\nu - f\bar{\nabla}_\nu \bar{\nabla}_\mu, \tag{7.67}$$

where a, b, c and f are free gauge parameters. These parameters can be chosen so as to cancel the nonminimal fourth-order terms $\bar{\nabla}_\mu \bar{\nabla}_\nu \bar{\nabla}_\alpha \bar{\nabla}_\beta$, $\bar{g}_{\mu\nu}\bar{\Box}\bar{\nabla}_\alpha \bar{\nabla}_\beta$ and $\bar{g}_{\nu\beta}\bar{\nabla}_\mu \bar{\Box}\bar{\nabla}_\alpha$ in the kinetic operator of $h_{\mu\nu}$. This can be achieved by the choice

$$a = \frac{1}{\beta + 4\gamma}, \quad b = \frac{4\alpha + \beta}{4(\gamma - \alpha)}, \quad c - f = \frac{2(\gamma - \alpha)}{\beta + 4\gamma} - 1. \tag{7.68}$$

In order to simplify the gauge-fixing term, we will further set $f = 1$.

Then, the quadratic terms in the action can be written in the form

$$S^{(2)} = \frac{1}{2} \int d^4 x \sqrt{\bar{g}}\, h_{\mu\nu} H^{\mu\nu\rho\sigma} h_{\rho\sigma} \tag{7.69}$$

where, suppressing indices, the Hessian has the form

$$H = K\bar{\Box}^2 + D^{\alpha\beta}\bar{\nabla}_\alpha \bar{\nabla}_\beta + W. \tag{7.70}$$

Here

$$K^{\mu\nu\rho\sigma} = \frac{\beta + 4\gamma}{4}\left(\bar{g}^{\mu\rho}\bar{g}^{\nu\sigma} + \frac{4\alpha + \beta}{4(\gamma - \alpha)}\bar{g}^{\mu\nu}\bar{g}^{\rho\sigma}\right), \tag{7.71}$$

and, for obvious dimensional reasons, $D_{\alpha\beta}$ is linear in the background curvature tensor and Z_N, and W is quadratic in the background curvature tensor, Z_N and Λ.

We now have to turn this quadratic form into a differential operator in the proper sense, namely a linear map of the space of covariant symmetric tensors into itself. As in section 3.4, this is achieved by factoring the tensor K, which can be viewed as a metric in the space of symmetric tensors:

$$H^{\mu\nu\rho\sigma} = K^{\mu\nu\alpha\beta}\Delta_{\alpha\beta}{}^{\rho\sigma} , \tag{7.72}$$

where

$$\Delta_{\alpha\beta}{}^{\rho\sigma} = \bar{\Box}^2 \mathbf{1}_{\alpha\beta}^{\rho\sigma} + V^{\lambda\tau}{}_{\alpha\beta}{}^{\rho\sigma}\bar{\nabla}_\lambda \bar{\nabla}_\tau + U_{\alpha\beta}{}^{\rho\sigma}. \tag{7.73}$$

7.4.3 *Derivation of beta functions*

In our quadratic action, we have the three operators: Δ acting on the graviton $h_{\mu\nu}$, the ghost operator Δ_{gh} and the operator $Y^{\mu\nu}$ in the gauge fixing. The one-loop EA of the theory is then given by

$$\Gamma = S + \frac{1}{2}\text{Tr}\log\left(\frac{\Delta}{\mu^2}\right) - \text{Tr}\log\left(\frac{\Delta_{gh}}{\mu^2}\right) - \frac{1}{2}\text{Tr}\log\left(\frac{Y}{\mu^2}\right) . \tag{7.74}$$

The last term is there to remove the determinant of Y coming from the gauge-fixing. It can be thought of as a Gaussian integral over a real anticommuting "third ghost".

In order to write the one-loop ERGE we choose cutoffs for the graviton, ghost and third ghost to be functions of the respective kinetic operators: $K R_k(\Delta)$ for the graviton, $R_k(\Delta_{gh})$ for the ghosts and $R_k(Y)$ for the third ghost. Since the kinetic operators contain the couplings, this is what we called a "spectrally adjusted" cutoff. Insofar as in the one-loop calculation these couplings are kept fixed, there is no drawback in this choice.

With this cutoff the ERGE reads

$$\partial_t \Gamma_k = \frac{1}{2} \text{Tr} \frac{\partial_t R_k(\Delta)}{P_k(\Delta)} - \text{Tr} \frac{\partial_t R_k(\Delta_{gh})}{P_k(\Delta_{gh})} - \frac{1}{2} \text{Tr} \frac{\partial_t R_k(Y)}{P_k(Y)} . \tag{7.75}$$

The traces can be evalated by means of (5.159), which gives the following expansion

$$\partial_t \Gamma_k = \frac{1}{2} B_0(\Delta) Q\left(4, \frac{d}{4}\right) + \frac{1}{2} B_2(\Delta) Q\left(4, \frac{d-2}{4}\right) + \frac{1}{2} B_4(\Delta) Q\left(4, \frac{d-4}{4}\right)$$

$$- B_0(\Delta_{gh}) Q\left(2, \frac{d}{2}\right) - B_2(\Delta_{gh}) Q\left(2, \frac{d-2}{2}\right) - B_4(\Delta_{gh}) Q\left(2, \frac{d-4}{2}\right)$$

$$- \frac{1}{2} B_0(Y) Q\left(2, \frac{d}{2}\right) - \frac{1}{2} B_2(Y) Q\left(2, \frac{d-2}{2}\right) - \frac{1}{2} B_4(Y) Q\left(2, \frac{d-4}{2}\right). \tag{7.76}$$

where the Q-functionals $Q(p, m)$, for an operator of order p, have been defined in (6.161) and calculated in (6.166) using the cutoff profile $R_k(z) = (k^p - z)\theta(k^p - z)$.

The last ingredient we need are the heat kernel coefficients. For the fourth order operator Δ they are given in [222–224] as

$$B_0(\Delta_h) = \frac{1}{(4\pi)^{d/2}} \frac{\Gamma(d/4)}{2\Gamma(d/2)} \frac{d(d+1)}{2} \int d^d x \sqrt{\bar{g}} \tag{7.77}$$

$$B_2(\Delta_h) = \frac{1}{(4\pi)^{d/2}} \frac{\Gamma((d-2)/4)}{2\Gamma((d-2)/2)} \int d^d x \sqrt{\bar{g}} \, \text{tr} \left[\frac{\bar{R}}{6} + \frac{1}{2d} V^\mu_\mu\right], \tag{7.78}$$

$$B_4(\Delta_h) = \frac{1}{(4\pi)^{d/2}} \frac{\Gamma(d/4)}{2\Gamma((d-2)/2)} \int d^d x \sqrt{\bar{g}} \left[\left(\frac{1}{90} \bar{R}^2_{\rho\lambda\sigma\tau} - \frac{1}{90} \bar{R}^2_{\rho\lambda} + \frac{1}{36} \bar{R}^2\right) \text{tr} \mathbf{1}\right.$$

$$+ \frac{1}{6} \text{tr} \, \Omega_{\rho\lambda} \Omega^{\rho\lambda} - \frac{2}{d-2} \text{tr} \, U - \frac{1}{6(d-2)} (2\bar{R}_{\rho\lambda} \text{tr} \, V^{\rho\lambda} - \bar{R} \text{tr} \, V^\rho_\rho)$$

$$\left. + \frac{1}{4(d^2-4)} (\text{tr} \, V^\rho_\rho V^\lambda_\lambda + 2\text{tr} \, V_{\rho\lambda} V^{\rho\lambda})\right], \tag{7.79}$$

where all the traces are on the indices $(\mu\nu)(\alpha\beta)$.

The ghost operator is nonminimal:

$$(\Delta_{gh})_{\mu\nu} = -\bar{g}_{\mu\nu}\bar{\Box} + \sigma_g \bar{\nabla}_\mu \bar{\nabla}_\nu - \bar{R}_{\mu\nu}, \tag{7.80}$$

with $\sigma_g = -(1 + 2b) = -\left(1 + 2\frac{\beta + 4\alpha}{4(\gamma - \alpha)}\right)$. The heat kernel coefficients for this type of operators are given in [225]:

$$B_0(\Delta_{gh}) = \frac{1}{(4\pi)^{d/2}}(1-\sigma_g)^{-d/2}\left(1+(d-1)(1-\sigma_g)^{d/2}\right)\int d^dx\,\sqrt{-\bar{g}}\,, \qquad (7.81)$$

$$B_2(\Delta_{gh}) = \frac{1}{(4\pi)^{d/2}}\frac{d+5+(1-\sigma_g)^{1-d/2}+\frac{12}{d}((1-\sigma_g)^{-d/2}-1)}{6}\int d^dx\,\sqrt{-\bar{g}}\,\bar{R}, \qquad$$

$$\begin{aligned}
B_4(\Delta_{gh}) = \frac{1}{(4\pi)^{d/2}}\int d^dx\sqrt{-\bar{g}}\Big[& \frac{d-16+(1-\sigma_g)^{\frac{4-d}{2}}}{180}\bar{R}^2_{\mu\nu\rho\lambda} \\
& -\frac{(1-\sigma_g)^{-d/2}}{180d(d^2-4)\sigma_g}\Big(d(d^2-4)\sigma_g^3 - 2d(d+2)(d+58)\sigma_g^2 \\
& +(d+2)(d^2+118d+720)\sigma_g - 1440d \\
& +(1-\sigma_g)^{d/2}\{(d^4-91d^3+596d^2-596d-1440)\sigma_g+1440d\}\Big)\bar{R}^2_{\mu\nu} \\
& +\frac{(1-\sigma_g)^{-d/2}}{72d(d^2-4)\sigma_g}\Big(d(d^2-4)\sigma_g^3 - 2d(d+2)(d+10)\sigma_g^2 \\
& +(d+2)(d^2+22d+144)\sigma_g - 576 \\
& +(1-\sigma_g)^{d/2}\{(d^4+11d^3-28d^2+52d-288)\sigma_g+576\}\Big)\bar{R}^2\Big], \quad (7.82)
\end{aligned}$$

Similarly for the third ghost operator

$$Y_{\mu\nu} = -\bar{g}_{\mu\nu}\Box + \sigma_Y\bar{\nabla}_\mu\bar{\nabla}_\nu + \bar{R}_{\mu\nu}, \qquad (7.83)$$

with $\sigma_Y = 1 - 2\frac{\gamma-\alpha}{\beta+4\gamma}$, we have

$$B_0(Y) = \frac{1}{(4\pi)^{d/2}}(1-\sigma_Y)^{-d/2}\left(1+(d-1)(1-\sigma_Y)^{d/2}\right)\int d^dx\,\sqrt{-\bar{g}}\,. \qquad (7.84)$$

$$B_2(Y) = \frac{1}{(4\pi)^{d/2}}\frac{d-7+(1-\sigma_Y)^{1-d/2}}{6}\int d^dx\,\sqrt{-\bar{g}}\,\bar{R}, \qquad (7.85)$$

$$\begin{aligned}
B_4(Y) = \frac{1}{(4\pi)^{d/2}}\int d^dx\sqrt{-\bar{g}}\Big[& \frac{d-16+(1-\sigma_Y)^{\frac{4-d}{2}}}{180}\bar{R}^2_{\mu\nu\rho\lambda} \\
& -\frac{d-91+(1-\sigma_Y)^{\frac{4-d}{2}}}{180}\bar{R}^2_{\mu\nu} + \frac{d-13+(1-\sigma_Y)^{\frac{4-d}{2}}}{72}\bar{R}^2\Big], \quad (7.86)
\end{aligned}$$

The calculation of the traces contained in the heat kernel coefficients is quite cumbersome and is best done by computer. Substituting the heat kernel coefficients in Eq. (7.76), and extracting the coefficients of 1, $\bar{R}$, $\bar{R}^2$, $\bar{R}_{\mu\nu}\bar{R}^{\mu\nu}$, and $\bar{R}_{\mu\nu\rho\sigma}\bar{R}^{\mu\nu\rho\sigma}$, we obtain the beta functions of Λ, G, α, β, γ.

7.4.4   Beta functions in $d = 4$

The beta functions of the higher-derivative couplings form a closed subsystem:

$$\beta_\alpha = \frac{1}{(4\pi)^2} \frac{90\alpha^2 + 15\alpha\beta - 120\alpha\gamma - 23\beta^2 - 199\beta\gamma - 338\gamma^2}{9(\beta + 4\gamma)^2} , \tag{7.87}$$

$$\beta_\beta = \frac{1}{(4\pi)^2} \frac{371}{90} , \tag{7.88}$$

$$\beta_\gamma = \frac{1}{(4\pi)^2} \frac{413}{180} . \tag{7.89}$$

or, changing variables,

$$\beta_\lambda = -\frac{1}{(4\pi)^2} \frac{133}{10} \lambda^2, \tag{7.90}$$

$$\beta_\omega = -\frac{1}{(4\pi)^2} \frac{25 + 1098\,\omega + 200\,\omega^2}{60} \lambda, \tag{7.91}$$

$$\beta_\theta = \frac{1}{(4\pi)^2} \frac{7(56 - 171\,\theta)}{90} \lambda . \tag{7.92}$$

The coupling λ is seen to have a logarithmic approach to asymptotic freedom, as in Yang-Mills theory. The condition $\lambda = 0$ also guarantees the vanishing of the other two beta functions. In order to find preferred values for ω and θ, it is customary to define a rescaled renormalization group time $\bar\tau$ such that $d\bar\tau = \lambda\, dk/k$. With this variable, one finds

$$\frac{d\omega}{d\bar\tau} = -\frac{1}{(4\pi)^2} \frac{25 + 1098\,\omega + 200\,\omega^2}{60} , \tag{7.93}$$

$$\frac{d\theta}{d\bar\tau} = \frac{1}{(4\pi)^2} \frac{7(56 - 171\,\theta)}{90} , \tag{7.94}$$

which have real fixed points:

$$\text{FP}_1 : (\omega_*, \theta_*) \approx (-5.46714, 0.327485) ; \qquad \text{FP}_2 : (\omega_*, \theta_*) \approx (-0.0228639, 0.327485) .$$

This flow is shown in Fig. 7.6. Note that $\frac{d\lambda}{d\bar\tau} = -\frac{1}{(4\pi)^2} \frac{133}{10} \lambda$, so also with this variable one sees the FP at $\lambda = 0$.

The beta functions for $\tilde{G}$ and $\tilde\Lambda$ can be written as

$$\beta_{\tilde\Lambda} = -2\tilde\Lambda + p(\lambda, \omega)\tilde\Lambda - q(\omega)\tilde{G}\tilde\Lambda + r(\omega)\tilde{G} + s(\lambda, \omega) + \frac{t(\lambda, \omega)}{\tilde{G}} , \tag{7.95}$$

$$\beta_{\tilde{G}} = 2\tilde{G} - u(\lambda, \omega)\tilde{G} - q(\omega)\tilde{G}^2 , \tag{7.96}$$

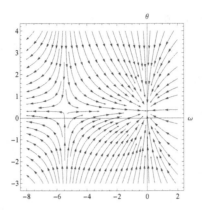

Fig. 7.6   The flow in the ω-θ plane in $d = 4$ with the fixed points FP$_1$ (left) and FP$_2$ (right).

where

$$p(\lambda, \omega) = \frac{1}{(4\pi)^2} \frac{1 + 86\omega + 40\omega^2}{12\omega} \lambda, \tag{7.97}$$

$$q(\omega) = \frac{171 + 298\omega + 152\omega^2 + 16\omega^3}{36\pi\sigma(1 + \omega)}, \tag{7.98}$$

$$r(\omega) = \frac{283 + 664\omega + 204\omega^2 - 128\omega^3 - 32\omega^4}{144\pi(1 + \omega)^2}, \tag{7.99}$$

$$s(\lambda, \omega) = -\frac{\sigma}{(4\pi)^2} \frac{1 + 10\omega}{4\omega} \lambda, \tag{7.100}$$

$$t(\lambda, \omega) = \frac{\sigma^2}{(4\pi)^2} \frac{1 + 20\omega^2}{256\pi\omega^2} \lambda^2, \tag{7.101}$$

$$u(\lambda, \omega) = \frac{1}{(4\pi)^2} \frac{3 + 26\omega - 40\omega^2}{12\omega} \lambda. \tag{7.102}$$

To picture the flow of $\tilde{\Lambda}$ and $\tilde{G}$, we set the remaining variables to their FP values $\omega = \omega_*$, $\theta = \theta_*$, and $\lambda = \lambda_* = 0$. Then, defining $r_* = r(\omega_*)$, $q_* = q(\omega_*)$, the flow equations (7.95) and (7.96) become very simple:

$$\beta_{\tilde{\Lambda}} = -2\tilde{\Lambda} + r_*\tilde{G} - q_*\tilde{G}\tilde{\Lambda}, \tag{7.103}$$

$$\beta_{\tilde{G}} = 2\tilde{G} - q_*\tilde{G}^2. \tag{7.104}$$

Except for unimportant numerical differences of the coefficients, which are anyway within the variability due to the choice of scheme, the resulting flow in the $(\tilde{\Lambda}, \tilde{G})$-plane, shown in Fig. 7.7, is exactly of the same form as the simple Einstein-Hilbert flow (7.26). It has two FPs for each FP of ω and θ. At FP$_1$, we have $r_* \approx -0.545$, $q_* \approx -0.931$. Then there is the Gaussian FP at $\tilde{\Lambda} = \tilde{G} = 0$ and another nontrivial FP at

$$\tilde{\Lambda}_* = \frac{r_*}{2q_*} \approx 0.293, \qquad \tilde{G}_* = \frac{2}{q_*} \approx -2.148. \tag{7.105}$$

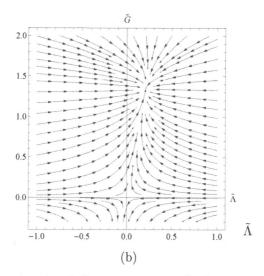

(b)

Fig. 7.7 The flow in the $(\tilde{\Lambda}, \tilde{G})$-plane for the FP $(\tilde{\Lambda}_*, \tilde{G}_*) \approx (0.293, -2.148)$.

At FP$_2$, one has $r_* \approx 0.620$, $q_* \approx 1.486$ and again there is a Gaussian FP and a non-Gaussian one at

$$\tilde{\Lambda}_* = \frac{r_*}{2q_*} \approx 0.209 \, , \qquad \tilde{G}_* = \frac{2}{q_*} \approx 1.346 \, . \tag{7.106}$$

We note that the FP occurs for positive G in the case of FP$_2$ and for negative G in the case of FP$_1$. The fact that in this truncation the coefficients of the beta functions are independent of $\tilde{\Lambda}$, concurs with the conclusion of section 7.3.

The attractivity properties of these FPs are determined by the stability matrix

$$M_{ij} = \frac{\partial \tilde{\beta}_i}{\partial \tilde{g}_j} = \begin{pmatrix} -2 - q_* \tilde{G}_* & r_* - q_* \tilde{\Lambda} \\ 0 & 2 - 2q_* \tilde{G}_* \end{pmatrix} \, .$$

At the Gaussian FP the eigenvalues of M are $(-2, 2)$; the attractive eigenvector points along the $\tilde{\Lambda}$ axis and the repulsive eigenvector has components $(r_*/4, 1)$. At the non-Gaussian FP the eigenvalues of M are $(-4, -2)$ with the same eigenvectors as before.

Note that the beta functions of λ, ω and θ, are universal, in the sense that they do not depend on the choice of the cutoff function R_k. This follows from the fact that they are proportional to the universal coefficients $Q(p, 0)$. The beta functions of $\tilde{\Lambda}$ and $\tilde{G}$ are not universal. There are however some contributions that are proportional to $Q(p, 0)$ times B_4 coefficients, which are universal. These are the functions p, s, t, u. On the other hand, the functions q and r are proportional to the scheme-dependent coefficients, $Q(p, 1)$ and $Q(p, 2)$, multiplying heat kernel coefficients B_2 and B_0. This can be confirmed also by looking at equations (3.2) in [226].[8] It is important to stress that although the value of $Q(p, 1)$ and $Q(p, 2)$

---

[8] A factor $1 + 10\omega^2$ in Eq. (8a) in [217] contains a misprint and should read $1 + 10\omega$.

can be changed by changing the function R_k, they always remain positive, so that the existence of a nontrivial FP for positive $\tilde{G}$ is universal.

We conclude with two comments. First, it is noteworthy that in this calculation we could keep the background metric $\bar{g}_{\mu\nu}$ completely arbitrary. This is an explicit manifestation that in this type of calculation it is possible to maintain the property of background independence.

Second, just like how the coupling $\tilde{G}$, which is asymptotically free in $d = 2$, develops a nontrivial fixed point when one moves a little away from $d = 2$, so the coupling $\tilde{\lambda} = \lambda k^{d-4}$ also develops a nontrivial fixed point away from $d = 4$. This has been studied in [219], where the RG flow has also been studied in $d = 3$, $d = 5$, $d = 6$, besides the case $d = 4 + \epsilon$. The case $d = 3$ is interesting because there are no propagating degrees of freedom in three-dimensional GR, but there are if one admits higher derivative couplings. It has been shown in [227] that the special case $8\alpha + 3\beta = 0$ describes massive gravitons and is free of ghosts. This particular combination is not a fixed point of the flow [228]. We now turn to another way of having gravitons in three dimensions, which involves an action with three rather than four derivatives.

7.5 Topologically massive gravity at one loop

Einstein gravity in three dimensions is a rather trivial theory, because it has no propagating degrees of freedom. Still, from (7.27) and using for example the type II cutoff coefficients (6.121), we see that in the simple Einstein-Hilbert approximation it has a nontrivial fixed point at

$$\tilde{\Lambda} = 0 , \qquad \tilde{G} = \frac{\pi}{20} \approx 0.157 . \tag{7.107}$$

This is due to the fact that the calculation is performed off-shell, and there are nonphysical degrees of freedom that give rise to a scale dependence of the EAA.

It is interesting to consider different actions for gravity that give propagating physical degrees of freedom. For this one has to consider actions containing more than two derivatives. We have already mentioned four-derivative actions in the preceding section. Here we consider an action containing three derivatives that is peculiar to three dimensions. In addition to the Hilbert term, it contains the Chern-Simons (CS) form constructed with the Levi-Civita connection:

$$S = Z_N \int d^3 x \sqrt{g} \left(2\Lambda - R + \frac{i}{2\mu} \frac{1}{\sqrt{g}} \epsilon^{\lambda\mu\nu} \Gamma^\rho_{\lambda\sigma} \left(\partial_\mu \Gamma^\sigma_{\nu\rho} + \frac{2}{3} \Gamma^\sigma_{\mu\tau} \Gamma^\tau_{\nu\rho} \right) \right) , \tag{7.108}$$

where $\epsilon^{\lambda\mu\nu}$ is the totally antisymmetric tensor density with entries $\pm 1, 0$. Since the action is independent of the metric (aside from the Christoffel symbols) the rules spelled out in section 5.2 imply that the Euclidean action must be imaginary. This accounts for the factor i.

The theory of gravity with this action is called "Topologically Massive Gravity" or TMG and μ is called the "topological mass" [229]. This theory is of considerable interest, as the presence of the cosmological term makes it possible to have a black hole solution, while the Chern-Simons term is responsible for the presence of a single propagating massive graviton. The properties of this theory have been the subject of intense scrutiny. For a generic value of the CS coupling, either black hole states (if $G < 0$) or graviton states (if $G > 0$) will have negative mass.

The perturbative renormalization of TMG has been studied by [230–232], where it was concluded that it is renormalizable. Here we will follow [233][9] and calculate the beta functions of the dimensionless couplings $\tilde\Lambda$, $\tilde G$ and $\tilde\mu$ using the ERGE. Actually, it will be useful to define the dimensionless combinations

$$\nu = \mu G \; ; \qquad \tau = \Lambda G^2 \; ; \qquad \phi = \mu / \sqrt{|\Lambda|} \; , \qquad (7.109)$$

not involving the RG scale k, and to look for their flow. Note in particular that $\frac{1}{32\pi\nu}$ is the coefficient of the CS term. We will see that there is a nontrivial fixed point that could be used to construct a UV completion of three-dimensional gravity.

The calculations in this section provide an opportunity to introduce an alternative method for the evaluation of the functional traces in the flow equations. This is because the general formulas for the heat kernel coefficients of the kinetic operator in this theory are not available, so that the general formulas discussed in section 5.7 cannot be used. As in section 5.6, one could use the method of spectral sums to compute the heat kernel coefficients and then use them in the general formula, but it is better to use the spectral sums to calculate directly the r.h.s. of the flow equation.

7.5.1   The quadratic action, gauge fixing and cutoff

As usual, we will not consider (7.108) as the action to be used in the definition of a functional integral, but rather as an ansatz for a truncation of the EAA. We expand this action to second order using the formulae of section 3.4 and add the standard gauge fixing term (5.82), (5.83) and ghost term (5.91).[10]

The beta functions will be obtained from the Wetterich equation (6.15), which for this theory has the form

$$\frac{d\Gamma_k^{(1)}}{dt} = \frac{1}{2}\text{Tr}\left(\frac{1}{\sqrt{\bar g}}\frac{\delta^2\,(S + S_{GF})}{\delta h_{\mu\nu}\delta h_{\rho\sigma}} + \mathcal{R}^{\mu\nu\rho\sigma}\right)^{-1}\frac{d\mathcal{R}_{\rho\sigma\mu\nu}}{dt}$$
$$-\text{Tr}\left(\frac{1}{\sqrt{\bar g}}\frac{\delta^2 S_{gh}}{\delta\bar C^\mu\delta C_\nu} + \mathcal{R}_\mu{}^\nu\right)^{-1}\frac{d\mathcal{R}^\mu{}_\nu}{dt} \; . \qquad (7.110)$$

Expanding in powers of $h_{\mu\nu}$, and discarding total derivative terms, the quadratic

---

[9]The related supergravity has been studied in [234].

[10]In [233] the gauge condition was written in terms of a gauge parameter $\rho = (4\beta + 1)/3$.

part of the action is given by

$$S^{(2)} + S_{GF} = \frac{1}{4} Z_N \int d^3 x \sqrt{\bar{g}} \left[h_{\mu\nu} \left(-\bar{\Box} + \frac{2\bar{R}}{3} - 2\Lambda \right) h^{\mu\nu} - \frac{2(1-\alpha)}{\alpha} h_{\mu\nu} \bar{\nabla}^\mu \bar{\nabla}_\rho h^{\rho\nu} \right.$$

$$- \left(2 - \frac{4(\beta+1)}{3\alpha} \right) h \bar{\nabla}^\mu \bar{\nabla}^\nu h_{\mu\nu} + \left(1 - \frac{2(\beta+1)^2}{9\alpha} \right) h \bar{\Box} h - \frac{1}{6} h (\bar{R} - 6\Lambda) h$$

$$\left. + \frac{i}{\mu} \varepsilon^{\lambda\mu\nu} h_{\lambda\sigma} \left(\bar{\nabla}_\mu \left(\bar{\Box} - \frac{\bar{R}}{3} \right) h^\sigma{}_\nu - \bar{\nabla}_\mu \bar{\nabla}^\sigma \bar{\nabla}^\rho h_{\rho\nu} \right) \right] . \tag{7.111}$$

For our purposes it will be sufficient to consider maximally symmetric backgrounds, for which

$$\bar{R}_{\mu\nu\rho\sigma} = \frac{\bar{R}}{6} (\bar{g}_{\mu\rho} \bar{g}_{\nu\sigma} - \bar{g}_{\mu\sigma} \bar{g}_{\nu\rho}) , \qquad \bar{R}_{\mu\nu} = \frac{\bar{R}}{3} \bar{g}_{\mu\nu} . \tag{7.112}$$

In order to achieve partial diagonalization of the inverse propagator we use the York decomposition. In addition to relations (5.25),(5.26),(5.27), we need also the following relations involving the gravitational CS terms:

$$\int d^3 x h_{\lambda\sigma} \epsilon^{\lambda\mu\nu} \bar{\nabla}_\mu \bar{\Box} h_\nu{}^\sigma = \int d^3 x \left[h_{\lambda\sigma}^T \epsilon^{\lambda\mu\nu} \bar{\nabla}_\mu \bar{\Box} h_\nu^{T\sigma} \right.$$

$$\left. - \xi_\lambda \epsilon^{\lambda\mu\nu} \bar{\nabla}_\mu \left(\bar{\Box} + \frac{\bar{R}}{3} \right) \left(\bar{\Box} + \frac{2\bar{R}}{3} \right) \xi_\nu \right]$$

$$\int d^3 x h_{\lambda\sigma} \epsilon^{\lambda\mu\nu} \bar{\nabla}_\mu h_\nu{}^\sigma = \int d^3 x \left[h_{\lambda\sigma}^T \epsilon^{\lambda\mu\nu} \bar{\nabla}_\mu h_\nu^{T\sigma} - \xi_\lambda \epsilon^{\lambda\mu\nu} \bar{\nabla}_\mu \left(\bar{\Box} + \frac{\bar{R}}{3} \right) \xi_\nu \right]$$

$$\int d^3 x h_{\lambda\sigma} \epsilon^{\lambda\mu\nu} \bar{\nabla}_\mu \bar{\nabla}^\sigma \bar{\nabla}^\rho h_{\rho\nu} = \int d^3 x \left[- \xi_\lambda \epsilon^{\lambda\mu\nu} \bar{\nabla}_\mu \left(\bar{\Box} + \frac{\bar{R}}{3} \right)^2 \xi_\nu \right] . \tag{7.113}$$

It is convenient to choose the gauge parameter $\beta = \alpha/2$, to eliminate the mixing between σ and h. Then, the quadratic action becomes

$$S^{(2)} + S_{GF} = \frac{Z_N}{2} \int d^3 x \sqrt{\bar{g}} \left[c_2 h_{\mu\nu}^{TT} \Delta_2^{\mu\nu\rho\sigma} h_{\rho\sigma}^{TT} + c_1 \hat{\xi}^\mu \Delta_{1\mu}{}^\nu \hat{\xi}_\nu + c_\sigma \hat{\sigma} \Delta_\sigma \hat{\sigma} + c_h h \Delta_h h \right] , \tag{7.114}$$

where we have defined the operators

$$\Delta_{2\mu\nu}{}^{\rho\sigma} = \left(-\bar{\Box} + \frac{2\bar{R}}{3} - 2\Lambda \right) \delta_{(\mu}^{(\rho} \delta_{\nu)}^{\sigma)} + \frac{i}{\mu} \varepsilon_{(\mu}{}^{\lambda(\rho} \delta_{\nu)}^{\sigma)} \bar{\nabla}_\lambda \left(\bar{\Box} - \frac{\bar{R}}{3} \right) ,$$

$$\Delta_{1\mu}{}^\nu = \left(-\bar{\Box} - \frac{1-\alpha}{3} \bar{R} - 2\alpha\Lambda \right) \delta_\mu^\nu ,$$

$$\Delta_\sigma = -\bar{\Box} - \frac{2-\alpha}{4-\alpha} \bar{R} - \frac{6\alpha\Lambda}{4-\alpha} , \tag{7.115}$$

$$\Delta_h = -\bar{\Box} - \frac{\bar{R}}{4-\alpha} - \frac{6\Lambda}{4-\alpha} ,$$

and coefficients

$$c_2 = \frac{1}{2} , \qquad c_1 = -\frac{1}{\alpha} , \qquad c_\sigma = \frac{4-\alpha}{9\alpha} , \qquad c_h = -\frac{4-\alpha}{36} . \tag{7.116}$$

Note that the CS term affects only the propagation of $h_{\mu\nu}^{TT}$.

The ghost action in the diagonal gauge becomes

$$S_{\text{ghost}}^{(2)} = \int d^3x \sqrt{\bar{g}} \left[\bar{V}^\mu \Delta_{V\mu}^{\ \nu} V_\nu + c_S \hat{\bar{S}} \Delta_S \hat{S} \right] , \tag{7.117}$$

where $c_S = (\alpha - 4)/3$ and

$$\Delta_{V\mu}^{\ \nu} = \left(-\bar{\Box} - \frac{\bar{R}}{3} \right) \delta_\mu^\nu , \qquad \Delta_S = -\bar{\Box} - \frac{2}{4-\alpha} \bar{R} . \tag{7.118}$$

We note that in the gauge $\alpha = 4$ the σ^2, h^2, $\hat{S}^2$ terms do not contain $\bar{\Box}$ and the method we shall use below to compute the beta functions will turn out to be not suitable to deal with this case. Therefore, in the sequel we will assume that $\alpha \neq 4$.

For each spin component we choose the cutoff to be a function of the corresponding operator given in (7.115). The operators contain the couplings Λ and μ, but since the bare couplings do not run in the one-loop approximation that we are using here, this cannot be called a spectrally adjusted cutoff in the present context. We follow the procedure for the type IIb cutoff discussed in section 6.8.3.

Since the bare couplings appearing in the second variation of the action do not depend on k, in the numerator of the first term in the r.h.s. of (7.110) we can replace

$$\frac{d}{dt} (Z_N c_i R_k(\Delta_i)) = Z_N c_i \partial_t R_k(\Delta_i) .$$

Then the overall factors Z_N and c_i cancel between numerator and denominator and (7.110) reduces to

$$k\frac{d\Gamma_k}{dk} = \frac{1}{2} \left[\text{Tr}_2 W(\Delta_2) + \text{Tr}_1 W(\Delta_1) + \text{Tr}_0 W(\Delta_\sigma) + \text{Tr}_0 W(\Delta_h) \right] \\ - \left[\text{Tr}_1 W(\Delta_V) + \text{Tr}_0 W(\Delta_S) \right] , \tag{7.119}$$

where $W(z) = \frac{\partial_t R_k}{P_k}$. We will use the cutoff

$$R_k(z) = (k^2 - |z|)\theta(k^2 - |z|) .$$

Note the absolute value, which is necessary because the operator acting on spin-2 modes is not positive.

Before addressing the full problem, let us briefly consider the case when the Chern-Simons term is absent. In this case all the operators appearing in the functional traces (7.119) are minimal Laplace-type operators of the form $\Delta = -\bar{\nabla}^2 1 + \mathbf{E}$ and the evaluation of the traces in (7.119) can be done using (5.159). For the function $W = \frac{\partial_t R_k}{P_k}$, the relevant Q-functionals are

$$Q_{3/2}(W) = \frac{8}{3\sqrt{\pi}} k^3 ; \qquad Q_{1/2}(W) = \frac{4}{\sqrt{\pi}} k .$$

The heat kernel coefficients b_0 and b_2 of the operators listed in (7.115) can be evaluated using standard methods and are listed in the following table.

	h^{TT}	$\hat{\xi}$	$\hat{\sigma}$	h	V	$\hat{S}$
$\mathrm{trb_0}$	2	2	1	1	2	1
$\mathrm{trb_2}$	$4\Lambda - 3\bar{R}$	$\frac{2(1-\alpha)\bar{R}+12\alpha\Lambda}{3}$	$\frac{(16-7\alpha)\bar{R}+36\alpha\Lambda}{6(4-\alpha)}$	$\frac{(10-\alpha)\bar{R}+36\alpha\Lambda}{6(4-\alpha)}$	$\frac{2}{3}$	$\frac{16-\alpha}{6(4-\alpha)}\bar{R}$

The result is

$$\partial_t \Gamma_k = \int d^3x \sqrt{\bar{g}}\, \frac{6(11 - 9\alpha - 2\alpha^2)\Lambda - (47 - 2\alpha^2)\bar{R}}{12\pi^2(4-\alpha)} k \;, \qquad (7.120)$$

from which we read off

$$A = \frac{8(11 + 9\alpha - 2\alpha^2)\tilde{\Lambda}}{\pi(4-\alpha)} \;; \qquad B = -\frac{4(47 - 2\alpha^2)}{3\pi(4-\alpha)} \;. \qquad (7.121)$$

These results agree, for $\alpha = 1$, and for $\tilde{\Lambda} \to 0$, with those of (6.121). Thus the fixed point occurs at (7.107) for $\alpha = 1$, and at $\tilde{\Lambda} = 0$, $\tilde{G} = \frac{3\pi}{47} \approx 0.2005$ for $\alpha = 0$.

7.5.2   *Evaluation of the beta functions*

In the presence of the Chern-Simons term the operator acting on the spin-2 field is of third order and its heat kernel coefficients are not available. The r.h.s. of the ERGE has to be evaluated directly by performing the spectral sums.

We have $\partial_t R_k(z) = 2k^2\theta(k^2 - |z|)$ and for $z < k^2$, $P_k(z) = k^2$, so that $W(z) = 2\theta(k^2 - |z|)$. Dividing numerator and denominator by k^2, we have

$$
\begin{aligned}
k\frac{d\Gamma_k}{dk} &= \sum_{\pm}\sum_n m_n^{T\pm}\theta(1 - |\tilde{\lambda}_n^{TT\pm}|) + \sum_n m_n^{\xi}\theta(1 - \tilde{\lambda}_n^{\xi}) \\
&+ \sum_n m_n^{\sigma}\theta(1 - \tilde{\lambda}_n^{\sigma}) + \sum_n m_n^{h}\theta(1 - \tilde{\lambda}_n^{h}) \\
&- 2\sum_n m_n^{V}\theta(1 - \tilde{\lambda}_n^{V}) - 2\sum_n m_n^{S}\theta(1 - \tilde{\lambda}_n^{S}) \;.
\end{aligned}
\qquad (7.122)
$$

Here $\tilde{\lambda}_n^{(i)} = \lambda_n^{(i)}/k^2$ are the distinct dimensionless eigenvalues of the operator Δ_i, and $m_n^{(i)}$ their multiplicities.

The spectrum of the Bochner Laplacian on S^3 has been given in section 5.6.1. The same basis of hyperspherical harmonics serve also as eigenfunctions for the operator coming from the variation of the Chern-Simons term. In fact, one has

$$\bar{\nabla}_{[\mu}Y^{(n,\pm 1)}_{\nu],q}(x) = \pm\frac{1}{2}\sqrt{\frac{\bar{R}}{6}}(n+1)\bar{\varepsilon}_{\mu\nu}{}^{\rho}\, Y^{(n,\pm 1)}_{\rho,q}(x) \;,$$

$$\bar{\nabla}_{[\mu}Y^{(n,\pm 2)}_{\nu]\rho,q}(x) = \pm i\sqrt{\frac{\bar{R}}{6}}(n+1)\bar{\varepsilon}_{\mu\nu}{}^{\sigma}\, Y^{(n,\pm 2)}_{\rho\sigma,q}(x) \;.$$

Using these results, the eigenvalues of the operators (7.115) are given by

$$\lambda_n^{TT\pm} = \frac{\bar{R}}{6}(n^2 + 2n + 2) - 2\Lambda \pm \frac{1}{\mu}\left(\frac{\bar{R}}{6}\right)^{3/2} n(n+1)(n+2) , \quad n \geq 2 ,$$

$$\lambda_n^{\xi} = \frac{\bar{R}}{6}\left(n^2 + 2n - 3 + 2\alpha\right) - 2\alpha\Lambda , \quad n \geq 2 ,$$

$$\lambda_n^{\sigma} = \frac{\bar{R}}{6}\left(n^2 + 2n - \frac{6(2-\alpha)}{4-\alpha}\right) - \frac{6\alpha\Lambda}{4-\alpha} , \quad n \geq 2 ,$$

$$\lambda_n^{h} = \frac{\bar{R}}{6}\left(n^2 + 2n - \frac{6}{4-\alpha}\right) - \frac{6\Lambda}{4-\alpha} , \quad n \geq 0 , \quad (7.123)$$

$$\lambda_n^{V} = \frac{\bar{R}}{6}\left(n^2 + 2n - 3\right) , \quad n \geq 1 ,$$

$$\lambda_n^{S} = \frac{\bar{R}}{6}\left(n^2 + 2n - \frac{12}{4-\alpha}\right) , \quad n \geq 1 ,$$

and the respective multiplicities are

$$m_n^{TT+} = m_n^{T-} = n^2 + 2n - 3 ,$$
$$m_n^{\xi} = m_n^{V} = 2(n^2 + 2n) , \quad (7.124)$$
$$m_n^{\sigma} = m_n^{h} = m_n^{S} = n^2 + 2n + 1 .$$

We see that the effect of the cutoff function is simply to terminate each sum at some maximal value $n_{\max}$.

The evaluation of the sums can be done using the Euler-Maclaurin formula. There are some subtleties that need to be emphasized. For each type of field, the effect of the step function is that the integral extends up to a finite value of n. Thus, in any spin sector in (7.122) we have to evaluate, schematically

$$\int_{n_0}^{\infty} dx\, m(x)\theta(1 - \tilde{\lambda}(x)) = \int_{n_0}^{n_{\max}} dx\, m(x) , \quad (7.125)$$

where $n_{\max}$ is a (real and positive) root of the equation

$$\tilde{\lambda}(x) = 1 . \quad (7.126)$$

When the eigenvalues are quadratic in n, Eq. (7.126) has two roots. Since the cutoff must be positive, it is always clear which one to choose.

Things are more complicated in the presence of the CS term. The eigenvalues of the TT field contain a term which is cubic in n, and this term occurs with opposite signs in the two chirality sectors. If we tried to define directly the one-loop effective action, this would lead to convergence problems similar to those discussed in [235]. Here, convergence is never a problem because we are only interested in the k-derivative of the one-loop effective action. Still, one has to deal with the fact that the $+$ and $-$ eigenvalues have different signs (at least for sufficiently large n) and grow at different rates.

Since Eq. (7.126) for $\tilde{\lambda}^{TT\pm}$ is cubic, one has to choose one among its roots

$$r - s + t ; \quad r + e^{i\pi/3}s - e^{-i\pi/3}t ; \quad r + e^{-i\pi/3}s - e^{i\pi/3}t ,$$

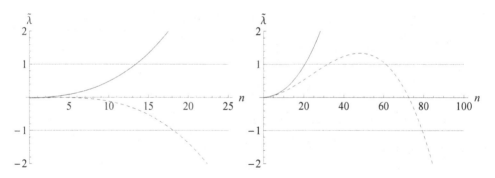

Fig. 7.8 The eigenvalues $\tilde{\lambda}_n^{TT+}$ (solid curve) and $\tilde{\lambda}_n^{TT-}$ (dashed curve) as functions of n, for $\tilde{R} = \tilde{\Lambda} = 0.01$. Left panel: small $\tilde{\mu}$ regime (here $\tilde{\mu} = 0.3$). Right panel: large $\tilde{\mu}$ regime (here $\tilde{\mu} = 3$).

where r, s, t depend on $\tilde{R} = \bar{R}/k^2$, $\tilde{\Lambda} = \Lambda/k^2$ and $\tilde{\mu} = \mu/k$. There is no root which is real for all parameter values.

One can understand this better by considering the plot of the functions $\tilde{\lambda}^{TT\pm}(x)$ for $x > 0$, see Figs. (7.8–7.10). Consider first the case when $\tilde{\mu} > 0$. The function $\tilde{\lambda}^{TT+}(x)$ is monotonically increasing, and for this function Eq. (7.126) has a single real root. Thus there is no ambiguity for the positive chirality modes. In the case of the negative chirality modes, however, for small x the function $\tilde{\lambda}^{TT-}(x)$ initially grows with x, until the cubic term prevails; hereafter it decreases. When $\tilde{R}$ and $\tilde{\Lambda}$ can be neglected, which is the situation we are interested in, the maximum is equal to $4\tilde{\mu}^2/27$. Thus if $\tilde{\mu} > \sqrt{27/4}$ Eq. (7.126) has two positive roots. It is clear that the smaller of the two has to be chosen, namely the one where the function is growing. For this reason we will call this an "ascending root cutoff". On the other hand if $0 < \tilde{\mu} < \sqrt{27/4}$, Eq. (7.126) has no positive root and the ascending root does not exist. However, we observe that the equation $\tilde{\lambda}^{TT-}(x) = -1$ has a real positive root for any $\tilde{\mu}$. As mentioned in Section 3, in the evaluation of the beta functions it is not important whether the modes have positive or negative eigenvalue. Therefore we can use this descending root to define the cutoff. We call this a "descending root cutoff". When $\tilde{\mu} < 0$, the discussion can be repeated interchanging the roles of $\tilde{\lambda}^{TT+}$ and $\tilde{\lambda}^{TT-}$.

Since the descending solution of $\tilde{\lambda}^{TT-} = -1$ always exists, while the ascending solution of $\tilde{\lambda}^{TT-} = 1$ only esists if $|\tilde{\mu}| > \sqrt{27/4}$, one wonders why not use always the former. The reason is that the information one can get from these different cutoffs is complementary. For example if we restrict ourselves to $\tilde{\mu} > 0$, when $\tilde{\mu}$ becomes very large, the descending solution of $\tilde{\lambda}_n^{TT-} = -1$ is much larger than the ascending solution of $\tilde{\lambda}_n^{TT+} = 1$. In fact the sum over negative chirality modes is divergent in the limit $\tilde{\mu} \to \infty$. The beta functions computed in this way will have fictitious singularities in this limit and it will not be possible to flow smoothly to the case when the CS term is absent. As we will see later, this also implies that the beta functions have fictitious singularities near $\tilde{G} = 0$, so it will not be possible to study the Gaussian FP in this cutoff scheme. On the other hand, the ascending solution has an imaginary part when $|\tilde{\mu}| < \sqrt{27/4}$ and therefore cannot be used to study the small $\tilde{\mu}$ region. The conclusion is that in order to properly understand the behavior of the theory over the whole range of values of $\tilde{\mu}$ one has to use both schemes: the "descending" cutoff for small $\tilde{\mu}$ and the "ascending" cutoff for large $\tilde{\mu}$.

Applying the Euler-Maclaurin formula to each of the sums in (7.122) and extracting from each of the integrals the appropriate powers of R, the r.h.s. of (7.122) can be written

$$\partial_t \Gamma_k = \sum \left[C_0 R^{-3/2} + C_2 R^{-1/2} + C_{3/2} + \frac{1}{2} F(n_0) - \frac{B_2}{2!} F'(n_0) \right] \qquad (7.127)$$

where the sum is over h^{TT}, $\hat{\xi}$, $\hat{\sigma}$, h, V, $\hat{S}$.

The contributions of each spin component to the $c_{3/2}$ term are independent of the cutoff scheme. They are listed in the following table:

	n_0	$C_{3/2}$	$F(n_0)$	$F'(n_0)$
$h^{TT}_{\mu\nu}$	2	12	20	24
$\hat{\xi}^\mu$	2	-24	32	24
$\hat{\sigma}$	2	-18	18	12
h	0	$-\frac{2}{3}$	2	4
V^μ	1	$-\frac{8}{3}$	12	16
$\hat{S}$	1	$-\frac{16}{3}$	8	8

In the second column, the lower end of integration in (7.125) is shown. Summing all the contributions we conclude that the coefficient of the R-independent term is exactly zero. More precisely, the sum $C_{3/2} + \frac{1}{2} F(n_0) - \frac{B_2}{2!} F'(n_0)$ is zero separately for the trace-free part of $h_{\mu\nu}$ (the sum of the first three lines), for the trace part h and for the ghosts (the sum of the last two lines). These are the components of the fields that can be defined by purely algebraic conditions. The cancellation does not occur for components defined by differential constraints, such as the transversality condition.

To some extent, the overall cancellation of the R-independent terms is expected. In the simpler setting of pure gravity without CS term, or any matter field coupled to gravity, the sums can be evaluated using the heat kernel expansion. On a manifold without boundary the trace of the heat kernel contains only integer powers of R, and in a 3-dimensional manifold the volume prefactor is proportional to $R^{-3/2}$, so that the expansion of $\partial_t \Gamma_k$ contains only odd powers of R, and there is no R-independent term. So the CS term will not be induced, if one starts without it.

To obtain the rest of the beta functional $\partial_t \Gamma_k$ there remains to sum the contributions of type C_0 and C_2 in (7.127); the final result has the following structure:

$$\partial_t \Gamma_k = \frac{V(S^3)}{16\pi} \left[k^3 A(\tilde{\Lambda}, \tilde{\mu}) + kB(\tilde{\Lambda}, \tilde{\mu})R + O(R^2) \right] , \tag{7.128}$$

where we have inserted powers of k such that the A- and B-coefficients are dimensionless. The volume of S^3 with radius ℓ is $V(S^3) = 2\pi^2 \ell^3$ with $\ell = \sqrt{\frac{6}{R}}$.

Equation (7.127) is an expansion in R, whose coefficients are functions of Λ, μ and k^2. As in section 7.3, we shall restrict ourselves to the parameter region where Λ and R are of the same order. Therefore we shall also expand the coefficients in (7.127) in powers of Λ, namely in C_0 we keep at most terms linear in Λ while in C_2 we only keep the Λ-independent terms. We will give explicit expressions for A and B later.

Evaluating the renormalized TMG action (7.108) on S^3 background, it can be written in the form

$$\Gamma_k = V(S^3) \left(\frac{2\Lambda}{16\pi G} - \frac{1}{16\pi G}\bar{R} + \frac{1}{12\sqrt{6}\pi G \mu}\bar{R}^{3/2} + O(\bar{R}^2) \right) , \tag{7.129}$$

where we have used that the integral of the CS term on S^3 is given by $\int \text{tr}(\omega d\omega + \frac{2}{3}\omega^3) = 32\pi^2$. The couplings Λ, G, μ are now (renormalized) running couplings evaluated at scale k. Rescaling the coupling constants as

$$G = \tilde{G}k^{-1} , \qquad \Lambda = \tilde{\Lambda}k^2 , \qquad \mu = \tilde{\mu}k , \tag{7.130}$$

so as to make them dimensionless, and comparing the t-derivative of (7.129) with (7.128), we obtain:

$$\frac{1}{8\pi\tilde{G}} \left(\partial_t \tilde{\Lambda} - \frac{\partial_t \tilde{G}}{\tilde{G}}\tilde{\Lambda} \right) = -\frac{3\tilde{\Lambda}}{8\pi\tilde{G}} + \frac{A}{16\pi} , \tag{7.131}$$

$$\frac{\partial_t \tilde{G}}{16\pi\tilde{G}^2} = \frac{1}{16\pi\tilde{G}} + \frac{B}{16\pi} , \tag{7.132}$$

$$\frac{1}{12\sqrt{6}\pi\tilde{\mu}\tilde{G}} \left(\frac{\partial_t \tilde{G}}{\tilde{G}} + \frac{\partial_t \tilde{\mu}}{\tilde{\mu}} \right) = 0 . \tag{7.133}$$

The last equation results from the fact that the terms of order $R^{3/2}$ in (7.128) cancel, and it implies that the dimensionless combination $\nu \equiv G\mu = \tilde{G}\tilde{\mu}$ does not run. From the other two equations one obtains the one-loop beta functions of $\tilde{G}$ and $\tilde{\Lambda}$:

$$\partial_t \tilde{G} = \tilde{G} + B(\tilde{\mu})\tilde{G}^2 ,$$

$$\partial_t \tilde{\Lambda} = -2\tilde{\Lambda} + \frac{1}{2}A(\tilde{\mu}, \tilde{\Lambda})\tilde{G} + B(\tilde{\mu})\tilde{\Lambda}\tilde{G} . \tag{7.134}$$

These equations have exactly the same form as in pure gravity with cosmological constant, except that the coefficients A and B are now $\tilde{\mu}$-dependent.

In order to solve for the RG flow, we use that ν does not run and substitute $\tilde{\mu} = \nu/\tilde{G}$. This yields two ordinary first order differential equations for $\tilde{G}(t)$ and $\tilde{\Lambda}(t)$, depending on the fixed constant value of the external parameter ν. Rather surprisingly, despite the very different functional form, the resulting flow is numerically quite similar to that of pure gravity with cosmological constant. Unlike in pure gravity, however, here we cannot give the solution in closed form. We will now describe the results for different cutoff schemes.

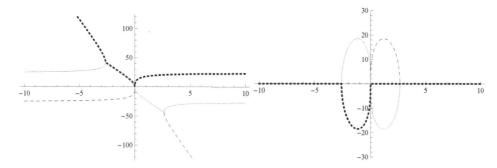

Fig. 7.9 The real (left panel) and imaginary (right panel) parts of the roots of the equation $\tilde{\lambda}_n^{TT+} = 1$, for $\tilde{R} = \tilde{\Lambda} = 0.01$, as functions of $\tilde{\mu}$. The first root (thick dotted line) always has positive real part and is complex for $-\sqrt{27/4} < \tilde{\mu} < 0$; the second root (dashed line) always has negative real part and is complex for $0 < \tilde{\mu} < \sqrt{27/4}$; the third root (solid line) is complex for $-\sqrt{27/4} < \tilde{\mu} < \sqrt{27/4}$. The solutions of the equation $\tilde{\lambda}_n^{TT-} = 1$ are obtained by the reflection $\tilde{\mu} \to -\tilde{\mu}$.

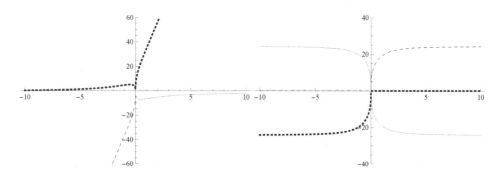

Fig. 7.10 The real (left panel) and imaginary (right panel) parts of the roots of the equation $\tilde{\lambda}_n^{TT-} = -1$, for $\tilde{R} = \tilde{\Lambda} = 0.01$, as functions of $\tilde{\mu}$. The first root (thick dotted) always has positive real part and is complex for $\tilde{\mu} < 0$; the second root (dashed) always has negative real part and is complex for $\tilde{\mu} > 0$; the third root (solid line) is always complex. The solutions of the equation $\tilde{\lambda}_n^{TT+} = -1$ are obtained by the reflection $\tilde{\mu} \to -\tilde{\mu}$.

7.5.3 *Ascending root cutoff*

For $|\tilde{\mu}| > \sqrt{27/4}$ we define the cutoff on the spin-2 modes as the smallest root of the equation $\tilde{\lambda}^{TT\pm} = 1$. In the case of positive $\tilde{\mu}$, the cutoff on $\tilde{\lambda}^{TT+}$ corresponds to the thick dotted line in Fig. (7.9), while the cutoff on $\tilde{\lambda}^{TT-}$ is obtained by reflecting the solid line. (Note that the two roots are different, so that the number of modes of positive and negative chirality is different.) Calculating the sums yields beta functions that are real and well defined for $0 < \tilde{G} < \sqrt{4/27\nu}$. For negative $\tilde{\mu}$ the roles of $\tilde{\lambda}^{TT+}$ and $\tilde{\lambda}^{TT-}$ are interchanged and one obtains beta functions that are real and well defined for $-\sqrt{4/27\nu} < \tilde{G} < 0$. The two calculations match smoothly along the line $\tilde{G} = 0$, so one can put them together to obtain a RG flow on the whole region $\tilde{G}^2 < 4\nu/27$.

With this prescription we can calculate the coefficients A and B, which arise as complicated functions involving cubic roots, but can be reduced to the following relatively simple form:

$$A(\tilde{\Lambda}, \tilde{\mu}) = -\frac{16}{3\pi} + \frac{9(2\sqrt{3}\cos 2\theta - \sqrt{3}\cos 4\theta + 8(\cos\theta)^3 \sin\theta)}{\pi(\cos 3\theta)^3}$$
$$+\frac{8(3 + 11\alpha - 2\alpha^2)}{\pi(4 - \alpha)}\tilde{\Lambda} + \frac{48(\cos\theta - \sqrt{3}\sin\theta)}{\pi \sin 6\theta}\tilde{\Lambda} , \qquad (7.135)$$

$$B(\tilde{\mu}) = -\frac{4(11 + 9\alpha - 2\alpha^2)}{3\pi(4 - \alpha)} - \frac{2(\sqrt{3}\sin\theta - \cos\theta) + 22(\sqrt{3}\sin 5\theta + \cos 5\theta)}{3\pi \sin 6\theta} ,$$

where we have introduced the angle

$$\theta = \frac{1}{3}\arctan\sqrt{\frac{4\tilde{\mu}^2}{27} - 1} . \qquad (7.136)$$

The beta functions admit a Taylor expansion around $\tilde{\Lambda} = \tilde{G} = 0$

$$\partial_t \tilde{G} = \tilde{G} - \frac{4(47 - 2\alpha^2)\tilde{G}^2}{3\pi(4 - \alpha)} - \frac{95\tilde{G}^4}{6\pi\nu^2} - \frac{2233\tilde{G}^6}{32\pi\nu^4} + O\left(\tilde{G}^7\right) , \qquad (7.137)$$

$$\partial_t \tilde{\Lambda} = -2\tilde{\Lambda} - \frac{4(14 - 27\alpha + 4\alpha^2)\tilde{G}\tilde{\Lambda}}{3\pi(4 - \alpha)} + \frac{\tilde{G}^3(42 + 115\tilde{\Lambda})}{6\pi\nu^2}$$
$$+\frac{11\tilde{G}^5(78 + 343\tilde{\Lambda})}{32\pi\nu^4} + O\left(\tilde{G}^6\right) .$$

The flow is shown, in the case $\nu = 5$, in Fig. 7.11. For any value of $\nu > 0$ there is a Gaussian FP, with scaling exponents 2 and -1, as expected. The eigenvectors coincide with the coordinate axes, as one can see from Fig. 7.11. In addition there is a nontrivial FP that for $\nu = 5$ occurs at $\tilde{\Lambda}_* = 0.000490471$ and $\tilde{G}_* = 0.200016$ which is UV attractive in both directions with scaling exponents 2.29401 and 1.00515.

The position of the FP and the eigenvalues of the stability matrix are given, for other values of ν, in Fig. 7.12. Note that $\tilde{\Lambda}$ is always positive but very small. This is due to the absence of a term of order $\tilde{G}^2$ in the expansion of $\partial_t \tilde{\Lambda}$ in (7.137). Actually if one plots the contours of the beta functions for $\tilde{\Lambda}$ one may see this as

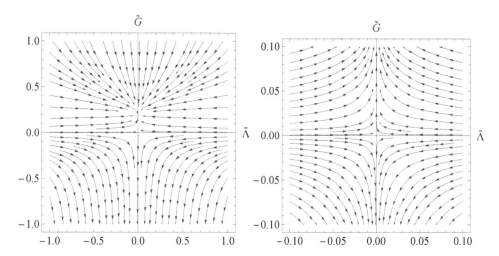

Fig. 7.11 The flow in the $\tilde{\Lambda}$-$\tilde{G}$ plane for $\alpha = 0$, $\nu = 5$, in the ascending root cutoff scheme. Right panel: enlargement of the region around the origin, showing the Gaussian FP. The beta functions become singular at $|\tilde{G}| = 1.9245$, outside the domain of the picture, but this singularity is an artifact of the scheme.

an effect of the deformation of the flow due to the presence of the boundary at $\tilde{G} = \nu\sqrt{4/27}$ (which occurs at $\tilde{G} = 1.924$ in Fig. 7.11). This can probably be regarded as a scheme artifact.

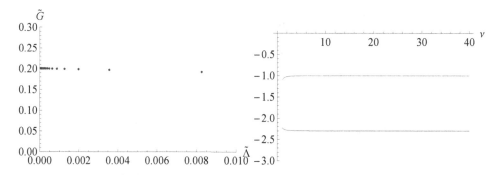

Fig. 7.12 Position of the FP (left panel) and eigenvalues of the stability matrix (right panel) for the nontrivial FP with $\alpha = 0$, $1 < \nu < 40$, in the ascending root cutoff scheme. Note that for this range of ν the singularity is always above the FP. In the left panel, ν grows from right to left. Note that $\tilde{\Lambda}_* > 0$ in this scheme. See Section 7.1 for a discussion. For large ν, $\tilde{G}_*$ tends to 0.2005 and the eigenvalues tend to -1 and -2.298.

Finally we observe that in the limit $\tilde{\mu} \to \infty$ the beta functions agree with the result for pure gravity without CS term derived in subsection 7.5.2.

7.5.4 *Descending root cutoff*

In this scheme the cutoff on the positive and negative chirality spin-2 modes is defined by two different equations, as follows:

$\tilde{\mu} > 0$	$\tilde{\mu} < 0$
$\tilde{\lambda}^{TT+} = 1$	$\tilde{\lambda}^{TT+} = -1$
$\tilde{\lambda}^{TT-} = -1$	$\tilde{\lambda}^{TT-} = 1$

Using the criteria described in the previous section, in the case of positive $\tilde{\mu}$, the cutoff on $\tilde{\lambda}^{TT+}$ corresponds once again to the thick dotted line in Fig. 7.9, while the cutoff on $\tilde{\lambda}^{TT-}$ is the thick dotted line in Fig. 7.10. For negative $\tilde{\mu}$ the roles of $\tilde{\lambda}^{TT+}$ and $\tilde{\lambda}^{TT-}$ are interchanged. Again the two roots are different, so that the number of modes of positive and negative chirality is different. In fact in this case the behavior is drastically different, since the cut on the descending mode grows linearly with $\tilde{\mu}$. This will give a singularity in the beta functions for $\tilde{\mu} \to \infty$, and therefore, when we make the substitution $\tilde{\mu} = \nu/\tilde{G}$, for $\tilde{G} \to 0$. Thus this scheme is really useful only for sufficiently small $\tilde{\mu}$.

With this prescription we find:

$$A(\tilde{\mu}, \tilde{\Lambda}) = -\frac{16}{3\pi} + \frac{\sqrt{3}}{\pi} \left[\left(\frac{2\cosh 2\eta - 1}{\cosh 3\eta} \right)^3 + \left(\frac{2\cosh 2\psi + 1}{\cosh 3\psi} \right)^3 \right] \tag{7.138}$$

$$+\frac{8(3 + 11\alpha - 2\alpha^2)}{\pi(4 - \alpha)}\tilde{\Lambda} + \frac{8\sqrt{3}}{\pi} \left[\frac{1}{2\cosh\eta + \cosh 3\eta} + \frac{1}{2\sinh\psi - \sinh 3\psi} \right] \tilde{\Lambda}$$

$$B(\tilde{\mu}) = -\frac{4(11 + 9\alpha - 2\alpha^2)}{3\pi(4 - \alpha)} - 20\sqrt{3}\frac{\cosh 2\eta + \cosh 2\psi}{3\pi \cosh 3\eta}$$

$$+\frac{2}{3\sqrt{3}\pi} \left(\frac{8\cosh 2\eta - 1}{\cosh 3\eta + 2\cosh\eta} + \frac{8\cosh 2\psi + 1}{\sinh 3\psi - 2\sinh\psi} \right)$$

where we have defined

$$\eta = \frac{1}{3}\operatorname{arctanh}\sqrt{1 - \frac{4\tilde{\mu}^2}{27}}\,, \qquad \psi = \frac{1}{3}\operatorname{arccoth}\sqrt{1 + \frac{4\tilde{\mu}^2}{27}}\,. \tag{7.139}$$

A representative flow is shown in Fig. 7.13 for $\nu = 0.1$. Superficially this may look similar to Fig. 7.11, but there are some important differences. First and foremost, there is no Gaussian FP, as is clear from the enlargement of the area around $\tilde{\Lambda} = \tilde{G} = 0$. The FP is wiped out by the singularity at $\tilde{G} = 0$, where the beta function of $\tilde{\Lambda}$ blows up. Another difference with Fig. 7.11 is that the flow lines near the $\tilde{G}$ axis are tilted in the opposite direction, and by a much larger amount.

For $\nu = 0.1$ the nontrivial FP occurs at $\tilde{\Lambda}_* = -0.0565337$ $\tilde{G}_* = 0.30036$ with critical exponents 2.76746 and 0.781453. The position of the FP and the eigenvalues, for $10^{-6} < \nu < 0.5$ are shown in Fig. 7.14. The FP occurs at positive $\tilde{\Lambda}$ for $\nu > 0.18$

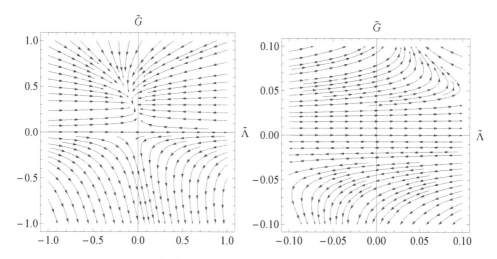

Fig. 7.13 The flow in the $\tilde{\Lambda}$-$\tilde{G}$ plane for $\alpha = 0$, $\nu = 0.1$, in the descending root cutoff scheme. Right panel: enlargement of the region around the origin, showing that there is no Gaussian FP. The beta functions diverge on the $\tilde{\Lambda}$ axis.

and negative $\tilde{\Lambda}$ for $\nu < 0.18$. Given that this cutoff scheme is more dependable for small $\tilde{\mu}$ (and hence, at fixed $\tilde{G}$, for small ν), it is again possible that the positive values of $\tilde{\Lambda}$ are a scheme artifact.

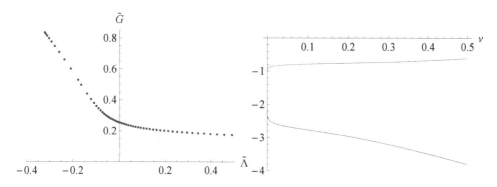

Fig. 7.14 Position of the FP (left panel) and eigenvalues of the stability matrix (right panel) for the nontrivial FP with $\alpha = 0$, $10^{-6} < \nu < 0.5$, in the descending root cutoff scheme. In the left panel, ν decreases from right to left. The cosmological constant changes sign for $\nu = 0.18$. The rightmost point ($\nu = 0.5$) has $\tilde{\mu} \approx 3 > \sqrt{27/4}$ and therefore is in the region where the scheme becomes unreliable.

Just as the beta functions in the ascending root scheme tend to a very simple form in the limit $\tilde{\mu} \to \infty$, the beta functions in the descending root scheme tend to a very simple form in the limit $\tilde{\mu} \to 0$:

$$\partial_t \tilde{G} = \tilde{G} - \frac{4}{3\pi} \frac{11 + 9\alpha - 2\alpha^2}{4 - \alpha} \tilde{G}^2 \, ,$$

$$\partial_t \tilde{\Lambda} = -2\tilde{\Lambda} - \frac{8}{3\pi} \left(1 + \frac{1 - 12\alpha + 2\alpha^2}{4 - \alpha} \tilde{\Lambda} \right) \tilde{G} \, . \tag{7.140}$$

These beta functions have a FP at $\tilde{\Lambda}_* = -1/3$, $\tilde{G}_* = \frac{3\pi(4-\alpha)}{4(11+9\alpha-2\alpha^2)}$, which for $\alpha = 0$ is $\tilde{G}_* \approx 0.8568$. The eigenvalues of the stability matrix are -2.182 and -1 corresponding to the directions $(1, 0)$ and $(-0.5499, 0.8352)$.

7.5.5 *Spectrally balanced cutoff*

The cutoff schemes discussed in the preceding two subsections are "spectrally unbalanced" in the sense that the summations over the spin-2 fields in (7.122) contained a different number of positive and negative chirality modes. Consideration of the behavior of the roots in Figs. (7.9) and (7.10) show that the ascending scheme becomes balanced for large $\tilde{\mu}$ while the descending scheme becomes balanced for small $\tilde{\mu}$. These are the regimes where these schemes are most reliable.

It seems natural to try a different cutoff method where the sums always contain equal numbers of positive and negative chirality modes, by construction. This can be achieved by cutting the sums $\sum_n m_n^{T+}$ and $\sum_n m_n^{T-}$ at the same value $n_{\max}$. Recalling that $m_n^{T+} = m_n^{T-}$, this amounts to replacing the sums on the spin-2 components in (7.122) by[11]

$$2 \sum_{n=2}^{n_{\max}} m_n^T \, . \tag{7.141}$$

We choose $n_{\max}$ to be the unique positive root of the equation $\tilde{\lambda}_n^{TT+} = 1$, if $\tilde{\mu} > 0$, and $\tilde{\lambda}_n^{TT-} = 1$ if $\tilde{\mu} < 0$.

With this prescription one gets the following results:

$$A(\tilde{\mu}, \tilde{\Lambda}) = -\frac{16}{3\pi} + \frac{2\sqrt{3}}{\pi(\cosh \eta)^3}$$

$$+ \frac{8(3 + 11\alpha - 2\alpha^2)}{\pi(4 - \alpha)} \tilde{\Lambda} + \frac{16\sqrt{3}}{\pi(\cosh 3\eta + 2\cosh \eta)} \tilde{\Lambda} \, , \tag{7.142}$$

$$B(\tilde{\mu}) = -\frac{4(11 + 9\alpha - 2\alpha^2)}{3\pi(4 - \alpha)} - \frac{8\sqrt{3}}{9\pi} \left(\frac{8 + 11\cosh 2\eta}{\cosh 3\eta + 2\cosh \eta} \right) \, ,$$

where η is given by (7.139). Thus, we observe that the apparent poles at $a = 0$, which correspond to $\eta = 0$, are actually absent as can be readily deduced from (7.142), and the coefficients A, B are real. Note also that $\eta = i\theta$ where θ is the angle that we have encountered in (7.136). Furthermore $\cosh 3\eta = 3\sqrt{3}/|\tilde{\mu}|$, and the results depend on the absolute value of $\tilde{\mu}$.

---

[11] Notice that here we are not deriving this formula from the ERGE, which would require defining different cutoff functions for the two classes of modes.

In the limit $\tilde{\mu} \to \infty$ we get

$$A \to \frac{8(11 + 9\alpha - 2\alpha^2)}{\pi(4 - \alpha)} \tilde{\Lambda}, \qquad B \to -\frac{4(47 - 2\alpha^2)}{3\pi(4 - \alpha)}. \tag{7.143}$$

This limit corresponds to neglecting the CS term, and we find agreement with the result of the heat kernel calculation in subsection 7.5.1.

The Taylor expansion of the beta functions around the Gaussian FP is

$$\partial_t \tilde{G} = \tilde{G} - \frac{4(47 - 2\alpha^2)\tilde{G}^2}{3\pi(4 - \alpha)} + \frac{28\tilde{G}^3}{3\pi\nu} - \frac{95\tilde{G}^4}{6\pi\nu^2} + O\left(\tilde{G}^5\right), \tag{7.144}$$

$$\partial_t \tilde{\Lambda} = -2\tilde{\Lambda} - \frac{4(14 - 27\alpha + 4\alpha^2)\tilde{G}\tilde{\Lambda}}{3\pi(4 - \alpha)} - \frac{4\tilde{G}^2(3 + 5\tilde{\Lambda})}{3\pi\nu} + \frac{\tilde{G}^3(42 + 115\tilde{\Lambda})}{6\pi\nu^2} + O\left(\tilde{G}^4\right).$$

In the limit $\tilde{\mu} \to 0$ the beta function of $\tilde{\Lambda}$ has the same expression as in (7.140), but the beta function of $\tilde{G}$ becomes singular. Still, the position of the FP seems to approximate closely the one that was found in the descending root cutoff. For $\nu = 10^{-6}$ we find $\tilde{\lambda}_* = -0.315$, $\tilde{G}_* = 0.830$. This should not come as a surprise, since in this limit the two positive roots of the equations $\tilde{\lambda}^{TT+} = 1$ and $\tilde{\lambda}^{TT-} = -1$ become equal, so the descending root cutoff becomes spectrally balanced.

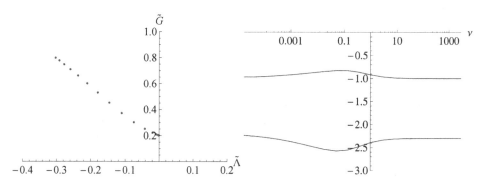

Fig. 7.15 Left panel: The position of the FP in the $\tilde{\Lambda}$-$\tilde{G}$ plane, varying ν from 0.002 (upper left) to 1000 (lower right). The point with coordinates (0,0.2005) is the limit $\nu \to \infty$. Right panel: The eigenvalues of M as functions of ν. For $\nu = 0.002$ they are -1 and -2.298 while for $\nu = 1000$ they are -0.969 and -2.238.

7.5.6 Summary

This was a long and rather technical section, so it is appropriate to pause and summarize the main points. The main technical novelty here has been the direct evaluation of the trace in the r.h.s. of the flow equation by spectral sums. The third order differential operator acting on the spin-2 degree of freedom poses considerable new challenges and has forced us to consider various types of cutoffs, that are reliable in different regimes.

The first and perhaps most striking result is the vanishing of the beta function of the dimensionless combination $\nu = \mu G$, which is the coefficient of the CS term.[12] One can interpret this as the consequence of some quantization condition. This issue has been discussed in [240] in the special case of flat space with asymptotically flat boundary conditions, where no quantization was found to be necessary. However, there will be topologies for which quantization is necessary. The vanishing of the beta function of ν is consistent with its universality in the UV.

Due to the fact that ν does not run, the remaining beta functions describe a flow in the $\tilde{\Lambda}$-$\tilde{G}$ plane, whose properties depend on the value of the fixed constant ν. The Einstein-Hilbert flow is recovered in the limit $\nu \to \infty$, but also for all values of ν the flow is qualitatively very similar to the Einstein-Hilbert flow (aside from cutoff pathologies). In particular, the flow is always governed by a Gaussian and a non-Gaussian FP. The Gaussian FP is UV attractive in the $\tilde{\Lambda}$ direction and UV repulsive in the $\tilde{G}$ direction, with scaling exponents equal to the canonical dimensions of Λ and G. The non-Gaussian FP is attractive in both directions and there exists a trajectory that connects the nontrivial FP in the UV to the Gaussian FP in the IR. The scaling exponents of the nontrivial FP depend weakly on ν but both for very large and very small ν they tend to similar values which are close to -1 and -2.3 (exactly -1, for large ν). As discussed in section 7.3.3, the combination $\sqrt{\Lambda}G$ is expected to have gauge-independent beta function. This beta function is proportional to

$$A + 6B\tilde{\Lambda} = -\frac{16}{3\pi}(1 + 3\tilde{\Lambda}) + \mu\text{-dependent terms} \qquad (7.145)$$

so the α-dependence cancels. The same is true for the beta function of $\phi = \mu/\sqrt{|\Lambda|}$. These findings confirm the robustness of the results, that we have already remarked in the four-dimensional case in connection with higher-derivative gravity.

Finally we note two advantages of this calculation over the four-dimensional one. The first is that for large $\tilde{\mu}$ the FP value of $\tilde{G}$ is relatively small and therefore more likely to be within the perturbative domain. The second is that in three dimensions the Riemann tensor is expressed entirely in terms of the Ricci tensor. As a result, it is possible to use field redefinitions to eliminate higher derivative terms, order by order in perturbation theory [241]. In three dimensions the higher derivative operators are technically redundant and therefore the truncation considered here is consistent. At least in perturbation theory, it is not necessary to consider an infinite set of operators. However, the redefinitions used to eliminate the higher derivative terms change the cosmological constant and Newton's constant. Therefore, the beta functions of $\tilde{\Lambda}$ and $\tilde{G}$ will receive corrections which have not been considered here.

---

[12]The analogous phenomenon in gauge theories has been investigated in [236, 237]. It was found that the coefficient of the CS term receives a finite renormalization. This however is an infrared effect [238]. In the context of the ERGE one can see that the coefficient of the CS term is constant for all finite k except for a finite jump in the limit $k \to 0$ [239]. Since our beta functions are only valid in the case $k \gg R$, we do not see such effect and we cannot exclude that such a finite renormalization happens.

7.6   A first peek beyond one loop

All the calculations reported so far were based on two independent approximations. On one hand, recall that the EAA is necessarily a function of two fields; the single-metric approximation consists in setting the VEV of the fluctuation field $h_{\mu\nu}$ to zero, or equivalently to set the VEV of the metric $g_{\mu\nu}$ equal to the background metric $\bar{g}_{\mu\nu}$. Then one remains with a functional of the background metric alone. On the other hand in the context of the ERGE the "one-loop" approximation consists in neglecting the running of any couplings that may appear in the r.h.s. of the equation. The preceding three sections were based on the single-field, one-loop approximation. In subsection 7.6.1 we will move a first step beyond one-loop, but remaining within the single-field approximation of the Einstein-Hilbert truncation. In subsection 7.6.2 we will move a first step beyond the single-metric approximation by considering the level-two Einstein-Hilbert truncation, both at one-loop and beyond one-loop.

7.6.1   *The background anomalous dimension*

In the single-metric Einstein-Hilbert truncation, the Hessian contains the prefactor $Z_N = \frac{1}{16\pi G}$, whose running would give the terms η_N in Eq. (6.120). The calculations in the preceding three sections were "one-loop" in the sense that we neglected η_N. To go beyond the one-loop approximation within the single-metric approximation one retains those additional terms. We thus return to the Einstein-Hilbert truncation as discussed in section 7.3. The flow equations are given by (7.25) and we now take into account the η_N-terms which were neglected there. The "anomalous dimension" η_N is defined by (6.116), or equivalently

$$\eta_N = \frac{\partial_t G}{G} = \frac{\partial_t \tilde{G}}{\tilde{G}} + 2 - d \ . \tag{7.146}$$

Inserting it in (7.25) and solving, we find

$$\frac{d\tilde{\Lambda}}{dt} = -2\tilde{\Lambda} + \tilde{G}\,\frac{A_1 + 2B_1\tilde{\Lambda} + \tilde{G}(A_2 B_1 - A_1 B_2)}{2(1 - B_2\tilde{G})} \ ,$$

$$\frac{d\tilde{G}}{dt} = (d-2)\tilde{G} + \frac{B_1 \tilde{G}^2}{1 - B_2\tilde{G}} \ . \tag{7.147}$$

We see that the beta functions are now much more nonlinear, due to the denominators containing $\tilde{G}$.

To proceed further we have to specify the coefficients A_1, A_2 etc. For coherence with section 7.3 we choose the type-II cutoff. The coefficients A_2 and B_2 are then given in (6.121). Restricting our attention to four dimensions, the flow equations become

$$\beta_{\tilde{\Lambda}} = -2\tilde{\Lambda} + \frac{1}{6\pi} \frac{(3 - 28\tilde{\Lambda} + 84\tilde{\Lambda}^2 - 80\tilde{\Lambda}^3)\tilde{G} + \frac{191 - 512\tilde{\Lambda}}{12\pi}\tilde{G}^2}{(1 - 2\tilde{\Lambda})(1 - 2\tilde{\Lambda} - \frac{13}{12\pi}\tilde{G})}$$

$$\beta_{\tilde{G}} = 2\tilde{G} - \frac{1}{3\pi} \frac{(23 - 20\tilde{\Lambda})\tilde{G}^2}{(1 - 2\tilde{\Lambda}) - \frac{13}{12\pi}\tilde{G}} \ . \tag{7.148}$$

This flow is shown in Fig. 7.16 and the properties of its non-Gaussian fixed point are listed in the third column of table 7.1. Note that there is still a singularity at $\tilde{\Lambda} = 1/2$ and that, compared to the fixed point of the equations (7.34), the imaginary part of the exponents is now even larger, resulting in a more pronounced spiralling.

Table 7.1   Fixed point of the Einstein-Hilbert truncation at level 0

	simple	one-loop	full
$\tilde{\Lambda}_*$	0.0326	0.0467	0.0924
$\tilde{G}_*$	0.8195	0.774	0.556
θ_1	4	$2.31 + 0.382i$	$2.425 + 1.27i$
θ_2	2	$2.31 - 0.382i$	$2.425 - 1.27i$
η_N	-	-	-2

Properties of the non-Gaussian fixed point in the Einstein-Hilbert truncation at level-zero (single-metric approximation), in three different approximations. The first column gives the fixed point of the simple Einstein-Hilbert flow (7.32), where the coefficients A_1 and B_1 are computed at $\tilde{\Lambda} = 0$. The second column gives the fixed point of the single-metric one-loop approximation, (7.34), where the $\tilde{\Lambda}$-dependence of the coefficients A_1 and B_1 is retained but η_N is still set to zero in the r.h.s. of the flow equation. The third column gives the fixed point of the flow (7.148). All results with type II cutoff in the gauge $\alpha = \beta = 1$.

We also notice from (6.116) that the "anomalous dimension" η_N is zero at the Gaussian fixed point and equal to $2 - d$ at a non-Gaussian fixed point. This is just the canonical mass dimension of G, so it may be a bit improper to call it an anomalous dimension, but this terminology is widely used in the literature on asymptotic safety. We shall encounter another notion of anomalous dimension in the next subsection. Insofar as it is justified to identify the cutoff k with an external momentum p, the propagator behaves effectively like $1/p^{2-\eta}$, so in four dimensions this would suggest that the propagator at a non-Gaussian fixed point behaves at high energy like $1/p^4$.

The main problem with this calculation is that already the starting Eq. (6.109) is not exact. Compared to the exact Eq. (6.106), the second functional derivatives with respect to the fluctuation field have been replaced with second functional derivatives with respect to the background field. We shall now do a different calculation that corrects this.

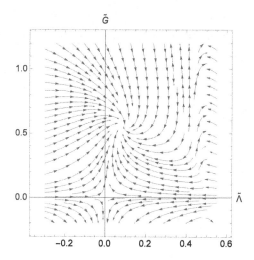

Fig. 7.16   The Einstein-Hilbert flow in the $\tilde{\Lambda}$-$\tilde{G}$ plane in $d = 4$, in the single-metric truncation.

7.6.2   *Level-two Einstein-Hilbert truncation*

We now make a first step beyond the single-field approximation by taking into account the anomalous dimensions of the graviton and of the ghosts. In the language of section 7.2.2 we will take into account five different operators and the respective couplings, which we now list: the background operators $\mathcal{O}_{(0,0,0,0)}$, $\mathcal{O}_{(2,0,0,0)}$ with the associated "level zero" cosmological constant and Newton constant; the level-two operators $\mathcal{O}_{(0,2,0,0)}$, $\mathcal{O}_{(2,2,0,0)}$ corresponding to the kinetic operator for the graviton. These are associated with the level two cosmological constant and Newton constant, which we shall reinterpret as wave function renormalization and mass of the graviton. Finally $\mathcal{O}_{(2,0,1,1)}$, the kinetic operator for the ghost with the associated ghost wave function renormalization.

To motivate the form of the ansatz for the EAA, we start from the second order terms in the expansion of the Hilbert action.

$$S_H(\bar{g} + h) + S_{GF}(h; \bar{g}) = Z_N \int d^d x \sqrt{\bar{g}} \left[2\Lambda - \bar{R} \right.$$

$$\left. -h_{\mu\nu}\left(\bar{R}^{\mu\nu} - \frac{1}{2}\bar{g}^{\mu\nu}\bar{R} + \bar{g}^{\mu\nu}\Lambda \right) + \frac{1}{2}\,h_{\mu\nu}\mathcal{H}^{\mu\nu\rho\sigma}h_{\rho\sigma} \right]. \quad (7.149)$$

The first, or level-zero, term is just the action of the background field; the second, or level-one, is proportional to the equation of motion; the third, or level-two, term, contains the Hessian $\mathcal{H}^{\mu\nu\rho\sigma}$, that for the Feynman-de Donder gauge was given in (3.135). Since these terms all come from the variation of the same action S_H, they all have the same prefactor $Z_N = \frac{1}{2\kappa^2} = \frac{1}{16\pi G}$. In a bi-field truncation, each of these operators will come with an independent coupling. We retain the terminology G and Λ for the two level-zero couplings. We will not be interested in the level-one couplings. For the level-two couplings we shall first redefine the quantum field as in (2.18) to give it canonical dimension, then we put a dimensionless wave function renormalization constant Z_h where previously there was the prefactor Z_N.

Now we recall that the gauge fixing term had a coefficient $Z_N/2\alpha$, and that in order to remove the non-minimal terms from the differential operator $Q^{\mu\nu\rho\sigma}$, we had chosen the Feynman gauge $\alpha = 1$. In order to achieve the same effect with the newly defined graviton kinetic term, we write the prefactor of the gauge fixing term $Z_h/2\alpha$ and we choose again the gauge $\alpha = 1$. Finally in the kinetic term for the graviton, as written in (3.144), we replace -2Λ by an independent mass squared parameter M^2.

Then our truncation for the EAA reads

$$\Gamma_k = \Gamma_{kH} + \Gamma_{k\,\text{gh}} \tag{7.150}$$

where

$$\Gamma_{kH} = \int d^dx\sqrt{\bar{g}}\left[Z_N(2\Lambda - \bar{R}) + \frac{1}{2}Z_h\phi_{\mu\nu}K^{\mu\nu\alpha\beta}((-\bar{\nabla}^2 + M^2)\mathbf{1}^{\rho\sigma}_{\alpha\beta} + W^{\rho\sigma}_{\alpha\beta})\phi_{\rho\sigma}\right], \tag{7.151}$$

is the level-two Einstein-Hilbert truncation, including the gauge fixing term, while the ghost action for the chosen gauge reads

$$\Gamma_{k\,\text{gh}} = Z_C\int d^dx\sqrt{\bar{g}}\,\bar{C}_\mu\left(\bar{\nabla}^\rho\bar{g}^{\mu\kappa}g_{\kappa\nu}\nabla_\rho + \bar{\nabla}^\rho\bar{g}^{\mu\kappa}g_{\rho\nu}\nabla_\kappa - \bar{\nabla}^\mu\bar{g}^{\rho\sigma}g_{\rho\nu}\nabla_\sigma\right)C^\nu, \tag{7.152}$$

with a wave-function renormalization $Z_C(k)$. Altogether we have five running couplings: the two background couplings Z_N (or equivalently G) and Λ, the wave function renormalizations Z_h, Z_C and the mass M.

7.6.3 The anomalous dimensions η_h and η_C

The beta functions of $\tilde{\Lambda}$ and $\tilde{G}$ are nearly the same as calculated in section 6.8.1, except for the replacement of η_N by

$$\eta_h = -\frac{\partial_t Z_h}{Z_h} \tag{7.153}$$

and a new contribution from the ghost anomalous dimension. Equation (6.119) is modified in

$$\frac{d\Gamma_k}{dt} = \frac{1}{2}\text{Tr}\frac{\partial_t R_k(\Delta_2) - \eta_h R_k(\Delta_2)}{P_k(\Delta_2) - 2\Lambda} - \text{Tr}\frac{\partial_t R_k(\Delta_{(gh)}) - \eta_C R_k(\Delta_{(gh)})}{P_k(\Delta_{(gh)})}$$

$$= \frac{1}{(4\pi)^{d/2}}\int dx\,\sqrt{g}\left\{\frac{d(d+1)}{4}Q_{\frac{d}{2}}\left(\frac{\partial_t R_k - \eta_h R_k}{P_k - 2\Lambda}\right) - dQ_{\frac{d}{2}}\left(\frac{\partial_t R_k - \eta_C R_k}{P_k}\right)\right.$$

$$+ \left[\frac{d(7-5d)}{24}Q_{\frac{d}{2}-1}\left(\frac{\partial_t R_k - \eta_h R_k}{P_k - 2\Lambda}\right) - \frac{d+6}{6}Q_{\frac{d}{2}-1}\left(\frac{\partial_t R_k - \eta_C R_k}{P_k}\right)\right]\bar{R}$$

$$\left. + O(\bar{R}^2)\right\}, \tag{7.154}$$

and consequently also Eq. (6.120) is replaced by

$$
\frac{d}{dt}\left(\frac{2\Lambda}{16\pi G}\right) = \frac{k^d}{16\pi}(A_1 + A_2\eta_h + A_3\eta_C)
$$
$$
-\frac{d}{dt}\left(\frac{1}{16\pi G}\right) = \frac{k^{d-2}}{16\pi}(B_1 + B_2\eta_h + B_3\eta_C) \ .
$$

(7.155)

The coefficients A_1, A_2, B_1, B_2 are still given by (6.121), and

$$
A_3 = \frac{64\pi}{(4\pi)^{d/2}(d+2)\Gamma(d/2)} \ ,
$$
$$
B_3 = \frac{16\pi(d+6)}{(4\pi)^{d/2}3d\Gamma(d/2)} \ .
$$

(7.156)

The anomalous dimension η_h and the beta function of M^2 can be extracted from the flow of the two point function on a flat background. Taking two functional derivatives of the ERGE with respect to the fluctuation field, using the ansatz (7.151), one finds for the l.h.s., in momentum space

$$
\frac{\delta^2 \partial_t \Gamma_k}{\delta\phi_{\mu\nu}(-q)\delta\phi_{\rho\sigma}(q)} = -\eta_h Z_h K^{\mu\nu\rho\sigma}(q^2 + M^2) \ .
$$

(7.157)

To describe the calculation that is needed for the r.h.s., let us introduce some notation. We suppress all indices and denote $\Gamma^{(n_1,n_2,n_3)}$ the derivative of the EAA n_1 times with respect to ϕ, n_2 times with respect to $\bar{C}$ and n_3 times with respect to C. The full graviton propagator is $G_h = (\Gamma^{(2,0,0)} + \mathcal{R}_h)^{-1}$ and the full ghost propagator is $G_c = (\Gamma^{(0,1,1)} + \mathcal{R}_c)^{-1}$. Then we have

$$
\frac{\delta^2\partial_t\Gamma_k}{\delta\phi\delta\phi} = \frac{\delta^2}{\delta\phi\delta\phi}\left[\frac{1}{2}\mathrm{Tr}\left(G_h\partial_t\mathcal{R}_h\right) - \mathrm{Tr}\left(G_c\partial_t\mathcal{R}_c\right)\right]
$$
$$
= \mathrm{Tr}\left(G_h\Gamma^{(3,0,0)}G_h\Gamma^{(3,0,0)}G_h\partial_t\mathcal{R}_h\right) - \frac{1}{2}\mathrm{Tr}\left(G_h\Gamma^{(4,0,0)}\partial_t\mathcal{R}_h\right)
$$
$$
-2\mathrm{Tr}\left(G_c\Gamma^{(1,1,1)}G_c\Gamma^{(1,1,1)}G_c\partial_t\mathcal{R}_c\right) + \mathrm{Tr}\left(G_c\Gamma^{(2,1,1)}G_c\partial_t\mathcal{R}_c\right) \ . \quad (7.158)
$$

The four terms on the r.h.s. can be represented graphically as in Fig. 7.17.

For the calculation of these diagrams, one needs the three- and four-graviton vertices, as well as two ghosts-one graviton and two ghosts-two graviton vertices. In the terminology introduced earlier, the vertices $\Gamma^{(n,0,0)}$, with two derivatives, correspond to level-three and level-four Newton constants. We have not introduced these couplings in the truncation (7.151),(7.152), so in order to close the system of flow equations we replace them by the level-zero Newton constant. Thus, the vertices $\Gamma^{(3)}$ and $\Gamma^{(4)}$ are obtained from the third and fourth variation of the background Hilbert term. The two ghosts-one graviton and two ghosts-two graviton vertices are obtained from (7.152).

To extract a scalar from the two-point function, we contract it with $K_{\mu\nu\rho\sigma}$ and divide by $K_{\alpha\beta\gamma\delta}K^{\alpha\beta\gamma\delta}$. Here it is necessary to resort heavily to the computer. We only give the final result. For some additional details the reader is referred to [242].

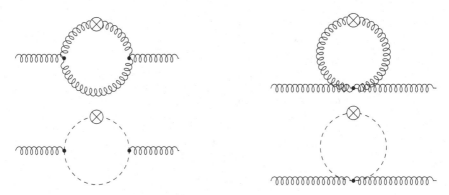

Fig. 7.17 Diagrams contributing to the graviton anomalous dimension. Springy lines are the graviton propagator and dashed lines are the ghost propagator. The cross is $\partial_t R_k$.

The anomalous dimension of the graviton can be written in the form

$$\eta_h = \left[a(\tilde{\Lambda}_k) + c(\tilde{\Lambda}_k)\eta_h + e(\tilde{\Lambda}_k)\eta_C \right] \tilde{G}_k \ , \tag{7.159}$$

with

$$a(\tilde{\Lambda}) = \frac{a_0 + a_1\tilde{\Lambda} + a_2\tilde{\Lambda}^2 + a_3\tilde{\Lambda}^3 + a_4\tilde{\Lambda}^4}{(4\pi)^{d/2}\Gamma(d/2)d^2(d^2 - 4)(3d - 2)(1 - 2\tilde{\Lambda})^4} \ , \tag{7.160}$$

$$a_0 = -4\pi(d - 2)\left(-896 + 264\,d + 1076\,d^2 - 434\,d^3 + 21\,d^4 + d^5\right) \ ,$$

$$a_1 = 16\pi(d - 2)\left(-2048 + 2552\,d - 318\,d^2 - 125\,d^3 + 2\,d^4 + d^5\right) \ ,$$

$$a_2 = -16\pi(12544 - 25760\,d + 16968\,d^2 - 4228\,d^3 + 354\,d^4 - 17\,d^5 + d^6) \ ,$$

$$a_3 = 4096\pi(d - 2)(-32 + 50\,d - 19\,d^2 + 2\,d^3) \ ,$$

$$a_4 = -2048\pi(d - 2)(-32 + 50\,d - 19\,d^2 + 2\,d^3) \ ;$$

$$c(\tilde{\Lambda}) = \frac{8\pi(d - 1)\left[128 + 720\,d - 350\,d^2 + 29\,d^3 + 32(d - 2)(d + 4)\tilde{\Lambda}\right]}{(4\pi)^{d/2}\Gamma(d/2)\,d^2(d + 2)(d + 4)(3d - 2)(1 - 2\tilde{\Lambda})^3} \tag{7.161}$$

$$e(\tilde{\Lambda}) = -\frac{128\pi\left(32 - 50\,d + 23\,d^2\right)}{(4\pi)^{d/2}\Gamma(d/2)d^2(d + 2)(d + 4)(3\,d - 2)} \ . \tag{7.162}$$

The ghost anomalous dimension is calculated in a similar way from the diagrams in Fig. 7.17. We obtain

$$\eta_C = \left[b(\tilde{\Lambda}_k) + d(\tilde{\Lambda}_k)\eta_h + f(\tilde{\Lambda}_k)\eta_C \right] \tilde{G}_k \ , \tag{7.163}$$

with

$$b(\tilde{\Lambda}) = \frac{64\pi\left[-8 + 4\,d + 18\,d^2 - 7\,d^3 + 2(4 - 9\,d^2 + 3\,d^3)\tilde{\Lambda}\right]}{(4\pi)^{d/2}\Gamma(d/2)d^2(d^2 - 4)(d + 4)(1 - 2\tilde{\Lambda})^2} \tag{7.164}$$

$$d(\tilde{\Lambda}) = \frac{-64\pi(4 - 4\,d - 9\,d^2 + 4\,d^3)}{(4\pi)^{d/2}\Gamma(d/2)d^2(d^2 - 4)(d + 4)(1 - 2\tilde{\Lambda})^2} \tag{7.165}$$

$$f(\tilde{\Lambda}) = \frac{-64\pi(4 - 9\,d^2 + 3\,d^3)}{(4\pi)^{d/2}\Gamma(d/2)d^2(d^2 - 4)(d + 4)(1 - 2\tilde{\Lambda})^2} \tag{7.166}$$

Note that the anomalous dimensions are not given explicitly. Rather, equations (7.159) and (7.163) form a system of linear equations for η_h and η_C whose solutions are the anomalous dimensions. The expressions for the solutions are complicated rational functions and it is easier to present the coefficients of the linear equations, out of which the anomalous dimensions can be obtained by linear algebra.

There is also another reason for presenting the results in this way. We recall that the one-loop approximation consists in neglecting the running of the couplings in the r.h.s. of the ERGE. Among the "couplings" are the wave function renormalization constants Z_h and Z_C, whose respective anomalous dimensions η_h and η_C appear in the r.h.s. of the equation. So, in this calculation, the one-loop approximation consists in neglecting $c(\tilde{\Lambda})$, $d(\tilde{\Lambda})$, $e(\tilde{\Lambda})$, $f(\tilde{\Lambda})$. The whole flow goes beyond one loop only insofar as the anomalous dimensions are given by the solutions of the full system (7.159), (7.163).[13]

Finally, the beta function of M^2 is

$$\partial_t \tilde{M}^2 = -2\tilde{M}^2 + \left(p(\tilde{\Lambda}, \tilde{M}) + q(\tilde{\Lambda}, \tilde{M})\eta_h + r(\tilde{\Lambda}, \tilde{M})\eta_C \right) \tilde{G} \tag{7.167}$$

where

$$p(\tilde{\Lambda}, \tilde{M}) = \frac{8\pi \left(p_0 + p_1 \tilde{M}^2 + p_2 \tilde{M}^4 + p_3 \tilde{M}^6 + p_4 \tilde{\Lambda} + p_5 \tilde{M}^2 \tilde{\Lambda} + p_6 \tilde{\Lambda}^2 \right)}{(4\pi)^{d/2} d(d^2 - 4)(d+4)(3d-2)(1+\tilde{M}^2)^3 \Gamma(d/2)} \tag{7.168}$$

$$q(\tilde{\Lambda}, \tilde{M}) = \frac{8\pi \left(q_0 + q_1 \tilde{M}^2 + q_2 \tilde{\Lambda} + q_3 \tilde{M}^2 \tilde{\Lambda} + q_4 \tilde{\Lambda}^2 \right)}{(4\pi)^{d/2} d(d^2 - 4)(d+4)(d+6)(3d-2)(1+\tilde{M}^2)^3 \Gamma(d/2)} \tag{7.169}$$

$$r(\tilde{\Lambda}, \tilde{M}) = \frac{256\pi d^2(d^2 - 4)(1+\tilde{M}^2)^3}{d+6} \tag{7.170}$$

$$
\begin{aligned}
p_0 &= 2d(d-2)\left(d^4 - 4d^3 - 77d^2 + 68d - 84\right) \\
p_1 &= d(d-2)\left(3d^4 - 29d^3 - 246d^2 - 16d - 288\right) \\
p_2 &= -192d^2\left(d^2 - 4\right) \\
p_3 &= -64d^2\left(d^2 - 4\right) \\
p_4 &= -2(d+4)\left(d^5 + 23d^4 - 242d^3 + 508d^2 - 168d - 32\right) \\
p_5 &= -2(d+2)(d+4)\left(3d^4 - 29d^3 + 44d^2 + 28d - 16\right) \\
p_6 &= -4(d+2)(d+4)\left(d^4 - 19d^3 + 40d^2 + 12d - 16\right)
\end{aligned}
\tag{7.171}
$$

---

[13]Note the difference with the single-field approximation, which can be the cause of confusion. In the Einstein-Hilbert truncation, and in the single-field approximation, the one-loop approximation consists in setting the anomalous dimension η_N to zero. Beyond level-zero, the wave-function renormalization is independent of Newton's coupling and the anomalous dimension can be nonzero also at one loop.

$$q_0 = -2d(d-1)(d-2)\left(d^3 - 86d + 156\right)$$
$$q_1 = -d(d-1)(d-2)(d+6)\left(3d^2 - 38d + 72\right)$$
$$q_2 = 2(d+6)\left(d^5 + 29d^4 - 300d^3 + 596d^2 - 112d - 64\right) \qquad (7.172)$$
$$q_3 = 2(d+4)(d+6)\left(3d^4 - 29d^3 + 44d^2 + 28d - 16\right)$$
$$q_4 = 4(d+4)(d+6)\left(d^4 - 19d^3 + 40d^2 + 12d - 16\right)$$

The flow is obtained by solving the equations for the anomalous dimensions as functions of $\tilde{G}$ and $\tilde{\Lambda}$, introducing the resulting expressions in the beta functions of $\tilde{G}$, $\tilde{\Lambda}$ and $\tilde{M}^2$. The flow given by these beta functions and anomalous dimensions is shown in Fig. 7.18.

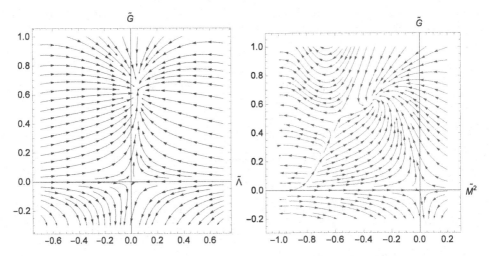

Fig. 7.18   Left: The full flow in the $\tilde{\Lambda}$-$\tilde{G}$ plane. Note the similarity to the simple Einstein-Hilbert flow. Right: The full flow in the $\tilde{M}^2$-$\tilde{G}$ plane. Note the singular line beginning at $\tilde{M}^2 = -1$ and a second (most likely spurious) fixed point with one attractive and one repulsive direction.

Since there are three couplings we give the sections in two different planes, which illustrates very well the different behavior of the cosmological constant $\tilde{\Lambda}$ and of the mass $\tilde{M}^2$. Note the close similarity of the figure on the left to Fig. 7.4, showing the one-loop, single field Einstein-Hilbert flow. Also, using the identification $M^2 \sim -2\Lambda$, one sees that there is a close similarity between the figure on the right and (7.16), showing the flow where the anomalous dimension of the graviton is identified with η_N. In particular, the singularity at $\tilde{\Lambda} = 1/2$ appears here at $\tilde{M}^2 = -1$.

Table 7.2 gives the properties of the nontrivial fixed point in the level-two approximation. The results are for type II cutoff, so they can be compared directly with the ones in table (7.1) Further comparisons can be found in [242] for the simplified case when M^2 is identified with -2Λ. It is shown there that the results change little when the cutoff of type Ia is used, and are also very similar to those of [243], where the anomalous dimension was obtained by contracting with the spin-2 projector (2.47) instead of the tensor K.

Table 7.2 Fixed point of the Einstein-Hilbert truncation at level 2

	one-loop	full flow
$\tilde{\Lambda}_*$	0.0508	0.0514
$\tilde{M}_*^2$	-0.565	-0.341
$\tilde{G}_*$	0.624	0.623
θ_1	3.94	3.92
$\theta_{2,3}$	$1.39 \pm 0.91i$	$1.37 \pm 0.93i$
η_h	0.500	0.530
η_C	-1.26	-1.26

Properties of the non-Gaussian fixed point in the Einstein-Hilbert truncation at level-two, taking into account the anomalous dimensions of the fluctuation fields. The couplings are defined in the action (7.151,7.152). The first column gives the one-loop approximation, the second gives the result of the full flow.

Notice that there is very little difference between the full system and the one-loop approximation. There is a more significant difference with the fixed point of the single-field approximation. In particular, the graviton anomalous dimension is considerably smaller in this calculation. Furthermore, there are now three scaling exponents, two of which form a complex conjugate pair. The corresponding eigenvectors lie approximately in the $\tilde{M}^2$-$\tilde{G}$ plane. The real exponent is close to the canonical dimension of Λ and its eigenvector is approximately in the $\tilde{\Lambda}$-direction.

7.7 Truncation to polynomials in R

We now come to a class of truncations where the Lagrangian density is a function of the Ricci scalar only. There is a vast literature on applications of these theories to cosmology. At one loop, the quantization of these theories has been discussed in [244]. Here we analyze the RG flow of this type of theory using the ERGE. The one-loop approximation does not seem to work too well, and we shall consider the RG flow improved with the background anomalous dimension as in section 7.6.1. We thus remain within the realm of single-field truncations.

The Euclidean action is

$$\Gamma_k[h_{\mu\nu}, C_\mu, \bar{C}_\nu; \bar{g}_{\mu\nu}] = \int d^d x \sqrt{g} f(R) + S_{GF}[h_{\mu\nu}; \bar{g}_{\mu\nu}] + S_{gh}[C_\mu, \bar{C}_\nu; \bar{g}_{\mu\nu}] , \quad (7.173)$$

the last two terms correspond to the gauge fixing and the ghost terms. The function f will be approximated by a polynomial later but can be kept general for now. We choose a simple second order gauge fixing

$$S_{GF} = \frac{k^{d-2}}{2\alpha} \int d^d x \sqrt{\bar{g}} \, F_\mu \bar{g}^{\mu\nu} F_\nu \quad (7.174)$$

where $F_\nu = \bar\nabla^\mu h_{\mu\nu} - \frac{1}{d}\bar\nabla_\nu h^\mu_\mu$ and α is a dimensionless gauge parameter. The choice of the gauge parameter $\beta = 0$ simplifies the subsequent formulas significantly. Note that unlike earlier treatment of the Einstein-Hilbert truncation, where the gauge fixing term had a prefactor Z_N, here we have put simply a power of k for dimensional reasons. Since the action contains four derivatives we could also use a four-derivative gauge fixing term of the type $F_\mu Y^{\mu\nu} F_\nu$ and $Y_{\mu\nu} = \bar g_{\mu\nu}\bar\nabla^2$ as in (7.67). These alternative choices lead only to minimal differences in the results. We give here an outline of the calculation, as presented originally in [245]. The interested reader is referred to [246] for more details.

7.7.1 *Hessian and gauge choice*

Expanding $\int \mathrm{d}^d x \sqrt{g} f(R)$ to second order in $h_{\mu\nu}$ gives

$$\frac{1}{2}\int \mathrm{d}^d x \sqrt{\bar g}\Big\{ \frac{1}{2}f''\left(h_{\alpha\beta}\bar\nabla^\alpha\bar\nabla^\beta\bar\nabla^\mu\bar\nabla^\nu h_{\mu\nu} - 2h\bar\nabla^2\bar\nabla^\alpha\bar\nabla^\beta h_{\alpha\beta}\right.$$

$$+ h(\bar\nabla^2)^2 h - 2\bar\nabla^\alpha\bar\nabla^\beta h_{\alpha\beta}\bar R_{\mu\nu}h^{\mu\nu} + 2\bar\nabla^2 h\bar R_{\mu\nu}h^{\mu\nu} + \left.\bar R_{\mu\nu}\bar R_{\alpha\beta}h^{\mu\nu}h^{\alpha\beta}\right)$$

$$+\frac{1}{2}f'\left(h_{\alpha\beta}\bar\nabla^2 h^{\alpha\beta} + h\bar\nabla_\mu\bar\nabla_\nu h^{\mu\nu} - 4h_{\alpha\beta}\bar\nabla^\alpha\bar\nabla^\mu h^\beta_\mu + 2h_{\alpha\beta}\bar\nabla^\mu\bar\nabla^\alpha h^\beta_\mu\right.$$

$$\left.-h\bar\nabla^2 h + 4\bar R_{\mu\nu}h^{\mu\beta}h^\nu_\beta - 2\bar R_{\mu\nu}h^{\mu\nu}h\right) + f\left(-\frac{1}{4}h_{\mu\nu}h^{\mu\nu} + \frac{1}{8}h^2\right)\Big\}, \qquad (7.175)$$

where the argument of f and its derivatives (denoted by a prime) is the background scalar curvature $\bar R$, which is assumed to be constant. To this one has to add the gauge fixing terms (7.174). In order to diagonalize the complete expression, we choose a maximally symmetric (Euclidean) background and decompose the metric fluctuation into tensor, vector and scalar parts, as we did in sections 6.8.4 for the cutoff of type Ib. Then one obtains for the tensor part

$$\Gamma^{(2)}_{h^T_{\mu\nu}h^T_{\alpha\beta}} = -\frac{1}{2}\left[f'\left(-\bar\nabla^2 - \frac{2(d-2)}{d(d-1)}\bar R\right) + f\right]\delta^{\mu\nu,\alpha\beta}, \qquad (7.176)$$

for the vector part

$$\Gamma^{(2)}_{\xi_\mu\xi_\nu} = \left(-\bar\nabla^2 - \frac{\bar R}{d}\right)\left[\frac{k^{d-2}}{\alpha}\left(-\bar\nabla^2 - \frac{\bar R}{d}\right) + \frac{2\bar R}{d}f' - f\right]\bar g^{\mu\nu}, \qquad (7.177)$$

and for the scalar part (which contains a nontrivial mixing between h and σ)

$$\Gamma^{(2)}_{hh} = \frac{d-2}{4d}\left[\frac{4(d-1)^2}{d(d-2)}f''\left(-\bar\nabla^2 - \frac{\bar R}{d-1}\right)^2 + \frac{2(d-1)}{d}f'\left(-\bar\nabla^2 - \frac{\bar R}{d-1}\right) - \frac{2\bar R}{d}f' + f\right]$$

$$\Gamma^{(2)}_{h\sigma} = \frac{d-1}{d^2}\left[(d-1)f''\left(-\bar\nabla^2 - \frac{\bar R}{d-1}\right) + \frac{d-2}{2}f'\right]\bar\nabla^2\left(\bar\nabla^2 + \frac{\bar R}{d-1}\right)$$

$$\Gamma^{(2)}_{\sigma\sigma} = \frac{d-1}{2d}\left[\frac{2(d-1)}{d}f''\bar\nabla^2\left(\bar\nabla^2 + \frac{\bar R}{d-1}\right) - \frac{d-2}{d}f'\bar\nabla^2 + \frac{2\bar R}{d}f' - f\right.$$

$$\left.+\frac{2(d-1)k^{d-2}}{d\alpha}\left(-\bar\nabla^2 - \frac{\bar R}{d-1}\right)\right]\bar\nabla^2\left(\bar\nabla^2 + \frac{\bar R}{d-1}\right). \qquad (7.178)$$

The standard Fadeev–Popov procedure gives a ghost action

$$S_c = \int d^d x \sqrt{g}\, \bar{C}_\mu \left[\delta^\mu_\nu \bar{\nabla}^2 + \bar{R}^\mu{}_\nu + \frac{(d-2)}{d} \bar{\nabla}^\mu \bar{\nabla}_\nu \right] C^\nu \qquad (7.179)$$

where $\bar{C}_\mu$ and C^μ are the ghost and anti-ghost fields.

The ghost and anti-ghost are also decomposed into transverse and longitudinal parts. The operators acting on these fields are:

$$\Gamma^{(2)}_{\bar{c}^T_\mu c^T_\nu} = \left(\bar{\nabla}^2 + \frac{\bar{R}}{d} \right) g^{\mu\nu}$$

$$\Gamma^{(2)}_{\bar{c}c} = -\frac{2(d-1)}{d} \left(\bar{\nabla}^2 + \frac{1}{d-1}\bar{R} \right) \bar{\nabla}^2 \qquad (7.180)$$

Finally, the decomposition of the ghosts gives rise to Jacobians involving the operators $J_c = -\bar{\nabla}^2$.

7.7.2 *Inserting into the ERGE*

We choose a cutoff of type Ib. The inverse propagators (7.176), (7.177), (7.178), (7.180) are all functions of $-\bar{\nabla}^2$. Then, for each type of tensor components, the (generally matrix-valued) cutoff function $\mathbf{R}_k$ is chosen to be a function of $-\bar{\nabla}^2$ such that

$$\mathbf{\Gamma}^{(2)}(-\bar{\nabla}^2) + \mathbf{R}_k(-\bar{\nabla}^2) = \mathbf{\Gamma}^{(2)}(P_k)\,, \qquad (7.181)$$

where P_k is defined as in (6.83) for some profile function R_k. Inserting everything into (6.109) gives:

$$\begin{aligned}
\frac{d\Gamma_k}{dt} ={}& \frac{1}{2}\mathrm{Tr}_{(2)}\, \frac{\frac{d}{dt}\mathbf{R}_{h^T h^T}}{\mathbf{\Gamma}^{(2)}_{h^T h^T} + \mathbf{R}_{h^T h^T}} + \frac{1}{2}\mathrm{Tr}'_{(1)}\, \frac{\frac{d}{dt}R_{\xi\xi}}{\Gamma^{(2)}_{\xi\xi} + R_{\xi\xi}} \\[2ex]
&+ \frac{1}{2}\mathrm{Tr}''_{(0)}\, \begin{pmatrix} \Gamma^{(2)}_{hh} + R_{hh} & \Gamma^{(2)}_{h\sigma} + R_{h\sigma} \\ \Gamma^{(2)}_{\sigma h} + R_{\sigma h} & \Gamma^{(2)}_{\sigma\sigma} + R_{\sigma\sigma} \end{pmatrix}^{-1} \begin{pmatrix} \frac{d}{dt}R_{hh} & \frac{d}{dt}R_{h\sigma} \\ \frac{d}{dt}R_{\sigma h} & \frac{d}{dt}R_{\sigma\sigma} \end{pmatrix} \\[2ex]
&+ \frac{1}{2} \sum_{j=0,1} \frac{\frac{d}{dt}R_{hh}(\lambda_j)}{\Gamma^{(2)}_{hh}(\lambda_j) + R_{hh}(\lambda_j)} \\[2ex]
&- \mathrm{Tr}_{(1)}\, \frac{\frac{d}{dt}R_{\bar{c}^T c^T}}{\Gamma^{(2)}_{\bar{c}^T c^T} + R_{\bar{c}^T c^T}} - \mathrm{Tr}'_{(0)}\, \frac{\frac{d}{dt}R_{\bar{c}c}}{\Gamma^{(2)}_{\bar{c}c} + R_{\bar{c}c}} \\[2ex]
&- \frac{1}{2}\mathrm{Tr}'_{(1)}\, \frac{\frac{d}{dt}R_{J_V}}{J_V + R_{J_V}} - \frac{1}{2}\mathrm{Tr}''_{(0)}\, \frac{\frac{d}{dt}R_{J_S}}{J_S + R_{J_S}} + \mathrm{Tr}'_{(0)}\, \frac{\frac{d}{dt}R_{J_c}}{J_c + R_{J_c}}\,. \qquad (7.182)
\end{aligned}$$

The first three lines contain the contribution from the metric fluctuation $h_{\mu\nu}$, which has been decomposed into its irreducible parts according to the York decomposition (5.8). Note that the trace over the scalar components is doubly primed, since the first two modes of the σ field do not contribute to $h_{\mu\nu}$. However, the first two modes of h do contribute, and their contribution to the trace is added separately in the third line. The fourth line contains the contributions of the ghosts, again decomposed into transverse and longitudinal parts. Note that the first mode of the scalar (longitudinal) parts is omitted, as it does not contribute to C^μ. The last line is the contribution of the Jacobians. Eliminating the Jacobians by a further field redefinition would produce some technically undesirable poles in the heat kernel expansion. For this reason we shall retain them explicitly.

In the gauge $\alpha = 0$ there are several simplifications in the structure of the equation, so that collecting the tensor, vector and scalar contributions one obtains

$$\frac{d\Gamma_k}{dt} = \frac{1}{2}\mathrm{Tr}_{(2)}\frac{f'\partial_t P_k + (P_k - \Delta)\partial_t f'}{(P_k - \frac{R}{3})f' + f} - \frac{1}{2}\mathrm{Tr}'_{(1)}\frac{\partial_t R_k}{P_k - \frac{R}{4}} - \frac{1}{2}\mathrm{Tr}''_{(0)}\frac{\partial_t R_k}{P_k - \frac{R}{3}}$$
$$+ \frac{1}{2}\mathrm{Tr}''_{(0)}\frac{(f' + 6(P_k - \frac{R}{3})f'')\partial_t P_k + (P_k - \Delta)(\partial_t f' + 3(P_k + \Delta - \frac{2}{3}R)\partial_t f'')}{\frac{2}{3}f + (P_k - \frac{2}{3}R)f' - 3f''(P_k - \frac{R}{3})^2}$$

$$(7.183)$$

From here on we specialize to $d = 4$. The traces are evaluated using the master formula (5.159). With the optimized cutoff, it turns out that terms up to b_8 in the heat kernel expansion are needed. These are given in table 5.2. The relevant Q-functionals have been evaluated in section 6.9.

Going to the dimensionless variables $\tilde{R} = k^{-2}R$ and $\tilde{f} = k^{-4}f$, $\tilde{f}' = k^{-2}f'$, $\tilde{f}'' = f''$, the flow equation becomes

$$\frac{d\Gamma_k}{dt} = \frac{384\pi^2}{30240\tilde{R}^2}\left\{ -\frac{1008(511\tilde{R}^2 - 360\tilde{R} - 1080)}{\tilde{R} - 3} - \frac{2016(607\tilde{R}^2 - 360\tilde{R} - 2160)}{\tilde{R} - 4} \right.$$

$$+ 20\frac{(311\tilde{R}^3 - 126\tilde{R}^2 - 22680\tilde{R} + 45360)(\partial_t\tilde{f}' + 2\tilde{f}' - 2\tilde{R}\tilde{f}'') - 252(\tilde{R}^2 + 360\tilde{R} - 1080)\tilde{f}'}{3\tilde{f} - (\tilde{R} - 3)\tilde{f}'}$$

$$+ \frac{1}{\tilde{f}''(\tilde{R} - 3)^2 + 2\tilde{f} + (3 - 2\tilde{R})\tilde{f}'}\left[1008(29\tilde{R}^2 + 360\tilde{R} + 1080)\tilde{f}' \right.$$

$$+ 4(185\tilde{R}^3 + 3654\tilde{R}^2 + 22680\tilde{R} + 45360)(\partial_t\tilde{f}' + 2\tilde{f}' - 2\tilde{R}\tilde{f}'')$$

$$- 2016(29\tilde{R}^3 + 273\tilde{R}^2 - 3240)\tilde{f}''$$

$$\left. \left. -9(181\tilde{R}^4 + 3248\tilde{R}^3 + 15288\tilde{R}^2 - 90720)(\partial_t\tilde{f}'' - 2\tilde{R}\tilde{f}''')\right] \right\}. \qquad (7.184)$$

Note the appearance of the third derivative. This is due to the fact that $\tilde{f}$ is a function of $\tilde{R}$, so $\partial_t f'' = \partial_t\tilde{f}'' - 2\tilde{R}\tilde{f}'''$. This is a nuisance if one tries to solve the differential equation for the fixed point $\partial_t\tilde{f} = 0$, but this equation should only be used within the domain of validity of the heat kernel expansion, *i.e.* for small $\tilde{R}$. Then one may as well use a Taylor expansion for $\tilde{f}$ and the presence of the third derivative is not a serious obstacle.

7.7.3   *Polynomial truncation*

If we assume that

$$f(R) = \sum_{i=0}^{n} g_i(k) R^i \tag{7.185}$$

then for the dimensionless function $\tilde{f}$

$$\tilde{f}(R) = \sum_{i=0}^{n} \tilde{g}_i(k) \tilde{R}^i \tag{7.186}$$

where $\tilde{g}_i = g_i k^{2i-4}$. Thus, the beta functions of the dimensionless couplings $\tilde{g}_i$ can be obtained by taking derivatives of Eq. (7.184):

$$\frac{d\tilde{g}_i}{dt} = \frac{1}{i!} \frac{\partial^i}{\partial \tilde{R}^i} \frac{1}{\tilde{V}} \frac{d\Gamma_k}{dt}\bigg|_{\tilde{R}=0}, \tag{7.187}$$

where $\tilde{V} = k^{-4} \int d^4x \sqrt{g} = \frac{384\pi^2}{\tilde{R}^2}$. These beta functions are very complicated expressions and can only be generated and manipulated by computer. Starting from $n = 1$, one looks for solutions of the complete system of beta functions. The case $n = 1$ is the Einstein-Hilbert truncation and the fixed point is already known to exist, being the one discussed in section 7.6.1 (apart from the different choice of gauge that leads only to slight numerical differences). Increasing the order of the truncation, one obtains systems of equations of higher order, which generally have more and more solutions. Many of these are complex, and others are unphysical for other reasons.[14] However, among the FP's it has always been possible to find one for which the lower couplings and critical exponents have values that are close to those of the previous truncation. That FP is then identified as the nontrivial FP in the new truncation. Table 7.3 gives the position of the nontrivial FP and table 7.4 gives the critical exponents, for truncations ranging from $n = 1$ (the Einstein-Hilbert truncation) to $n = 8$.

Looking at the columns of Tables 7.3 and 7.4 we see that in general the properties of the FP are remarkably stable under improvement of the truncation. In particular the projection of the flow in the $\tilde{\Lambda}$-$\tilde{G}$ plane agrees well with the case $n = 1$. The greatest deviations seem to occur in the row $n = 2$, and in the columns g_2 and ϑ_2. The value of g_{2*} decreases steadily with the truncation. The critical exponent ϑ_2 appears for the first time in the truncation $n = 2$ with a very large value, but it decreases quickly and seems to converge around 1.5. This behavior may be related to the fact that g_2 is classically a marginal variable.

In all truncations only three operators are relevant. One can conclude that in this class of truncations the UV critical surface is three-dimensional. Its tangent space at the FP is spanned by the three eigenvectors corresponding to the eigenvalues

---

[14]A similar phenomenon is known to occur in scalar theory in the local potential approximation [187, 191].

Table 7.3   Position of the fixed point

n	$10^3 \times$								
	$\tilde{g}_{0*}$	$\tilde{g}_{1*}$	$\tilde{g}_{2*}$	$\tilde{g}_{3*}$	$\tilde{g}_{4*}$	$\tilde{g}_{5*}$	$\tilde{g}_{6*}$	$\tilde{g}_{7*}$	$\tilde{g}_{8*}$
1	5.226	-20.14							
2	3.292	-12.73	1.514						
3	5.184	-19.60	0.702	-9.682					
4	5.059	-20.58	0.270	-10.97	-8.646				
5	5.071	-20.54	0.269	-9.687	-8.034	-3.349			
6	5.051	-20.76	0.141	-10.20	-9.567	-3.590	2.460		
7	5.042	-20.97	-0.034	-9.784	-10.521	-6.048	3.421	5.905	
8	5.066	-20.75	0.088	-8.581	-8.926	-6.808	1.165	6.196	4.695

Position of the FP for increasing order n of the truncation.   To avoid writing too many decimals, the values of $\tilde{g}_{i*}$ have been multiplied by 1000.

Table 7.4   Scalig exponents

n	$Re\vartheta_1$	$Im\vartheta_1$	ϑ_2	ϑ_3	$Re\vartheta_4$	$Im\vartheta_4$	ϑ_6	ϑ_7	ϑ_8
1	2.382	2.168							
2	1.376	2.325	26.86						
3	2.711	2.275	2.068	-4.231					
4	2.864	2.446	1.546	-3.911	-5.216				
5	2.527	2.688	1.783	-4.359	-3.761	-4.880			
6	2.414	2.418	1.500	-4.106	-4.418	-5.975	-8.583		
7	2.507	2.435	1.239	-3.967	-4.568	-4.931	-7.572	-11.076	
8	2.407	2.545	1.398	-4.167	-3.519	-5.153	-7.464	-10.242	-12.30

Critical exponents for increasing order n of the truncation.   The first two critical exponents ϑ_0 and ϑ_1 are a complex conjugate pair. The critical exponent ϑ_4 is real in the truncation $n = 4$ but for $n \geq 5$ it becomes complex and we have set $\vartheta_5 = \vartheta_4^*$.

with negative real part. In the parametrization (7.186), it is the three-dimensional subspace in $\mathbf{R}^9$ defined by the equation:

$$\tilde{g}_3 = 0.0006 + 0.0682\,\tilde{g}_0 + 0.4635\,\tilde{g}_1 + 0.8950\,\tilde{g}_2$$

$$\tilde{g}_4 = -0.0092 - 0.8365\,\tilde{g}_0 - 0.2089\,\tilde{g}_1 + 1.6208\,\tilde{g}_2$$

$$\tilde{g}_5 = -0.0157 - 1.2349\,\tilde{g}_0 - 0.7254\,\tilde{g}_1 + 1.0175\,\tilde{g}_2$$

$$\tilde{g}_6 = -0.0127 - 0.6226\,\tilde{g}_0 - 0.8240\,\tilde{g}_1 - 0.6468\,\tilde{g}_2 \qquad (7.188)$$

$$\tilde{g}_7 = -0.0008 + 0.8139\,\tilde{g}_0 - 0.1484\,\tilde{g}_1 - 2.0181\,\tilde{g}_2$$

$$\tilde{g}_8 = 0.0091 + 1.2543\,\tilde{g}_0 + 0.5085\,\tilde{g}_1 - 1.9012\,\tilde{g}_2$$

There is a clear trend for the eigenvalues to grow with the power of R. In fact, in the best available truncation, the real parts of the critical exponents differ from their classical values d_i by at most 2.1, and there is no tendency for this difference to grow for higher powers of R.

With a finite dimensional critical surface, one can make definite predictions in quantum gravity. The real world must correspond to one of the trajectories that emanate from the FP, in the direction of a relevant perturbation. Such trajectories lie entirely in the critical surface. Thus, at some sufficiently large but finite value of k one can choose arbitrarily three couplings, for example $\tilde{g}_0$, $\tilde{g}_1$, $\tilde{g}_2$ and the remaining six are then determined by (7.188). These couplings could then be used to compute the probabilities of physical processes, and the relations (7.188), in principle, could be tested by experiments. The linear approximation is valid only at very high energies, but it should be possible to numerically solve the flow equations and study the critical surface further away from the FP.

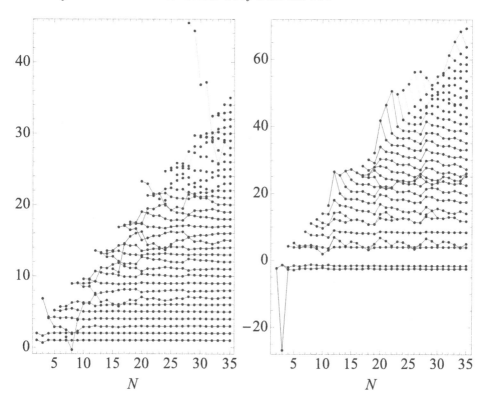

Fig. 7.19   Left: convergence of the couplings at the fixed point with the order of the truncation. Here $N = n + 1$ is the number of couplings in the truncation. For visual convenience the values have been displaced according to the formula $i + 1 + g_i(N)/g_i(35)$ and $0 \leq i \leq N - 1$. Right: the real part of the scaling exponents as functions of N. Reproduced with permission from [247].

The stability of these results has been checked in [246]. In the truncation $n = 3$ it has been found that the results are very stable over a large range of values for the gauge parameter α. The results also change very little if one uses a fourth-order gauge fixing of the form

$$S_{GF} = \frac{1}{2\alpha} \int d^d x \sqrt{\bar{g}} \, F_\mu \bar{g}^{\mu\nu} \bar{\nabla}^2 F_\nu \tag{7.189}$$

which is analogous of the one used in section 7.4 for higher derivative gravity. In both cases, they also change little when one makes the choice $\beta = 1$ instead of $\beta = 0$.

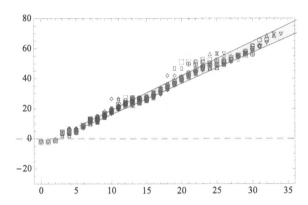

Fig. 7.20 All the scaling exponents, for all truncations $2 \leq N \leq 35$. The shaded area is the estimate of the theoretical errors (7.190,7.191). Reproduced with permission from [247].

Using a more efficient recursive technique, these results have been extended up to $n = 34$ in [247]. Instead of giving long tables of values it is more convenient to display the results for the fixed point couplings and scaling exponents graphically. One can see from figure (7.19) that some couplings and critical exponents converge very rapidly, while others tend to oscillate longer. These results confirm the picture described above, of a fixed point with three relevant directions. One nice property of the fixed point is that the critical exponents do not deviate very much from the canonical dimensions. This can be made quantitative by performing a least-square linear fit of the form

$$\theta_i = ai - b \tag{7.190}$$

from which one gets the values

$$a = 2.17 \pm 5\% \; ; \qquad b = 4.06 \pm 10\% \, , \tag{7.191}$$

which are remarkably close to the classical values $a = 2$ and $b = 4$. This fit is illustrated in Fig. (7.20).

7.8   Effect of matter

Gravity does not exist in isolation, so an asymptotically safe theory of gravity alone is not sufficient: matter should be UV safe too. We shall ask here the following question: does the addition of minimally coupled, non-self-interacting matter, spoil the fixed point that we have found for pure gravity? There are several good reasons to consider this issue even before having a full and satisfactory understanding of the situation of pure gravity.

The first comes from the analogy with Yang-Mills theories, where asymptotic freedom is only possible if the number of fermion fields is not too large. It is reasonable to expect that too many matter fields of one type or another destroy the gravitational fixed point (if one exists for pure gravity). If we found compelling theoretical evidence that the kind of matter that exists in the universe is incompatible with a gravitational fixed point, then the whole program would lose much appeal.

The second reason is that matter might actually make things easier instead of more complicated. There may exist some kind of large N limit, where N is related in some way to the number of matter fields, in which the matter contributions dominate and the complicated graviton contributions can be neglected. Something of this type has already been discussed in the literature, in simple cases [248, 249]. Recently, precisely such a limit has been used to prove the existence of a nontrivial fixed point in certain gauge theories coupled to scalars and fermions [250].

Third, from a conceptual point of view, it is much easier to think of suitable physical observables in the presence of matter than for pure gravity. In fact, as is well-known, pure gravity does not have *any* local observable. We have already discussed the challenges posed by the detection of single gravitons. The gravitational scattering of standard model particles, however negligible it may be at ordinary energies, is certainly closer to well-understood experimental situations than the gravitational scattering of gravitons. More generally, a theory of gravity and matter opens more possibilities of experimental verification.

Given that we already have a candidate fixed point for pure gravity, it is reasonable to start by adding matter a little at the time and seeing what it does to the fixed point. In fact, from a mathematical point of view, we can treat the numbers of matter fields N_S, N_D and N_V as continuous parameters. Starting from the origin in this parameter space, we will look for the envelope within which the fixed point still exists with the same properties.

One should keep in mind that this does not exhaust the logical possibilities, since there may exist a viable fixed point for gravity coupled to certain types of matter that is not in any sense "the same" as the fixed point of pure gravity. This would be very interesting, because it would constitute a vindication of the logic that led to supergravity, albeit in a very different context. In any case, here we shall limit ourselves to the question posed above. We shall consider first the single-metric one-loop results and then extend them by considering the effect of the anomalous dimensions of the fluctuation fields, which can themselves be calculated in a one-loop approximation or in the full truncated RG flow, as explained in section 7.6.3.

7.8.1 *Single-metric, one-loop results*

We have already computed in section 6.6 the contribution of minimally coupled massless matter fields to the gravitational beta functions. Here we shall put those results together with the one-loop gravitational beta functions of section 6.8.3. In $d = 4$ and using the type-II cutoff, the beta functions become

$$\beta_{\tilde{G}} = 2\tilde{G} + \frac{\tilde{G}^2}{6\pi} \left(N_S + 2N_D - 4N_V - 46 \right), \tag{7.192}$$

$$\beta_{\tilde{\Lambda}} = -2\tilde{\Lambda} + \frac{\tilde{G}}{4\pi} \left(N_S - 4N_D + 2N_V + 2 \right)$$
$$+ \frac{\tilde{G}\tilde{\Lambda}}{6\pi} \left(N_S + 2N_D - 4N_V - 16 \right). \tag{7.193}$$

The numbers -46, 2 and -16 represent the contributions of gravitons and ghosts to the beta functions. These contributions are such that the RG flow admits a nontrivial fixed point when matter is absent. Let us see what effect matter has, in this approximation. The beta functions have a nontrivial fixed point at

$$\tilde{\Lambda}_* = -\frac{3}{4} \frac{N_S - 4N_D + 2N_V + 2}{N_S + 2N_D - 4N_V - 31}, \tag{7.194}$$

$$\tilde{G}_* = -\frac{12\pi}{N_S + 2N_D - 4N_V - 46}. \tag{7.195}$$

Since the beta functions vanish for $\tilde{G} = 0$, flow lines cannot cross from negative to positive $\tilde{G}$. Since the low-energy Newton's coupling is experimentally bound to be positive, we require that also the fixed point occurs at positive $\tilde{G}$. This puts a bound on the matter content. In the following we shall find it convenient to present the results in the N_S-N_D-plane, treating the number of gauge fields as a fixed parameter. Positivity of $\tilde{G}_*$ demands that

$$N_D < 23 + 2N_V - \frac{1}{2}N_S. \tag{7.196}$$

Notice that gauge fields contribute with the same sign as gravity, so they facilitate the existence of the fixed point, whereas scalars and fermions tend to destroy it. When their number increases, the fixed-point value of $\tilde{G}_*$ increases and reaches a singularity on the line $N_D = 23 + 2N_V - \frac{1}{2}N_S$. On the other side of the singularity $\tilde{G}_*$ is negative. Fig. (7.21) shows the existence region of a positive fixed point for $\tilde{G}_*$ for various numbers of gauge fields. We see that the existence region grows with the number of gauge fields, while for a given number of gauge fields, only a finite number of combinations of scalar and Dirac fields is allowed.

The behavior of the cosmological constant is shown in figure 7.22. It has a singularity on the line

$$N_D = \frac{31}{2} + 2N_V - \frac{1}{2}N_S. \tag{7.197}$$

This singularity in $\tilde{\Lambda}_*$ is parallel to the singularity in $\tilde{G}_*$ and is shifted downwards by $N_D = 7.5$. There are fixed points in the intermediate region between these singularities, but they are disconnected from the one in the origin. For "phenomenological" applications we will restrict our attention to points that are below the singularity in the cosmological constant. The allowed region is therefore somewhat smaller than the one shown in fig. 7.21. In the absence of gauge fields this leaves only the area

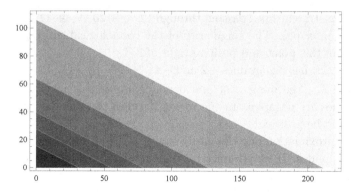

Fig. 7.21 The shaded triangles are the areas in the N_S-N_D plane compatible with a gravitational fixed point for $N_V = 0$ (darkest triangle in bottom left corner), $N_V = 6, 12, 24, 45$ (from bottom to top). The last three numbers correspond to the dimensions of the gauge groups in the standard model, in $SU(5)$ and $SO(10)$ GUTs.

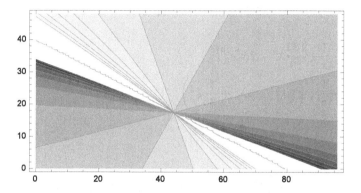

Fig. 7.22 Level lines of $\tilde{\Lambda}$ at the fixed point, for $N_V = 12$. $\tilde{\Lambda} = 0$ on the straight line intersecting the axes at $N_D \approx 6$. More negative values correspond to darker color and more positive values to lighter color. There is a singularity on the line joining the points $(0, 40)$ and $(80, 0)$. For smaller N_V the singularity is closer to the origin, for larger N_V further away, roughly coinciding with the lines in the preceding figure.

$N_D < \frac{31}{2} - \frac{1}{2}N_S$, which means that at most 15 Weyl spinors or 31 scalars, or a combinations thereof, are admissible.

When we restrict ourselves to the allowed region, the sign of the cosmological constant at the fixed point is determined by the numerator in (7.194): above (or left) of the line

$$N_D = \frac{2 + 2N_V + N_S}{4} \qquad (7.198)$$

the cosmological constant is negative, whereas below (or right) of this line it is positive. In the beta function for $\tilde{\Lambda}$ the contribution of each field is weighted with the number of degrees of freedom it carries, with a plus sign for bosons and a minus sign for fermions. The line (7.198) is where any supersymmetric theory would lie. The contours of constant $\tilde{\Lambda}_*$ are straight lines passing through $(2N_V + 20, N_V + 11/2)$, where (7.197) and (7.198) intersect. The singularity of the cosmological constant on (7.197) is negative left of this point and positive right of it.

The stability matrix (7.28) has eigenvalues -2 and $-4\frac{N_S+2N_D-4N_V-31}{N_S+2N_D-4N_V-46}$. Below the singularites of $\tilde{\Lambda}_*$ and $\tilde{G}_*$, the numerator and denominator of this ratio are positive, so both eigenvalues are negative. In the region between the singularities the second eigenvalue would be positive.

In this perturbative approximation one can easily examine the effect of matter on higher gravitational couplings. If we parametrize the curvature squared terms, up to total derivatives, as

$$\int d^4x \sqrt{g} \left[\frac{1}{2\lambda}C^2 + \frac{1}{\xi}R^2 \right] , \tag{7.199}$$

where C is the Weyl tensor, the beta functions of the couplings are

$$\beta_\lambda = -\frac{1}{(4\pi)^2} \frac{133}{10}\lambda^2 - 2\lambda^2 a_\lambda^{(4)} ,$$

$$\beta_\xi = -\frac{1}{(4\pi)^2} \left(10\lambda^2 - 5\lambda\xi + \frac{5}{36}\xi^2 \right) - \xi^2 a_\xi^{(4)} ,$$

where

$$a_\lambda^{(4)} = \frac{1}{2880\pi^2} \left(\frac{3}{2}N_S + 9N_D + 18N_V \right) , \tag{7.200}$$

$$a_\xi^{(4)} = \frac{1}{2880\pi^2} \frac{5}{2} N_S . \tag{7.201}$$

It is remarkable that all types of matter contribute with the same sign to the running of these couplings, which are therefore always asymptotically free.

The coefficients (7.200) are universal and, as noticed in [249], with type II cutoff and with the optimized shape function (6.44), the contribution of matter to the running of all couplings multiplying terms with six or more derivatives is identically zero.

7.8.2 *Inclusion of anomalous dimensions*

Following [242], we will now include the effect of the anomalous dimension of all the fields. We will use the same approximations and techniques of section 7.6.3, except that here we will not consider an independent ("level-two") graviton mass M^2. The truncation is given by (7.150), with M^2 replaced by -2Λ, plus the matter action

$$\Gamma_{\text{matter}} = S_S + S_D + S_V$$

$$S_S = \frac{Z_S}{2} \int d^d x \sqrt{g}\, g^{\mu\nu} \sum_{i=1}^{N_S} \partial_\mu \phi^i \partial_\nu \phi^i$$

$$S_D = i Z_D \int d^d x \sqrt{g} \sum_{i=1}^{N_D} \bar{\psi}^i \slashed{D} \nabla \psi^i,$$

$$S_V = \frac{Z_V}{4} \int d^d x \sqrt{g} \sum_{i=1}^{N_V} g^{\mu\nu} g^{\kappa\lambda} F^i_{\mu\kappa} F^i_{\nu\lambda}$$

$$+ \frac{Z_V}{2\xi} \int d^d x \sqrt{\bar{g}} \sum_{i=1}^{N_F} \left(\bar{g}^{\mu\nu} \bar{D}_\mu A^i_\nu \right)^2$$

$$+ \frac{1}{2} \int d^d x \sqrt{\bar{g}} \sum_{i=1}^{N_V} \bar{C}_i (-\bar{D}^2) C_i \,. \tag{7.202}$$

In each case, i is a summation index over matter species. In the Dirac action the covariant derivative is $\nabla_\mu = \partial_\mu + \frac{1}{8}[\gamma^a, \gamma^b]\omega^{ab}_\mu$, where the spin-connection ω^{ab}_μ can be expressed in terms of the vierbeins. This introduces an additional $O(d)$ local gauge invariance. We adopt a symmetric gauge-fixing of $O(d)$, such that vierbein fluctuations can be re-expressed completely in terms of the metric fluctuations [251, 252]. We will therefore not rewrite the gravitational part of the action in terms of vierbeins; full details of the procedure can be found in [206].

There are no gauge interactions, so the fermions (as well as the scalars) are uncharged and there are no gauge covariant derivatives. In the Abelian gauge field action the second term is a gauge fixing term with gauge-fixing parameter ξ and the third term represents the abelian ghosts. The ghosts are decoupled from the metric and gauge field fluctuations and therefore do not contribute to the running of Z_h and Z_V, but they are coupled to the gravitational background and therefore contribute to the beta functions of G and Λ, as we have seen in section 3.2.4.

The truncation for the gravitational and matter action contains two essential couplings G and Λ, and five inessential wave-function renormalization constants Z_Ψ with $\Psi = (h, C, S, D, V)$. Being inessential means that the Z_Ψ can be eliminated from the action by field rescalings and do not appear explicitly in any beta function. However, they enter the beta-functions of the essential couplings in a nontrivial way via the anomalous dimensions

$$\eta_\Psi = -\partial_t \ln Z_\Psi \,. \tag{7.203}$$

Taking this into account, the beta functions for $\tilde{G}$ and $\tilde{\Lambda}$ have the following form:

$$\frac{d\tilde{\Lambda}}{dt} = -2\tilde{\Lambda} + \frac{8\pi\tilde{G}}{(4\pi)^{d/2}d(d+2)\Gamma[d/2]}\left[\frac{d(d+1)(d+2-\eta_h)}{1-2\tilde{\Lambda}} - 4d(d+2-\eta_c)\right.$$

$$\left. +2N_S(2+d-\eta_S) - 2N_D2^{[d/2]}(2+d-\eta_D) + 2N_V(d^2-4-d\,\eta_V)\right]$$

$$-\frac{4\pi\tilde{G}\tilde{\Lambda}}{3d(4\pi)^{d/2}\Gamma[d/2]}\left[\frac{d(5d-7)(d-\eta_h)}{1-2\tilde{\Lambda}} + 4(d+6)(d-\eta_C)\right.$$

$$\left. -2N_S(d-\eta_S) - N_D2^{[d/2]}(d-\eta_D) + 2N_V(d(8-d)-(6-d)\eta_V)\right] \quad (7.204)$$

$$\frac{d\tilde{G}}{dt} = (d-2)\tilde{G} - \frac{4\pi\tilde{G}^2}{3d(4\pi)^{d/2}\Gamma(d/2)}\left[\frac{d(5d-7)(d-\eta_h)}{1-2\tilde{\Lambda}} + 4(d+6)(d-\eta_C)\right.$$

$$\left. -2N_S(d-\eta_S) - N_D2^{[d/2]}(d-\eta_D) + 2N_V(d(8-d)-(6-d)\eta_V)\right] . \quad (7.205)$$

To evaluate the anomalous dimensions η_Ψ we expand around flat space and extract from the r.h.s. of the flow equation terms quadratic in momentum and in the fluctuation field Ψ. The terms in this calculation have a natural diagrammatic expression as one-loop corrections to the running of the two point function of the field Ψ. The graviton anomalous dimension can be written as

$$\eta_h = \eta_h\big|_{\text{gravity}} + \eta_h\big|_{\text{matter}},$$

where the first term has already been calculated in section 7.6.3 and the second is due to the diagrams of Fig. 7.23. As we neglect matter-ghost couplings at this order of the approximation, the ghost anomalous dimension will be the same as in the pure gravity case, and there is no ghost contribution to the matter anomalous dimensions.

Fig. 7.23 Matter loops contributing to the graviton anomalous dimension. Springy lines are the graviton propagator and continuous lines are the matter propagator. The cross is $\partial_t R_k$. The momentum structure of the vertices in the present truncation implies that the tadpole diagrams on the right vanish.

Finally, graviton fluctuations induce nontrivial matter anomalous dimensions through the diagrams of Fig. 7.24.

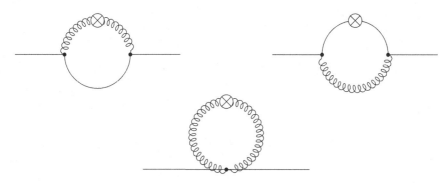

Fig. 7.24 Graviton loops contributing to the matter anomalous dimension. Springy lines are the graviton propagator and continuous lines are the matter propagator. The cross is $\partial_t R_k$.

The resulting anomalous dimensions of the matter fields are

$$
\eta_S = -\frac{32\pi\tilde{G}}{(4\pi)^{d/2}\Gamma(d/2)}\left[\frac{2}{d+2}\frac{1}{(1-2\tilde{\Lambda})^2}\left(1-\frac{\eta_h}{d+4}\right)+\frac{2}{d+2}\frac{1}{1-2\tilde{\Lambda}}\left(1-\frac{\eta_S}{d+4}\right)\right.
$$
$$
\left.+\frac{(d+1)(d-4)}{2d(1-2\tilde{\Lambda})^2}\left(1-\frac{\eta_h}{d+2}\right)\right],
\tag{7.206}
$$

$$
\eta_D = \frac{32\pi\tilde{G}}{(4\pi)^{d/2}\Gamma(d/2)}\left[\frac{(d-1)(d^2+9\,d-8)}{8d\,(d-2)(d+1)(1-2\tilde{\Lambda})^2}\left(1-\frac{\eta_h}{d+3}\right)\right.
$$
$$
\left.+\frac{(d-1)^2}{2d(d+1)(d-2)}\frac{1}{1-2\tilde{\Lambda}}\left(1-\frac{\eta_D}{d+2}\right)-\frac{(d-1)(2d^2-3d-4)}{4d(d-2)(1-2\tilde{\Lambda})^2}\left(1-\frac{\eta_h}{d+2}\right)\right]
\tag{7.207}
$$

$$
\eta_V = -\frac{32\pi\tilde{G}}{(4\pi)^{d/2}\Gamma(d/2)}\left[\frac{(d-1)(16+10\,d-9\,d^2+d^3)}{2d^2(d-2)(1-2\tilde{\Lambda})^2}\left(1-\frac{\eta_h}{d+2}\right)\right.
$$
$$
\left.+\frac{4(d-1)(2d-5)}{d(d^2-4)(1-2\tilde{\Lambda})}\left(1-\frac{\eta_V}{d+4}\right)+\frac{4(d-1)(2d-5)}{d(d^2-4)(1-2\tilde{\Lambda})^2}\left(1-\frac{\eta_h}{d+4}\right)\right].
\tag{7.208}
$$

The formulas (7.159), (7.163), (7.206), (7.207), (7.208) do not directly give the anomalous dimensions, rather they give a set of linear equations for the anomalous dimensions. If we denote $\vec{\eta} = (\eta_h, \eta_c, \eta_S, \eta_D, \eta_V)$, these equations can be written collectively in the form

$$
\vec{\eta} = \vec{\eta}_1(\tilde{\Lambda}, \tilde{G}) + A(\tilde{\Lambda}, \tilde{G})\vec{\eta} ,
\tag{7.209}
$$

where $\vec{\eta}_1$ is the leading one-loop term and A is a matrix of coefficients. The full anomalous dimension is obtained by solving this algebraic system of equations.

7.8.3   *Fixed points*

7.8.3.1   *No matter*

We start from the case $N_S = N_D = N_V = 0$. We list the properties of the fixed point in the following table. Recall that in the present bi-field approach, the one-loop approximation includes the anomalous dimensions, as discussed in section 7.6.3 (recall in particular footnote 10).

This is almost the same calculation that was discussed in sections 7.6.2–3, except that here we identify $M^2 = -2\Lambda$. Comparing with Tables 7.1 and 7.2, we see that without making this distinction, the value of $\tilde{G}_*$ is very close to that of the single-field one-loop approximation. Distinguishing between the two mass parameters has a more significant effect.

Table 7.5   Fixed point without matter

	1L-II	full-II	full-Ia	Ref. [243]
$\tilde{\Lambda}_*$	0.010	0.009	−0.049	−0.008
$\tilde{G}_*$	0.772	0.776	1.579	1.446
θ_1	3.298	3.317	3.991	3.323
θ_2	1.954	1.925	1.290	1.954
η_h	0.269	0.299	0.540	0.072
η_C	−0.806	−0.814	−1.390	−1.503

Properties of the non-Gaussian fixed point in the level-2 truncation used in the present section. The first column is the one-loop approximation, the second column refers to the full flow. Both use a type II cutoff. The third column gives the result obtained with a cutoff of type Ia instead of II. The last column gives the results of reference [243], who also used a cutoff of type Ia. It differs from the other three in the definition of the anomalous dimension, which is defined by contracting with the spin-2 projector instead of the tensor K. All columns are evaluated in the gauge $\alpha = \beta = 1$.

7.8.3.2   *Scalar matter*

Even though physically N_S must be an integer, mathematically one can treat N_S as a continuous parameter. For $N_S \leq 12$ the effect of scalars is to push $\tilde{\Lambda}_*$ towards larger values, while $\tilde{G}_*$ is almost stable. The product $\tilde{\Lambda}_* \tilde{G}_*$, which we have seen in section 7.3.3 to be gauge-independent at leading order in $\tilde{\Lambda}$, and is also known to be quite insensitive to the cutoff choice, increases slowly. In this regime the critical exponents change little while the anomalous dimensions increase in absolute value, maintaining the same sign ($\eta_h > 0$, $\eta_C < 0$ and $\eta_S < 0$). There is a sharp change of behavior of $\tilde{\Lambda}_*$ for $N_S \geq 12$. Beyond this value, the cosmological constant stops growing with N_S, while $\tilde{G}_*$ begins to grow and also the critical exponents become very large ($O(10^3)$).

The change of behavior occurs smoothly over the whole range, so one has a continuous deformation of the pure gravity fixed point up to $N_S \approx 27.7$ where

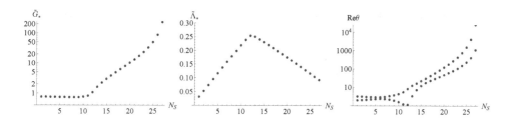

Fig. 7.25 Left and middle: Position of the fixed point as a function of the number of scalar fields. Right: critical exponents. Note the logarithmic scales. All with type II cutoff and one-loop anomalous dimensions.

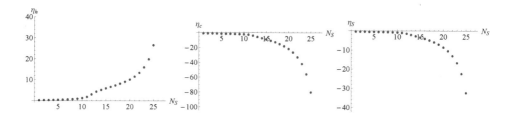

Fig. 7.26 The graviton (left), ghost (middle) and scalar (right) anomalous dimensions as functions of the number of scalar fields.

$\tilde{G}_*$ diverges. As in the perturbative approximation, there is therefore a maximal number of scalar fields that is compatible with the existence of the fixed point. The effect of scalar fields on the position of the fixed point, on the critical exponents and on the anomalous dimensions is shown in figs. (7.25), (7.26), at one loop and with type II cutoff.

With a type Ia cutoff the results are different for $N_S > 12$, and the fixed point becomes complex at $N_S \approx 17$. This, together with the fact that the anomalous dimensions become rather large, makes the full RG improved equations unreliable.

In the future we may understand better which truncation gives physically reliable results but for the time being the scheme dependence has to be taken as a measure of the theoretical uncertainties. We can say that in the present approximation the fixed point ceases to exist when the number of scalars becomes of the order of 20. Larger truncations may be needed to be sure of the fate of the fixed point.

7.8.3.3 *Fermionic matter*

The effect of fermions is to push $\tilde{G}_*$ to larger values and $\tilde{\Lambda}_*$ to more negative values, see Fig. 7.27. At a critical number of fermions $N_D \approx 10.1$, $\tilde{G}_*$ goes to $+\infty$ and $\tilde{\Lambda}_*$ goes to $-\infty$. This is similar to the behavior seen in the perturbative analysis. Accordingly fermions have a destabilizing effect on asymptotic safety in gravity, reminiscent of a similar effect of fermions on asymptotic freedom in gauge theories.

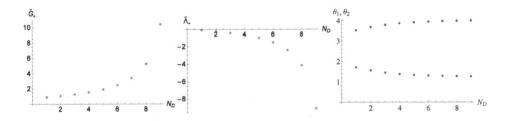

Fig. 7.27   Left: The values of $\tilde{G}_*$ (left) and $\tilde{\Lambda}_*$ (middle) as functions of the number of Dirac fields. Right: The critical exponents $\theta_{1,2}$ as functions of the number of Dirac fields.

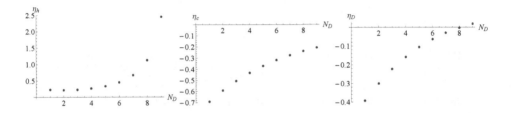

Fig. 7.28   The graviton (left), ghost (middle) and fermion (right) anomalous dimensions as functions of the number of fermion fields, all at one loop and with type II cutoff.

Fermionic fluctuations have only a small effect on the values of the critical exponents. In contrast, the graviton anomalous dimension grows, cf. Fig. (7.28). These results show only a very weak scheme-dependence. The main difference in the RG improved case lies in the fact that the fermionic anomalous dimension remains negative up to the critical value of N_D.

7.8.3.4   *Scalars and fermions*

The main result up to this point lies in the existence of a maximum number of fermions and scalars compatible with the gravitational fixed point within our truncation. This is true also for combinations of scalars and fermions, as seen in Fig. 7.29, which shows the existence region of the fixed point in the N_S-N_D-plane for $N_V = 0$. Note the qualitative agreement with the analysis of the perturbative approximation in section 7.8.1. We conclude that the inclusion of dynamical matter can fundamentally change a quantum theory of gravity, or even make it inconsistent. It is thus crucial to include realistic matter degrees of freedom in the investigation of the asymptotic-safety scenario for quantum gravity.

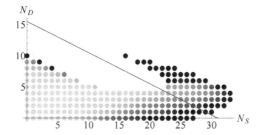

Fig. 7.29 The points in the N_S-N_D plane compatible with a gravitational fixed point with two relevant directions for $N_V = 0$. The line represents the perturbative bound (7.197). Lighter shades of gray mean smaller η_h; black means $\eta_h > 10$.

7.8.3.5 *Vector fields*

In contrast to scalars and fermions, there is no bound on the number of vector fields compatible with a viable gravitational fixed point. The effect of vector degrees of freedom is always to decrease $\tilde{G}_*$ and to increase $\tilde{\Lambda}_*$. The position of the fixed point and the values of the critical exponents and anomalous dimensions are shown in figs. 7.30 and 7.31, for $0 \leq N_V \leq 50$, covering all phenomenologically interesting models. The behavior is very smooth. From the point of view of the N_V-dependence, however, this is still a transient range. For very large N_V all quantities reach the following asymptotic values:

	$\tilde{G}_*$	$\tilde{\Lambda}_*$	θ_1	θ_2	η_h	η_C	η_V
$\lim_{N_V \to \infty}$	0	3/8	4	2	9/10	0	0

This picture holds with small quantitative changes also when the RG improvement is taken into account, and with type Ia cutoff. The most significant difference lies in the fact that the vector anomalous dimension does not change sign even for large N_V, when the RG improvement is taken into account. This suggests that the existence of the fixed point is a true feature of gravity coupled to vector fields. It will be interesting to see whether the gauge coupling remains asymptotically free when Yang-Mills is coupled to gravity.

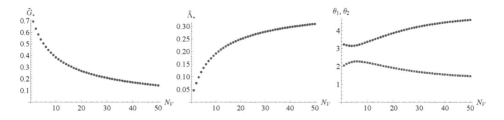

Fig. 7.30 Position of the fixed point (left and middle) and critical exponents (right) as a function of number of vector fields.

Table 7.6   Fixed point with standard model matter

	1L-II	full-II	1L-Ia	full-Ia
$\tilde{\Lambda}_*$	−2.399	−2.348	−3.591	−3.504
$\tilde{G}_*$	1.762	1.735	2.627	2.580
θ_1	3.961	3.922	3.964	3.919
θ_2	1.644	1.651	2.178	2.187
η_h	2.983	2.914	4.434	4.319
η_C	−0.139	−0.129	−0.137	−0.125
η_S	−0.076	−0.072	−0.076	−0.073
η_D	−0.015	0.004	−0.004	0.016
η_V	−0.133	−0.145	−0.144	−0.158

Properties of the fixed point with N_S = 4, N_D = 22.5, N_V = 12. The first two column refer to type II cutoff, the last two to type Ia cutoff. In the first and third column the anomalous dimension is computed at one loop, in the other two with the full formula (7.209).

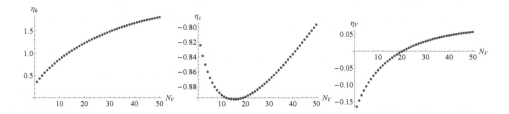

Fig. 7.31   The graviton (left), ghost (middle) and vector (right) anomalous dimensions as functions of the number of vector fields.

7.8.3.6   *Specific matter models*

We begin with the matter content of the standard model in its original form, excluding right-handed neutrinos: four scalars, 45 Weyl spinors, which in the present context are equivalent to 22.5 Dirac spinors, and 12 gauge fields. Table 7.6 reports the properties of the non-Gaussian fixed point with this matter content, with different cutoff schemes (type II or type Ia) and approximations for the anomalous dimensions (one-loop or full). In all these cases the matter content of the standard model seems to be compatible with the existence of a fixed point. By adding, one at a time, the vector fields, then the scalars, then the fermions, one can convince oneself that this fixed point is a continuous deformation of the one of pure gravity.

Theories that go beyond the standard model contain more fields. A very minimal extension is a single further scalar field, which can be viewed as a model of dark matter [253–256]. This has a small effect, as seen in the third row of table 7.7. In the fourth row we consider a model with three right-handed neutrinos, to account for neutrino masses. This has a somewhat larger effect but is still clearly compatible with asymptotic safety. In the fifth row we consider a model with three right-handed neutrinos and two scalars, one of which can be thought of as the axion [257–260], the other as dark matter. This model is still in the allowed region with the type II cutoff, but if one were to use the more stringent type Ia cutoff it would be quite close to the boundary. This model is therefore nearly as extended as one can get without adding further gauge fields. The extent of the allowed region with $N_V = 12$ is shown in figure 7.32.

Table 7.7 Fixed points with various matter contents

model	N_S	N_D	N_V	$\tilde{G}_*$	$\tilde{\Lambda}_*$	θ_1	θ_2	η_h
no matter	0	0	0	0.77	0.01	3.30	1.95	0.27
SM	4	45/2	12	1.76	−2.40	3.96	1.64	2.98
SM +dm scalar	5	45/2	12	1.87	−2.50	3.96	1.63	3.15
SM+ 3 ν's	4	24	12	2.15	−3.20	3.97	1.65	3.71
SM+3ν's + axion+dm	6	24	12	2.50	−3.62	3.96	1.63	4.28
MSSM	49	61/2	12	-	-	-	-	-
SU(5) GUT	124	24	24	-	-	-	-	-
SO(10) GUT	97	24	45	-	-	-	-	-

Properties of the non-Gaussian fixed point with type-II cutoff, gauge $\alpha = \beta = 1$ and one-loop anomalous dimensions, for various models.

Many popular models seem to be incompatible with asymptotic safety, at least within the present approximations. The MSSM has the same number of gauge fields as the SM, but too many fermions. GUTs have the same fermion content as the SM (typically with right-handed neutrinos included) and there are more gauge fields, so one may hope that they are compatible with a fixed point. In this case, however, it is the large number of scalars required for symmetry breaking that poses a severe challenge.[15] Technicolor-like models [262], which dispense with fundamental scalars, and instead introduce further fermions and gauge bosons, could very well be compatible with a fixed-point scenario for gravity, as larger numbers of vectors also imply a larger number of fermions compatible with the fixed point. Some of these theories actually do have fixed points, but here we restrict our attention to the fixed point that is a continuous deformation of the one that is known for pure gravity. This is clearly not a fundamental physical requirement, just a temporary selection criterion. It may be relaxed as our understanding progresses.

Fig. 7.32 shows the region in the N_S-N_D-plane where a fixed point exists with $\tilde{G}_* > 0$, $\theta_1, \theta_2 > 0$ for $N_V = 12$, at one loop and with type II cutoff. In comparison to the perturbative results, the inclusion of the anomalous dimensions leads to a more complicated shape of the boundary, but it remains true that continuous deformations of the fixed point without matter are only possible in a bounded domain of the plane. When one increases the number of scalars or fermions at fixed N_V one encounters a singularity, or the fixed point becomes complex. The fixed

---

[15]In the case of SO(10) a minimal scalar sector would contain the adjoint (45 real fields) the fundamental (10 complex fields) and one 16-dimensional complex spinor [261] leading to NS = 97, which is outside the permitted region.

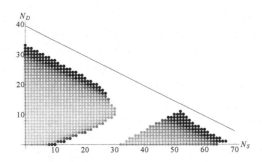

Fig. 7.32 The region of the N_S-N_D plane compatible with the existence of a gravitational fixed point with $\tilde{G}_* > 0$ and two attractive directions for $d = 4$, $N_V = 12$. The line represents the perturbative bound (7.197). Lighter shades of gray mean smaller η_h; black means $\eta_h > 10$.

points in the disconnected island on the right cannot be continuously deformed into the one without matter. Instead, they are the continuation of a fixed point that is complex in the permitted region connected to the origin, and becomes a pair of real fixed points for larger number of scalars. For small N_V the gap closes and there are combinations of matter fields such that the two fixed points are both real. We see here an example of the complicated phenomena that may happen just due to minimally coupled matter. The physical viability of the second fixed point requires a deeper investigation.

The shades of grey in figures (7.29) and (7.32) are related to the value of the graviton anomalous dimension, with darker tones indicating a larger anomalous dimension. We observe that η_h becomes very large $(O(10^3))$ at some points in the horn of Fig. (7.29) and near the boundary, at small N_S. The restriction $\eta_h > -2$ is automatically satisfied everywhere and does not add significant restrictions, however the dark dots indicate that the truncation used is unreliable. These graphs should therefore be taken only as a rough estimate, as the shape and position of the boundary could change in extended truncations.

7.8.3.7 *Higher-dimensional cases*

Large extra dimensions have a number of theoretical motivations, and have been shown to be compatible with asymptotically safe gravity in the Einstein-Hilbert truncation [263] and under the inclusion of fourth-order derivative operators [219]. While extra dimensions are not necessary for the consistency of the model, they seem compatible with it. Phenomenological implications have been studied in [264–268]. Experimentally, the upper bounds on their radius come from LHC results, see, e.g., [269, 270].

While the extra dimensions have to be compactified in a realistic setting, we can neglect the effect of compactifications here: in the UV limit, the compactification radius is much larger than the inverse cutoff scale. The allowed regions for $d = 5, 6$ and $N_V = 12$, at one loop and with type II cutoff, are shown in Fig. 7.33. They

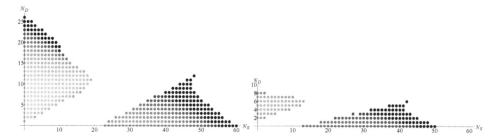

Fig. 7.33   The region of the N_S-N_D plane compatible with the existence of a gravitational fixed point with $\tilde{G}_* > 0$ and two attractive directions for $d = 5$ (left) and $d = 6$ (right) and $N_V = 12$. Lighter shades of gray mean smaller η_h; black means $\eta_h > 10$.

evidently become smaller with increasing dimension, so that the standard model would still be (barely) compatible with a fixed point in $d = 5$ but it is not in $d = 6$. It would be incompatible also in $d = 5$ if we used the type Ia cutoff.

Chapter 8

The asymptotic safety programme

The calculations in the preceding chapter illustrate various aspects of the search for a gravitational fixed point. These examples cover only a small part of the existing literature and have been presented without much commentary and with minimal references. Here we shall partly make up for these omissions by reviewing the existing literature and then discussing the main results, open problems and possible future developments.

8.1 Overview of the literature

8.1.1 *Pre-ERGE work*

The notion of nonperturbative renormalizability introduced in section 7.1, arose from the study of the possible behaviors a QFT may have in the IR and UV limit. This goes back to the work of Wilson [209] and others on phase transitions and critical phenomena. In the late 1970's Steven Weinberg got interested in these developments and suggested that nonperturbative renormalizability may solve the UV problems of gravity [38]. Having in mind the notion of asymptotic freedom, of which nonperturbative renormalizability is a generalization, he called "asymptotically safe" a theory that approaches a nontrivial fixed point in the UV.

To substantiate the conjecture that gravity is asymptotically safe, one needs a calculation of the beta functions in a theory of gravity. The first attempts were made by Christensen and Duff [271] and Gastmans et al. [272] in two dimensions, where Newton's coupling is dimensionless, or near two dimensions. The resulting beta function

$$\partial_t G = -\frac{38}{3} G^2$$

shows that G is asymptotically free in two dimensions. This implies that in $2 + \epsilon$ dimensions there is a nontrivial fixed point for $\tilde{G} = G k^\epsilon$. On the occasion of the centenary of Einstein's birth, a celebratory volume was published by Hawking and Israel. Weinberg's contribution [39] contained a discussion of quantum field theory approaches to quantum gravity and a more detailed presentation of the notion of asymptotic safety. The work of Christensen and Duff and Gastmans et al. was presented as preliminary evidence for asymptotic safety of gravity.

With the exception of a paper by Smolin [248], where a fixed point was found in the presence of a large number of matter fields, there followed about ten years when no progress was made along this line. Further calculations in $2 + \epsilon$ dimensions were performed by Jack and Jones [273] and by Kawai and others in Japan in the early 1990's, both at one loop [274, 275] and two loops [276]. An interesting feature of these calculations is that they did not use the linear background-quantum split $g_{\mu\nu} = \bar{g}_{\mu\nu} + h_{\mu\nu}$ but rather the exponential decomposition $g_{\mu\nu} = \bar{g}_{\mu\rho}(e^h)^\rho{}_\nu$ discussed in section 5.4.6. In any case it was not clear that the continuation to four dimensions ($\epsilon \to 2$) is possible, and further progress had to wait for a reliable method to compute beta functions in any dimension, in particular in $d = 4$.

Such a tool was introduced by Martin Reuter. He had been working on the notion of "Effective Average Action" and its application to gauge theories using the background field method. In the early 1990's it became clear that the EAA satisfies an exact equation [189–191]. This equation was initially used to study scalar models, but applications to gauge theories followed soon after [197]. Reuter's seminal work [207] applied the same techniques to gravity. After writing the exact equation (6.103), Reuter calculated the single-field beta functions for Λ and G with a type Ia cutoff, see equations (6.130). By identifying the cutoff scale k with the inverse of the distance from the origin, and inserting the scale-dependent Newton coupling in Newton's potential, he obtained an estimate of the quantum corrections to the Newtonian potential. This is very much in the spirit of Ühling's calculation in QED. Newton's constant was found to decrease with k, a behavior that can be described as antiscreening. This yields a correction to the Newtonian potential that has $\beta < 0$, in the notation of Eq. (4.17). This is actually opposite to the behavior that we discussed in section 4.5.4, but this was discovered only later. We shall return to this point.

8.1.2 *The first ten years of ERGE*

In reviewing the subsequent developments we shall roughly divide the work of the first ten years (1996–2005) from the work done in the following ten years (2006-2015). Reuter's paper was followed by [208], where the beta functions were calculated using the type Ib cutoff scheme, and furthermore the contribution of minimally coupled matter fields was calculated. Also [277] considered the beta functions of gravity coupled to matter, more precisely for $N = 4$ gauged supergravity. There were also early attempts to go beyond the Einstein-Hilbert truncation: in [278, 279] the effect of a term R^2 on the running of Newton's constant was calculated. In those early papers the standard choice for the cutoff profile was exponential, and it did not allow an exact evaluation of the Q-functionals. Thus, the beta functions of Λ and G were written as momentum integrals. A solution of the fixed point equations required numerical analysis. This was done by Souma [280], who proved the existence of a nontrivial fixed point and calculated its scaling exponents.

Later, Lauscher and Reuter [281] studied in greater detail the flow equations and its fixed points, with a type Ib cutoff. There followed several other papers that studied various properties of the Einstein-Hilbert truncation, such as dependence on gauge or cutoff scheme [282]. Litim used the so-called "optimized cutoff", which permits the Q-functionals to be integrated analytically and therefore to write the beta functions in closed form [283]. A conceptually different, approximate alternative to the ERGE, the proper time RG equation [284–286], was applied to gravity by Bonanno and Reuter [287] and found to lead to essentially the same results. In [288] Lauscher and Reuter considered the extension of the Einstein-Hilbert truncation by a term proportional to R^2 and found a fixed point where all three couplings are nonzero. The addition of nonlocal terms involving a function of the volume was considered by Reuter and Saueressig [289].

At about the same time Max Niedermaier developed an alternative approximation to the flow equations [290]: instead of keeping all the degrees of freedom of the metric and truncating the action, in these papers gravity is simplified by considering only metrics with two Killing vectors, while retaining the most general action. The resulting theory was shown to be asymptotically safe.

The effect of minimally coupled Gaussian matter on the Einstein-Hilbert fixed point was studied by Percacci and Perini [291], and in [292] this was extended by allowing a scalar to interact with gravity via a Lagrangian of the form

$$\frac{1}{2}(\nabla\phi)^2 + V(\phi^2) - F(\phi^2)R \; . \tag{8.1}$$

In these papers the fermions were treated with a type I cutoff, which we have seen in section 6.6.3 to be problematic.

Among the early ideas for the interpretation and possible phenomenology associated to asymptotic safety was the notion of fractality of spacetime. This would manifest itself in a scale-dependence of the effective metric, as discussed in [293]. In particular, it has been argued that there may be a notion of minimal length in asymptotically safe gravity [294, 295]. Direct consequences of a fixed point have been proposed from the very early days by Bonanno and Reuter, both for black holes [296], and for early [297] and late [298] cosmology (the latter based on the existence of a hypothetical IR fixed point). They are based on the identification of the cutoff scale k with a suitable function of radius, or time, respectively. Reuter and Weyer identified the physical world with a specific RG trajectory in the Einstein-Hilbert truncation, close to the critical line joining the Gaussian to the non-Gaussian fixed point [299]. Classical gravity prevails in a certain range of scales and quantum effects appear both at smaller and larger scales. They argued that such quantum effects may explain the galaxy rotation curves [300]. Further analysis of cosmological models with varying Λ and G was made by Reuter and Weyer [301] and Reuter and Saueressig [302].

The first ten years of applications of the ERGE to gravity can be closed with the appearance of several review papers: the "Living Review" by Niedermaier and Reuter [303], and then [304] and [305].

8.1.3 *The subsequent ten years*

One of the main challenges for the asymptotic safety programme is the extension of the truncation to include higher order terms. Up to 2006 the only serious attempt to go beyond the Einstein-Hilbert truncation was [288], where a single additional coupling was considered. In 2006, Codello and Percacci [217] applied the ERGE to compute the beta functions of higher derivative gravity at one loop in four dimensions. This calculation has been described in section 7.4. It established a link between the literature on the ERGE and the old works on the asymptotic freedom of higher derivative gravity. The same one-loop calculation has been also repeated, with small differences due to the choice of cutoff procedure, in [218], and much more recently has been generalized to arbitrary dimension by Ohta and Percacci [219]. The list of these one-loop calculations is completed by the work by Sezgin and Percacci [233] on Topologically Massive Gravity, described in section 7.5, and its extension to Topologically Massive Supergravity [234].

The calculation of beta functions of curvature squared gravity, but without making the one-loop approximation, has been done by Benedetti, Machado and Saueressig in [306], soon thereafter also in the presence of an additional scalar field in [307]. The striking difference with the one-loop calculations is that the dimensionless couplings in the four-derivative sector are no longer asymptotically free, but instead have nonzero limits. One issue with the "completely nontrivial" fixed point is that already in the truncation studied in [288], there seems to be no trajectory joining it to the Gaussian one [308]. Which result is physically correct is not known for sure at the moment.

Another generalization was to add higher powers of R. Codello, Percacci and Rahmede found that a fixed point exists for polynomials of order up to six in R, and that it has three relevant directions [245]. Machado and Saueressig repeated the calculation confirming and extending the results [309]. Among other things, they also wrote the flow equation for $f(R)$, although the function f was then truncated to a polynomial of order six. Later the order of the polynomial was raised to eight [246] and more recently, using recursive methods, Falls, Litim, Nikolakopoulos and Rahmede pushed the analysis to polynomials of order 34 (so the truncation contains 35 couplings) [247].

The paper [246] also contains a discussion of the fixed point of curvature squared gravity in the presence of matter. The existence of a fixed point for such a truncation shows that the fixed point does not cease to exist in the presence of terms that are non-renormalizable in perturbation theory. This result was extended beyond the one-loop approximation in [307].

The scalar-tensor theories of the form (8.1) have been studied again in more detail, in polynomial truncations, by Narain and Percacci [310]. This analysis was generalized to actions of the form

$$\frac{1}{2}(\nabla\phi)^2 - F(\phi^2, R) , \qquad (8.2)$$

by Narain and Rahmede [311]. Gravity coupled to QED has been studied in [312]. Also aside from the asymptotic safety literature, the effect of gravity on the running of gauge couplings has been the subject of a vigorous debate [313–320].

Much progress has also been made in better understanding the Einstein-Hilbert truncation. Nink and Reuter [321] have emphasized that the anti-screening behavior producing the nontrivial fixed point results from a competition between dia- and paramagnetic terms in the kinetic operators, in which the paramagnetic terms dominate.

Different, but classically equivalent formulations have been examined in detail. The tetrad formulation has been studied in [206, 322]. First-order formulations of gravity with an independent connection are only equivalent to Einstein gravity on-shell. It is nevertheless a meaningful question to ask how they differ at the quantum level. The RG flow for Einstein-Cartan gravity with or without the Immirzi parameter has been studied by Daum, Harst and Reuter [323,324]. The large number of fields makes for a very complicated calculation. A somewhat simpler, one-loop calculation of the same flow has been given by Benedetti and Speziale [325], with similar results. The one-loop flow for the most general action quadratic in torsion and nonmetricity has been calculated more recently by Pagani and Percacci [326]. The main issue with these calculations is that they do not contain a genuine kinetic term for the connection. Basically, the terms that have been considered so far, amount to mass terms for the difference between the dynamical connection and the Levi-Civita connection.[1] This is what makes them equivalent to Einstein theory on-shell: the equation of motion derived from a mass term just says that the field has to vanish. However, this is certainly not a suitable truncation to discuss the ultraviolet limit.

In the papers up to this point the ghost sector had always been treated as fixed, without any running coupling. The Einstein-Hilbert truncation has been extended by taking into account the ghost wave function renormalization in two papers by Groh and Saueressig [327] and by Eichhorn and Gies [328] which appeared simultaneously on the archive. The properties of the fixed point are not very different from earlier treatments, but the scheme dependence is less pronounced. Also, the different cutoff schemes of the two calculations produce only minor differences in the results. The ghost anomalous dimension has to be taken into account in the bi-field truncations based on the level expansion, that will be mentioned below. A separate nonminimal ghost coupling has also been considered in [329].

---

[1]For metric connections, this is known as the contortion tensor.

An important issue concerns the relation between the Euclidean beta functions computed in almost all papers on asymptotic safety, and Lorentzian quantum gravity. A direct calculation of the beta functions in Lorentzian spacetime has been given by Manrique, Rechenberger and Saueressig [330]. The fixed point is present with little change.

A different way of approximating the theory has been proposed by Reuter and Weyer in 2008: the "Conformally Reduced Einstein-Hilbert truncation" or CREH [331, 332]. In this case one freezes all degrees of freedom except for the conformal one, so that gravity reduces essentially to a scalar field, albeit with a rather peculiar action. This simplification allows one to study much more general truncations. Perhaps surprisingly, the analog of the Einstein-Hilbert truncation in this context has a nontrivial fixed point that is not too different from the one of the full theory.[2] This work was followed up by Machado and Percacci [333], who considered conformally reduced gravity with four derivative terms and found connections with earlier work by Antoniadis and Mottola [334]. Another study by Bonanno and Guarnieri [335] used the proper time flow equation to compare the beta functions obtained in the CREH from the potential or from the kinetic term. Falls recently observed that since the spin-0 degree of freedom is non-propagating in Einstein's theory, it may be more appropriate to treat the CREH as a topological field theory [336].

The role of conformal invariance in the gravitational RG flow has been discussed in a series of papers. Besides some calculations in [333], mentioned above, it has been shown in full generality in [337] that it is possible to preserve Weyl invariance along the flow, provided one chooses a suitable (position-dependent) cutoff. In the resulting flow, there is no fixed point, but the couplings nonetheless reach asymptotically finite limits. This has been shown to be true also when Weyl-invariance is guaranteed by the presence of a non-integrable connection [338]. On the other hand, with a more standard cutoff, a Wess-Zumino-like term will appear in the EAA [339]. In a related development, Codello, d'Odorico and Pagani have studied the existence of a c-function in the ERGE flows [340–342]. A different proposal for a c-function, closer in spirit to the "F-theorem" [343], has been made by Becker and Reuter [344].

On the theory side, there remain to discuss two very important research lines. The first are functional single-field truncations, the prime example being $f(R)$ truncation, without the assumption that f is a polynomial. Ordinary QFT methods are not well-suited to treat such problems, but this is precisely what asymptotic safety requires, and it is here that the full power of the ERGE is expected to make a difference. The application of the ERGE to a fully functional treatment of the Wilson-Fisher fixed point is now well understood and similar results are conceivable also for gravity. The second are bi-field truncations. Essentially all papers

---

[2]If, in the York decomposition, one only considers the contribution of the spin-2 degrees of freedom, they generate a fixed point too.

cited up to now treated gravity in the single-field approximation. As we have discussed in section 6.7, the EAA for gravity must necessarily be a functional of two fields. The treatment of bi-field approximations is technically much more involved and has gained momentum rather slowly. Since both topics are still not settled, we will discuss them in the next section.

Although most work has been devoted to establishing the existence of a fixed point with suitable properties, the derivation of observable consequences from asymptotic safety has always been seen as an equally important task. Short of direct experimental tests, which are very difficult for any theory of quantum gravity, the comparison of analytic results to numerical simulations is a very useful alternative. There has been significant exchange of ideas between the asymptotic safety community and the CDT community. Early papers on the fractal structure of quantum spacetime, already mentioned earlier, dealt with the scale dependence of the metric. A direct connection to CDT was proposed by Daum and Reuter [345]. It was soon understood that one aspect of this fractal structure would be a scale dependence of the spectral dimension. This has been discussed in detail by Reuter and Saueressig [346, 347] and also in [348] and [308]. The relation to the notion of spectral action of noncommutative geometry has been made by Alkofer et al. [349].

Work on direct applications to open problems of astrophysics and cosmology also continued. In one form or other, such applications are always based on the identification of the cutoff k with some characteristic scale of the problem under consideration. For spherically symmetric spacetimes this is generally a function of the radial coordinate, and for cosmological models a function of time. The spacetime-dependent cutoff is then inserted in the running couplings. This can be done in three different ways: at the level of the solution, at the level of the equations of motion or at the level of the action. When this "RG improvement" is performed at the level of the equations, one can choose to impose that the energy-momentum tensor of matter is conserved, in which case diffeomorphism invariance results in a relation between the gradients of the cutoff and the beta functions. Alternatively one may allow an exchange of energy between the matter and the varying couplings, which behave to some extent like fields. The general issues related to these procedures have been discussed by [301, 350], who note that the running couplings produce an effective Brans-Dicke-like dynamics. Other points of view have been expressed in [351, 352] and elsewhere.

In an application to , Bonanno and Reuter [353] identified the cutoff with the Hubble parameter and considered the case where one does not require separately the conservation of the energy momentum tensor. There is then an effective flow of energy from the time-varying cosmological constant to the matter degrees of freedom. They showed that this process can generate the right amount of entropy that is observed in the universe.

Hindmarsh, Litim and Rahmede [354] assumed instead that the energy momentum tensor is conserved. This puts constraints on the form of the cutoff. The field equations and the RG equations were written as a coupled autonomous system and various classes of solutions of these equations have been discussed.

Coming specifically to inflation, Weinberg [355] has given the conditions for a long almost de Sitter phase in the context of a general gravitational action near a fixed point. Tye and Xu [356] repeated this analysis and pointed out that inflation occurs sufficiently below the Planck scale that the couplings are not at their fixed point values.

Motivated by examples in QCD, Bonanno [357] argued in favor of RG improving the action, rather than the equations of motion. He assumed that the cutoff is proportional to $\sqrt{R}$, which, upon substitution in the Hilbert action leads to an $f(R)$ theory. In the vicinity of the fixed point, the flow can be solved by linearization and consists of spiralling trajectories. The resulting effective theory contains a term of the form $\cos\,(\log R)$. There exist infinitely many de Sitter solutions, some stable and others unstable. In particular there are unstable solutions with sufficiently many e-foldings to produce inflation.

Hindmarsh and Saltas [358] also identified the cutoff with $\sqrt{R}$ in the action, up to a numerical factor r, and then analyzed the theory going to the Einstein frame. Like Bonanno, they found infinitely many de Sitter solutions. In the Λ-G plane, the evolution of the universe would correspond to the piece of RG trajectory that starts from an "outer" de Sitter solution in the UV (i.e. in the past, producing inflation), passes near the Gaussian fixed point and then approaches another de Sitter point (accelerated expansion). Viability of the picture in the classical regime requires r near one, but this would lead to excessive primordial fluctuations. Viability of the inflationary phase requires a large r. The authors suggest that this discrepancy may be solved in the presence of additional degrees of freedom.

Copeland, Rahmede and Saltas [359] considered effective actions of the form $R + R^2$. The beta functions are shown to have a nontrivial fixed point where the R^2 term is asymptotically free (as in one-loop calculations, and unlike the fixed point of [288]). They show that there are RG trajectories that describe Starobinski inflation well. In particular the value of the R^2 coupling at the Planck scale is determined from CMB data to be of order 10^{-9}.

Other applications to cosmology have been discussed in [360–366].

Coming to black holes, the fate of the central singularity has been discussed by Cai and Easson [367], taking into account higher derivative terms, by Casadio, Hsu and Mirza [368] and by Kofinas and Zarikas [369]. Reuter and Tuiran consider the Kerr black hole [370]. Koch and Saueressig [371] considered the RG-improved Schwarzschild-de Sitter black hole solution and found interesting deviations from the RG-improved Schwarzschild solution, due to the role of the cosmological constant. As usual, they identified the cutoff with a (multiple of) radial distance from the origin and the improvement was made at the level of solution. At the nontrivial fixed point, the improved solution has exactly the same form as the classical one, but the role of the cosmological and Newton couplings are interchanged. As a consequence, the singularity in the origin is not removed. The entropy of the solution is shown to correspond to the effective average action evaluated at a self-consistent solution, suggesting that the microscopic origin of the black hole entropy is in the fluctuations of the geometry. The difference between RG-improving the black hole solution, and the black hole solution of the RG-improved equations, is discussed in [372].

Criticism to asymptotic safety is sometimes based on arguments involving the formation of black holes [373]. One way to avoid such arguments is to show that the weakening of gravity at short length scales prevents the formation of microscopic black holes. This issue has been discussed by Falls, Litim, Raghuraman [374, 375] and by Basu and Mattingly [376] and Ward [377].

The threshold for quantum gravity effects can be lowered in models with additional dimensions. This is therefore a natural context in which to explore possible experimental signatures of asymptotic safety. The existence of a fixed point in dimensions higher than four was implicit in earlier work such as [281], where the beta functions had been computed for arbitrary dimension, but the first detailed discussion of this aspect is due to Fischer and Litim [263]. Incidentally, the existence of a fixed point for any dimension is a very strange occurrence in the context of critical phenomena. The fact that the couplings become very large with increasing dimension suggest that this may be an artifact. We shall return to this later. The signatures of asymptotic safety in collider experiments such as Drell-Yan scattering have been discussed by Hewett and Rizzo [378] and Litim and Plehn [264–266]. Black hole formation has been discussed by Koch [267] while Döbrich and Eichhorn suggested that processes that are absent at tree level in the standard model, such as photon-photon scattering, may offer the best opportunities for detection [268].

Finally, it is worth recalling some particle physics-related developments. Shaposhnikov and Wetterich used the notion of asymptotic safety to argue that the Higgs self-coupling should be near zero at the Planck scale, and from there derived a prediction of the Higgs mass that proved remarkably close to the measured value [379]. More details of this argument are given in [380]. Even though the scalar self coupling could be near zero at the Planck scale for different reasons, this is definitely a very striking prediction.

Litim and Sannino showed that certain gauge theories with Yukawa interactions are asymptotically safe, with a nontrivial FP that is within the domain of perturbation theory [250]. This fills an important gap in the literature: previous known non-gravitational examples of asymptotic safety existed only in dimensions $2 < d < 4$. These models can be studied by perturbative methods only in a suitable large-N limit, but it is likely that the property of asymptotic safety persists also at finite N. Thus, in addition to providing a much-needed, reliable setting where the properties of an asymptotically safe theory can be studied, they may also become relevant to BSM physics. Although so far not conclusive, these developments show that experimental support for asymptotic safety may very well come from particle physics before cosmology.

8.2   State of the art

8.2.1   *Finite dimensional truncations*

As discussed in the preceding section, there is a lot of evidence for the existence of a fixed point in finite-dimensional single-field truncations. Before discussing the much more complicated bi-field and/or infinite dimensional truncations, let us stop for a moment to critically appraise this evidence. The fixed point seems to be remarkably robust, having shown up in essentially all conceivable approximations, but so far it has been impossible to make reliable quantitative predictions of its properties. One persistent problem is the dependence of the results on unphysical choices, such as gauge parameters or the shape of the cutoff function. This dependence is unavoidable, and one learns to live with it, but it is a sign that the quantities one is dealing with are not physically observable. As pointed out originally by Weinberg, it would be desirable to define asymptotic safety in terms of couplings that are closely related to measurable quantities, such as cross sections, or at least on-shell amplitudes. The ERGE allows us instead to calculate the running of couplings appearing in the EAA, off-shell. To understand the need to remain off-shell, let us focus on the Einstein-Hilbert truncation with a cosmological constant. The equations of motion are given by (3.127) and if we use them in the Hilbert action, both terms become proportional to the volume. It becomes then impossible to extricate the beta function of the cosmological constant from the beta function of Newton's constant. On the other hand, on the background of a four-sphere, the equation of motion $R = 4\Lambda$ implies that the on-shell Hilbert action is

$$-\frac{\Lambda}{8\pi G} \int d^4x \sqrt{g} = -\frac{3\pi}{\Lambda G} \ ,$$

where in the second step we used (5.135). The dimensionless combination ΛG should therefore be measurable. Several calculations have indeed demonstrated that the beta function of this quantity has a milder gauge- and cutoff-dependence than the beta functions of G and Λ separately [208, 281].[3] Still, different cutoff procedures can produce significantly different results. As discussed in section 7.2.1, this residual dependence can be taken as a measure of the errors due to the truncation.

The other quantities that should be measurable are the critical exponents. Here the positive news is that all evidence points towards a finite number of relevant deformations, between two and four for pure gravity. This means that the deviation from Gaussianity due to the loop corrections is not very large, and is a sign that the theory will be very predictive. However, also in this case it has not been possible so far to make any reliable quantitative estimates. In fact, it is not even clear at present (at least not to this author) whether the most relevant critical exponents are real or imaginary. As discussed in section 7.3, perturbative calculations yield real exponents. In the Einstein-Hilbert truncation the dependence of the beta

---

[3]See also sections 7.3.3 and 7.5.6.

functions on the cosmological term and the use of the background anomalous dimension to "RG improve" the beta functions produce a complex conjugate pair. As already emphasized in section 7.3, and like most things having to do with the cosmological constant, the dependence of the beta functions on the cosmological term is questionable, as we shall further discuss below.

The main source of uncertainty, however, is the use of the flow equation (6.109), which, as already mentioned, is not exact to start with. In practice, in the single-field approximation (which is what the use of the background anomalous dimensions amounts to) one replaces the Hessian with respect to the fluctuation field with the Hessian with respect to the background field, and this can potentially introduce spurious background dependence in the flow [381]. As discussed in section 7.6, an independent evaluation of the anomalous dimension yields again real critical exponents for the flow of the level-zero (background) couplings, but complex ones when the level-two cosmological constant is concerned. It appears that a satisfactory resolution of this issue will have to wait until the use of bi-field approximations and the related Ward identities are better developed.

Another issue that has been raised is that the beta functions obtained from the ERGE cannot be used in the calculation of observables in the same manner in which the logarithmic running of couplings is used to "RG improve" perturbative amplitudes [382]. In fact, from this point of view, one should not even speak of a beta function for Newton's constant. This is to some extent a semantic issue, but it highlights the difference between the beta functions of renormalizable theories, where the dependence of the coupling on renormalization scale can be directly translated into a dependence of amplitudes on external momenta, from the beta functions obtained from the ERGE, where this cannot generally be done. The ERGE gives the dependence of couplings appearing in the EAA on an unphysical cutoff scale, and to obtain the dependence of amplitudes on external momenta one should first integrate the flow down to $k = 0$ to obtain the full effective action.

It is possible that the situation could be improved even within the single-field truncation. As noted in [383], at least some of the issues arise because the calculations are performed too far off-shell. Another possible source of problems may lie in an inadequate parametrization of the degrees of freedom of the theory. Progress along these lines can be traced to a paper by Benedetti trying to derive some flow equations that are valid on-shell [384], and a paper by Eichhorn applying the flow equation to unimodular gravity [385], where one cannot use the standard linear background-quantum split but must use an exponential split (5.106). Independently of its use in unimodular gravity, the exponential parametrization has the conceptual virtue that when $h_{\mu\nu}$ is allowed to fluctuate freely, one covers the whole space of Riemannian metrics [386, 387]. In contrast, when one uses the standard linear split, large fluctuations can lead to degenerate metrics or even to metrics with different signatures.[4] Thus, if one takes seriously the notion of performing a

---

[4]This issue does obviously not arise in perturbative calculations, where one only integrates over

functional integral over Riemannian metrics on a given manifold, the exponential parametrization seems to be better suited than the linear one.[5] Another novelty introduced in [383] was the use of a "physical gauge", *i.e.* a gauge where one directly removes the unphysical degrees of freedom from the functional integral. This is possible if one uses the York decomposition, see section 5.4.5. The choice of gauge was to remove the trace h and the transverse vector ξ_μ from the degrees of freedom. In the exponential parametrization the expansion of the cosmological term only produces powers of the trace h, and if h is removed as a partial gauge choice, the Hessian for the remaining modes, $h_{\mu\nu}^{TT}$ and σ, is independent of Λ. Consequently, also the beta functions of any coupling will be independent of Λ. This has rather far reaching consequences. As we saw in section 7.3, the ubiquitous terms $1 - 2\tilde{\Lambda}$ in the denominators generate a singularity that precludes a smooth evolution of the flow towards the infrared. Furthermore, the Λ-dependence of the beta functions was also responsible for the complex critical exponents in the Einstein-Hilbert truncation. Both features are removed when one uses the exponential parametrization and the physical gauge, leaving a flow that resembles very closely the perturbative Einstein-Hilbert flow, pictured in Fig. 7.3. There is also another remarkable fact: unlike the standard flow equations described in chapter 7, the ones obtained by this procedure only admit a suitable fixed point if the dimension is below some upper bound [388, 389]. This is more in line with known critical phenomena and may be a hint that the exponential parametrization better reflects the physics.

The exponential parametrization has one last virtue that has been emphasized by Falls [389]. One can redefine the field h in such a way that it appears only linearly in the Taylor expansion of the action. Then, it would be completely absent from the Hessian, which would depend only on the two physical degrees of freedom $h_{\mu\nu}^{TT}$ and s. This Hessian is the same on- and off-shell. One can gauge fix in the standard Faddeev-Popov way and all the dependence on the gauge parameters cancels out. (This point has been mentioned in section 5.4.6.) With these choices Falls finds that the most relevant scaling exponent is close to 3 [390], which would be in accordance with Monte Carlo calculations of Hamber [391].

Recent systematic analyses of gauge and parametrization dependence within the Einstein-Hilbert truncation confirm the features described above [214, 392]. Thus, there are several hints that, all else being equal, the use of the exponential parameterization and of the physical gauge lead to more reliable results. The exploration of other truncations within this framework is under way.

---

small fluctuations around the background.

[5]Of course, one could restrict $h_{\mu\nu}$ in the linear decomposition to be such that $g_{\mu\nu}$ is always positive definite. Then, treating the transformation from linear to exponential split as a change of coordinates in field space, and taking into account the corresponding Jacobian, one would obtain completely equivalent results. Here we are instead assuming that the measure for h is the same, so that it makes a difference whether h is the one appearing in the linear or in the exponential split. In principle, these two choices define different quantum theories.

8.2.2 *Functional truncations*

Finite-dimensional truncations do not exploit the full power of the ERGE. For that one has to make an ansatz with a free function of the field, such as in $f(R)$ gravity or in the scalar-tensor theory (8.2). Although the flow equation for $f(R)$ gravity had already been written in [246, 309], the first serious attempt to find a functional solution was made by Benedetti and Caravelli four years later [393]. They wrote a slightly different equation, obtained by a spectral sum instead of the heat kernel expansion, but did not find a scaling solution. Later, Benedetti concluded that if a scaling solution exists, it must have a finite number of relevant deformations [394]. Dietz and Morris [395] made a systematic analysis of the equations in the literature and found that the number of solutions of the fixed point equation can be reliably determined by parameter counting. Given that the equation is third order, one would need to provide three conditions to reduce the number of solutions to a discrete set. Two such conditions are provided by the requirement that the solution continues past the two singularities of the equation for positive R. Accordingly, several lines of fixed points have been identified. Such fixed points would lead to a continuum of eigen-perturbations, a physically unacceptable situation. They tentatively suggested that the correct solutions must be valid also for negative R, where a further singularity in the equation provides an additional restriction. Numerical analyses support these conclusions, but a solution extending from minus to plus infinity was not found. Later, they found that all eigenperturbations of the solutions are redundant [396]. With these negative results it became clear that genuinely functional fixed points are much more sensitive to the details of the flow equation than the ones that had been hitherto found in finite dimensional truncations. Nevertheless, Demmel, Saueressig and Zanusso found a fixed functional in the three-dimensional CREH [397], see also [398]. Later, they introduced two free parameters in the cutoff scheme, by which they could control the number of fixed singularities of the equation [399] and a candidate fixed functional in $d = 4$ was found [400].

Percacci and Vacca looked for functional solutions in the scalar-tensor theory (8.2) using the exponential parametrization for the metric and a physical gauge [383]. The flow equations are much simpler than the ones obtained previously in [310], to the point that some nontrivial analytic solutions can be found. In three dimensions, using more sophisticated numerical techniques, Borchardt and Knorr found a solution that can be interpreted as a gravitationally dressed Wilson-Fisher fixed point [401]. In four dimensions there are various analytic nontrivial solutions, all with a constant potential, whose physical significance is not clear, and no solution with a nontrivial potential has been found, either analytically or numerically. This suggests that, at least within this truncation, the "triviality" problem of scalar fields persists also in the presence of gravity. These solutions have been generalized to the case of an $O(N)$-invariant scalar multiplet [388]. The existence of the solutions places upper bounds both on N, confirming the results of [242], and on the dimension.

The same type of approximation has been applied also to $f(R)$ gravity by Ohta, Percacci and Vacca [402]. Simple solutions have been shown to exist, although their physical significance is not completely clear. The existence of these solutions is pleasing, but the situation is altogether still not satisfactory. An important aspect that remains unclear is the physical meaning of the limit $\tilde{R} \equiv R/k^2 \to \infty$, i.e. of coarse-graining at length scales that are much larger than the diameter of the manifold. In fact it had been suggested in [399] that one may not even need to solve any equation beyond some finite value of $\tilde{R}$. More precisely, the solution would just scale as $\tilde{R}^2$, as demanded by classical dimensional arguments, for $\tilde{R}$ beyond a certain value. But then, if nontrivial, the solution would be non-analytic somewhere. The problem lies not just in the solution, but in the equations themselves. One can write the equation using the heat kernel method, but the heat kernel expansion is reliable only for $k^2 \gg R$. Alternatively, one can write the equation using a spectral sum. In order to be able to calculate the integrals in closed form one uses Litim's optimized cutoff, but this cutoff gives rise to a beta functional that is a non-differentiable staircase. To deal with it in practice, one has to approximate it by a smooth interpolating function. Either way, the behavior of the equation contains some arbitrary element.

Note that in finite-dimensional truncations one only considers the region of small $\tilde{R}$, which for fixed dimensionful R corresponds to the UV limit. In this case the heat kernel expansion is adequate, the result is expected to be independent of the topology and therefore the use of the sphere is justified. It is only in the functional case that one has to consider also large $\tilde{R}$. This means that one has to control also the infrared limit of the theory, where the results become sensitive to the topology. We note that from a conceptual point of view, the sphere also suffers from the problem of the Wick rotation, mentioned in section 5.2.

8.2.3 *Bi-field truncations and the shift Ward identity*

The other main open front is the extension beyond the single-field truncations, to functionals of two fields ("bi-field truncations"), and here the issue is to understand and bring under control the dependence of the EAA on its two arguments. This is probably the single most important issue with the use of the ERGE in gravity and more generally gauge theories, and a proper understanding of this may bring significant changes to the current picture.

The standard background EA depends on two variables $\Gamma(h; \bar{g})$ but the dependences on the two variables are not unrelated, so that effectively the information stored in the EA must be the same as if it depended on a single field. Ideally the

EA would simply obey the relation $\Gamma(h; \bar{g}) = \Gamma(\bar{g} + h)$, which holds for the classical action. This is equivalent to asking for invariance under the "shift symmetry",

$$\bar{g} \to \bar{g} + \epsilon \; ; \qquad h \to h - \epsilon \; . \tag{8.3}$$

Under these circumstances the Green functions defined by the functional derivatives with respect to the fluctuations

$$\left. \frac{\delta}{\delta h(x_1)} \cdots \frac{\delta}{\delta h(x_n)} \Gamma(h; \bar{g}) \right|_{h=0} \; , \tag{8.4}$$

would be equal to the Green functions of the background field calculated at zero fluctuation:

$$\frac{\delta}{\delta \bar{g}(x_1)} \cdots \frac{\delta}{\delta \bar{g}(x_n)} \Gamma(0; \bar{g}) \; . \tag{8.5}$$

This property is actually true when the background field method is applied to a linear scalar theory. It is not true in the case of a gauge theory, because the shift symmetry is broken by the background gauge fixing term, which is quadratic in h. Nevertheless, one can show that this difference is physically unimportant and that both systems of Green functions lead to the same S-matrix [403].

In the case of the EAA at non-zero cutoff k, even this weaker result is no longer true: the cutoff introduces an additional breaking of the shift symmetry that is not so tame as the one introduced by the gauge fixing. The two sets of Green functions are still related in some way, but the relation is much more complicated. It can be expressed by means of a "shift Ward identity", that has been discussed recently by Safari [404] (see also [405], where it is referred to as a Nielsen identity). The effect of the shift Ward identities would be to restrict the dependencies of $\Gamma_k(h; \bar{g})$ on its two arguments, reducing the information to that of a functional of a single field. Thus, the shift Ward identity would have to be solved together with the flow equation.

Let us see what has been done along these lines. As explained in section 7.2, there are two main points of view regarding the double field-dependence. One can view the EAA as a functional of two metrics ("comma notation") or a background metric and a symmetric tensor $h_{\mu\nu}$ ("semicolon notation"). In the former case one speaks of bi-metric truncations, in the latter one usually performs a "level expansion" in powers of $h_{\mu\nu}$. Bi-metric truncations have been discussed first by Manrique and Reuter in [406]. They subsequently considered the effect of matter fields [407] and the so-called "double Einstein-Hilbert truncation" [408], which consists of two separate Einstein-Hilbert actions for the full metric and for the background. It thus contains four couplings: two cosmological constants and two Newton constants. This truncation has been then studied in greater detail in the monumental paper by Becker and Reuter [213], where it has been shown how the requirement of background independence in the infrared effectively reduces the number of independent couplings to two. An interesting side-effect of this calculation is that the anti-screening that is necessary for the formation of a nontrivial fixed point only occurs at high energy. In the low-energy, semi-classical regime a screening behavior is found. This solves the issue mentioned in the end of section 8.1.1, reconciling the naive "RG improvement" of Newton's potential with the EFT calculations described in section 4.5.4.

The level expansion has been used instead by Christiansen et al. [409,410]. They calculated the full momentum dependence of the two point function of $h_{\mu\nu}$ and read off the beta function of the "level two" cosmological and Newton constant. More recently the results has been extended to the three-point function and the locality of the flow in momentum space has been demonstrated [411]. The calculation of the two-point function (though limited to the small-momentum regime) has been interpreted differently by Codello, d'Odorico and Pagani [243], who treat the level-two Newton coupling as the wave function renormalization of the graviton, and hence as a redundant coupling. This is essentially the calculation described in section 7.6.3. The same interpretation has been adopted also by Donà, Eichhorn and Percacci [242], as discussed in section 7.8. At present, the calculations show significant differences between the "same" coupling at different levels. It is possible that these differences are precisely those required by the Ward identities, but this will have to be checked when the identities themselves are made explicit.

Finally, a functional *and* bi-metric truncation has been discussed by Dietz and Morris [412]. The approximation used is the first order in the derivative expansion in the CREH. The main result is that the redundancy due to the presence of two scalar fields (corresponding to the background and dynamical conformal factors) can be eliminated by using the shift Ward identity. More precisely, it is shown that combining the four flow equations for the two potentials and the two coefficients of ∂^2, each a function of two variables, with the Ward identities, one obtains two flow equations for two functions of a single variable.

8.3 Outlook

It is often said that the two revolutionary theories of the twentieth century, General Relativity and quantum theory, have vastly different conceptual foundations that are hard to reconcile. This is certainly the reason for the difficulties that are encountered in the canonical approaches to quantum gravity. It is less evident that these deep issues have much to do with the narrow technical problem of renormalizability, which was the main stumbling block for the covariant approaches. After all, there are plenty of non-renormalizable theories that pose none of these issues. It is therefore not at all obvious that the formalism of QFT is in any way incompatible with gravity. In fact, EFT techniques can be successfully applied to gravity below the Planck scale. We have discussed in section 1.6 some reasons for trying to go beyond this EFT and we have argued that such a complete theory of quantum gravity is really a "quantum theory of spacetime".

Asymptotic safety is the most conservative extension of the low-energy EFT: it assumes that the same broad ideas and methods of QFT that successfully describe the other interactions, when suitably extended to a non-perturbative domain, can be used to construct a UV-complete theory of gravity. It has two main features that distinguish it from other approaches to quantum gravity and make it very appealing.

The first is a direct link with low-energy physics. Asymptotic safety is a "bottom up" approach, that starts from the tried and tested low-energy EFT formalism. This formalism describes well the neighborhood of the origin in Fig. 7.4, for example. In this picture, points with small $\tilde{\Lambda}$ and positive $\tilde{G}$ are pushed up for increasing cutoff, and if nothing happened there would be a divergence. Asymptotic safety implies that this will not happen and that the trajectories will end at some finite $\tilde{G}$ and $\tilde{\Lambda}$. The theory will just cross-over to a non-perturbative regime at the Planck scale, possibly without having to change the kinematical degrees of freedom. The non-trivial FP controls the behavior of the theory at trans-Planckian energies. But then, we are essentially guaranteed that starting from the UV fixed point there are trajectories that will end describing the correct physics at low energy. This should be contrasted with "top down" approaches to quantum gravity, where recovering low-energy physics is a very hard challenge.

The other main feature is predictivity. The requirement of UV completeness is not a goal in itself but rather the means by which — under favorable circumstances — all except for a few couplings will be determined.[6] It thus gives us a way, at least in principle, to calculate physical processes at the Planck scale. We are still very far from being able to do so, or for that matter to calculate any physical observable. Section 8.1 contains references to many papers that have tried to extract some physical predictions from the existing results on asymptotic safety. These give us useful hints as to what to expect, but are not yet fully reliable and in any case contain large theoretical errors. There are several reasons for this. The first is that most quantities that can be calculated show a more or less pronounced dependence on details of the cutoff scheme. This should somehow go away in the calculation of a physical observable. The situation will be even more complicated in the bi-field approach, where there are many ways of deriving the beta functions for what is essentially the same coupling. As discussed earlier, the way to address this issue is to solve simultaneously the flow equation and the shift Ward identity. How to efficiently do this is being actively investigated and one may reasonably expect that a better understanding will be reached within a few years.

The other source of uncertainties is in the so-called "cutoff identification". In all applications so far, the effect of asymptotic safety on physical processes is accounted for by replacing the fixed couplings (either in a solution, or in the equations, or in the action) by the running coupling, where the cutoff is identified with some

---

[6]In principle even quantities such as the fine structure constant may turn out to be calculable [312].

characteristic momentum scale of the process. The theoretical justification for using the EAA at some finite cutoff scale k is as follows: if k is really an infrared cutoff in the physical situation under consideration, then the RG flow at scales below k will stop and therefore the full EA will be well approximated by the EAA at scale k. This procedure may give roughly correct results, but is inherently ambiguous, because the characteristic scale is always defined up to factors of order one. When this type of procedure is applied in renormalizable theories, such ambiguity hardly matters, because the running of couplings is only logarithmic. In GR one encounters quadratic or even quartic running and the situation is much worse.

In principle, the proper procedure would be as follows:

- in the chosen approximations, find the UV fixed point;
- identify the RG trajectory that describes the real world;
- integrate the flow equation from some initial condition close to the fixed point, down to $k = 0$. This gives the EA. Because the fixed point is at finite couplings, the integration is not expected to produce divergences.
- use this EA to calculate the quantum effects on the process of interest.

This type of calculation has never been performed from a to z. Some covariant calculations of the EA using the ERGE have been described in [166, 413–416], but they do not involve an UV fixed point. For the case of a flow between two fixed points see [417]. Much more work will be needed to address this type of question.

Finally let us briefly discuss asymptotic safety as a "quantum theory of space-time" *i.e.* its implications for the short-distance structure of spacetime. Of course the fact that the ERGE is formulated in continuous spacetime does not imply, in itself, that spacetime *is* continuous. EFT's are also formulated in continuous space-time, but in an EFT with an UV cutoff Λ_{UV} one cannot resolve features shorter than Λ_{UV}^{-1}, so one cannot really say what happens at shorter length scales. In asymptotic safety, however, we really try to achieve the *continuum limit* $\Lambda_{UV} \to \infty$, and if this endeavour was successful one could really say that there is some sense in which spacetime is continuous.

Still, it is not completely clear what this would mean. To give this a meaning one would have to invent some experiment by which the measurement of arbitrarily short distances is possible. In other approaches to quantum gravity one often says that this is impossible because any measurement of a distance shorter than Planck's length would result in the formation of a black hole. We have already mentioned that in asymptotically safe gravity one cannot take this as a foregone conclusion, because the weakening of gravity at short distances (anti-screening) may prevent the formation of black holes. But even if a black hole did not form, the question is not completely settled. In a theory of pure gravity it is natural to take the Planck length as the unit of length. If we measure the cutoff in such units, then there are two possibilities. The first is that the unit is defined by the low-energy limit of the Planck length. In this case there is indeed no lower limit on the measurable distances. The second possibility is that one uses the Planck length at the scale k as a unit. This may be a more appropriate choice in some cases [294,295]. The statement that there is a fixed point for (the dimensionless!) Newton's constant, namely that there is a highest value for $k^2 G$, can be read as the statement that there is a maximal value for the cutoff in Planck units, or equivalently that there is a minimum value for the distances. Which one of the two interpretations is physically correct is something that cannot be decided abstractly but rather depends on the experiment. Some discussion of this in scattering experiments has been given in [418]. The provisional conclusion to be drawn from these considerations is that asymptotic safety need not be necessarily incompatible with some form of discreteness.

Numerical simulations show that gravity may have different phases. An extended phase resembling a four-dimensional manifold has only been found after adding the requirement of causality on the triangulations. So far, work on asymptotic safety has been done almost exclusively in Euclidean quantum gravity, and then mostly on the sphere. As long as only finite-dimensional truncations are involved, this should not make much difference, because any two manifolds look the same at short distances. Functional truncations, however, are sensitive to the global topology of the manifold and one will have to be more careful. In particular, as in CDT, it may be useful or perhaps even necessary to restrict the topology to manifolds that admit a Lorentzian metric. This would also eliminate much of the embarrassing profusion of four-dimensional topologies.

The existence of different phases of gravity, analogous for example to the "broken" and "unbroken" phases of scalar and gauge theories, can also be anticipated from the structure of the gravitational field variables. As mentioned in section 4.5.3, there is a very close analogy between GR and the chiral models, which describe the low-energy broken chiral symmetry phase of certain gauge theories. From this point of view, gravity as we know it must correspond to a "broken" phase of some fundamental theory. There have been some attempts, within the conformal reduction of the theory, to describe also an "unbroken" phase [332]. It is hard to see, however, how this can be reconciled with a formal path integral over metrics. This is particularly true if one makes use of the exponential parametrization (5.106). It is therefore prudent to assume that the type of calculations being performed in the literature on asymptotic safety refer only to the broken phase. The description of an unbroken phase "without spacetime" may require the use of drastically different variables [42, 107, 108].

One way of categorizing approaches to quantum gravity is to list all the structures that are present in the classical theory and to specify which ones of these are assumed to be fixed and which ones "fluctuate" [419]. In GR one has to specify

- a point set
- a topology
- a differentiable structure

- a causal/conformal structure
- a (pseudo-) Riemannian metric

In the radical approaches where one wants to have a vacuum that corresponds to "no spacetime" and then add "atoms of spacetime", one must necessarily give up all of the above. It is implicit in much of the work on asymptotic safety that everything up to the differentiable structure is fixed, and that the metric fluctuates. As long as the classical field corresponding to the metric is nondegenerate, one would always be in what was called above a broken phase. However, it may well be that in certain circumstances the theory only determines the conformal structure of the metric and not the metric itself. This may be the case sufficiently near the fixed point. Physics would be conformally invariant and it would be impossible, within such a medium, to measure any distance. Understanding the physics of such a situation already seems challenging enough.

Presso e lontano, lì, né pon né leva

Dante, Paradiso XXX, 121

Appendix A

Appendix

A.1 Units

The following table gives the values of the "fundamental constants" in MKS units:

speed of light	c	LT^{-1}	$2.99792458 \times 10^8 ms^{-1}$
Planck's constant	$\hbar$	$L^2 T^{-1} M$	$1.0545718 \times 10^{-34} m^2 kg s^{-1}$
Newton's constant	G	$L^3 T^{-2} M^{-1}$	$6.674 \times 10^{-11} m^3 kg^{-1} s^{-2}$

There is a unique way of combining these three constants to obtain a length, a time and a mass. These are the Planck length, time and mass:

Planck length	ℓ_P	$\sqrt{\frac{\hbar G}{c^3}}$	$1.616 \times 10^{-35} m$
Planck time	t_P	$\sqrt{\frac{\hbar G}{c^5}}$	$5.391 \times 10^{-44} s$
Planck mass	m_P	$\sqrt{\frac{\hbar c}{G}}$	$2.176 \times 10^{-8} kg$

It may be more convenient to set $\kappa^2 = 8\pi G$ as a unit, instead of G. This leads to the "reduced Planck units":

Planck length	ℓ_P	$\sqrt{\frac{8\pi\hbar G}{c^3}}$	$8.103 \times 10^{-35} m$
Planck time	t_P	$\sqrt{\frac{8\pi\hbar G}{c^5}}$	$2.703 \times 10^{-43} s$
Planck mass	m_P	$\sqrt{\frac{\hbar c}{8\pi G}}$	$4.341 \times 10^{-9} kg$

The Planck unit of energy is

$$E_P = m_P c^2 = 1.956 \times 10^9 J .$$

(This is a Planck unit that is comprehensible in daily life, being comparable to the kinetic energy of a train.) In particle physics, the standard unit of energy is the electron-Volt:

$$1eV = 1.602176 \times 10^{-19} J = 8.19061 \times 10^{-29} E_P .$$

This defines a unit of distance

$$\frac{\hbar c}{1 eV} = 1.97327 \times 10^{-7} m \ ,$$

so in natural units $1 m$ corresponds to $5.06773 \times 10^{6} eV^{-1}$. Furthermore, the Planck unit of energy is

$$E_P = 1.221 \times 10^{28} eV = 1.221 \times 10^{19} GeV$$

or

$$m_P = 1.221 \times 10^{19} GeV/c^2 \ ,$$

while the reduced Planck mass is

$$m_P/\sqrt{8\pi} = 2.435 \times 10^{18} GeV/c^2 \ .$$

Mostly we use natural units where $\hbar = 1$ and $c = 1$ but not Planck units. Then G remains indicated as such. Its low energy value is the Planck area or squared Planck mass

$$G \sim \ell_P^2 = 2.612 \times 10^{-70} m^2 \sim 6.709 \times 10^{-57} eV^{-2} \ ,$$

where "$\sim$" means "equal in natural units".

A.2 Notations

A.2.1 *Conventions*

In chapter 2 all formulas are written in Minkowski signature. In subsequent chapters all formulas are written in Euclidean signature, unless otherwise stated. Rules for translating between the two are given in section 5.2.

Minkowski metric:
$\eta_{\mu\nu}$ is diagonal with entries $-1, 1 \ldots 1$.

In any signature:

$$\Box = \partial^2 \ .$$

In Minkowski signature this is the d'Alembertian $-\partial_0^2 + \sum_i \partial_i^2$. The (positive) flat space Laplace operator is $-\Box$.

Symmetrization:

$$T_{[\mu_1 \ldots \nu_n]} = \frac{1}{n!} \sum_{\pi} T_{\pi(\mu_1) \ldots \pi(\mu_n)}$$

where π are elements of the permutation group of n elements.

Antisymmetrization:

$$T_{[\mu_1 \ldots \nu_n]} = \frac{1}{n!} \sum_{\pi} (-1)^{|\pi|} T_{\pi(\mu_1) \ldots \pi(\mu_n)}$$

where $|\pi|$ is the parity of the permutation π.

Covariant derivatives:

$$\nabla_\lambda T^\mu{}_\nu = \partial_\lambda T^\mu{}_\nu + \Gamma_\lambda{}^\mu{}_\rho T^\rho{}_\nu - \Gamma_\lambda{}^\rho{}_\nu T^\mu{}_\rho \ .$$

In any signature:

$$\Box = \nabla^2 = g^{\mu\nu}\nabla_\mu\nabla_\nu \ .$$

In Lorentzian signature, this is the d'Alambert operator. In positive definite signature there are several Laplace operators. The Bochner Laplacian is

$$\Delta_B = -\Box \ .$$

For other Laplacians see section 5.3.

Commutator of covariant derivatives:

$$[\nabla_\mu, \nabla_\nu]T^\rho{}_\sigma = R_{\mu\nu}{}^\rho{}_\lambda T^\lambda{}_\sigma - R_{\mu\nu}{}^\lambda{}_\sigma T^\mu{}_\lambda \ .$$

Riemann tensor:

$$R_{\mu\nu}{}^\rho{}_\sigma = \partial_\mu \Gamma_\nu{}^\rho{}_\sigma - \partial_\nu \Gamma_\mu{}^\rho{}_\sigma + \Gamma_\mu{}^\rho{}_\tau \Gamma_\nu{}^\tau{}_\sigma - \Gamma_\nu{}^\rho{}_\tau \Gamma_\mu{}^\tau{}_\sigma \ .$$

(This is the natural definition of curvature tensor in a gauge theory, which is a Lie algebra-valued two-form. The first pair of indices are the form indices and the second pair labels a basis in the Lie algebra of $GL(d)$, which consists of $d \times d$ matrices.)

Ricci tensor:

$$R_{\mu\nu} = R_{\rho\mu}{}^\rho{}_\nu$$

Einstein's equations

$$R_{\mu\nu} - \frac{1}{2}Rg_{\mu\nu} + \Lambda g_{\mu\nu} = 8\pi G T_{\mu\nu}.$$

The standard prefactor of the Einstein-Hilbert action is $\frac{1}{16\pi G}$. This is needed to produce the factor $8\pi G$ in the r.h.s. of Einstein's equations. This in turn comes from comparing Einstein's equations with the Poisson equation of Newtonian physics, which conventionally has a factor $4\pi G$. This is well-suited to describe gravity in three space dimensions, but is unnatural in other dimensions. Nevertheless it seems that all authors write the equations in the same form in all dimensions, and I will follow this convention here.

The factor $8\pi G$ in Einstein's equations is denoted κ^2. In this way the prefactor of the Hilbert action is

$$\frac{1}{16\pi G} = \frac{1}{2\kappa^2} = Z_N \ .$$

In the literature one often encounters different conventions for κ. From section 6.8 onward, the prefactor of the Hilbert action is denoted Z_N. This is motivated by the similarity with the wave function renormalization constant, when the equations are linearized.

The symbol Tr denotes a functional trace while tr denotes a finite dimensional trace. Similarly Det and det for determinants.

A cutoff kernel that is a matrix in some space (in this case the Lie algebra of the gauge group) is designated $\mathcal{R}_k$. The symbol R_k is reserved for real functions, such as the cutoff kernel of a real scalar field.

A.2.2 *Acronyms*

ADM: Arnowitt, Deser and Misner
BSM: Beyond the Standard Model
CDT: Causal Dynamical Triangulations
CREH: Conformally Reduced Einstein-Hilbert
CS: Chern Simons
EDT: Euclidean Dynamical Triangulations
EA: Effective Action
EAA: Effective Average Action
EFT: Effective Field Theory
EH: Einstein-Hilbert
ERGE: Exact Renormalization Group Equation
FP: Fixed Point
GR: General Relativity
GUT: Grand Unified Theory
HL: Hořava-Lifshitz
IR: InfraRed
LQG: Loop Quantum Gravity
MSSM: Minimally Supersymmetric Standard Model
QCD: Quantum ChromoDynamics
QED: Quantum ElectroDynamics
QFT: Quantum Field Theory
RG: Renormalization Group
SM: Standard Model
SUSY: SUperSYmmetry
SUGRA: SUperGRAvity
TOE: Theory Of Everything
UV: UltraViolet
YM: Yang Mills

Bibliography

[1] J. Stachel, Early history of quantum gravity (1916-1940), in "Black Holes, Gravitational radiation and the Universe", B.R. Iyer and B. Bhawal eds, (Kluwer Academic Publisher, Netherlands 1999), pp. 525-534.

[2] G. Gorelik, "Matvei Bronstein and quantum gravity: 70th anniversary of the unsolved problem", Physics-Uspekhi vol 48, no 10 (2005) pp. 1039-1053.

[3] C. Rovelli, arxiv:gr-qc/0006061.

[4] L. Rosenfeld, "Zur Quantelung der Wellenfelder", Ann der Physik 5 (1930) 113; "Uber die Gravitationswirkungen des Lichtes", Zeit fur Phy s 65 (1930) 589.

[5] M.P. Bronstein "Quantentheories schwacher Gravitationsfelder", Physikalische Zeitschrift der Sowietunion 9 (1936) 140.

[6] M. Fierz, "Force-free particles with any spin", Hel. Phys. Acta 12 (1939) 3-37; W. Pauli and M. Fierz, "On Relativistic Field Equations of Particles With Arbitrary Spin in an Electromagnetic Field", Hel. Phys. Acta 12 (1939) 297.

[7] S.N. Gupta, Proc. Phys. Soc. A65 (1952) 608-619.

[8] R. Arnowitt, S. Deser and C.W. Misner, "The dynamics of general relativity", in Gravitation: An Introduction to Current Research, ed. L. Witten, p 227 (Wiley, New York, 1962).

[9] B. S. DeWitt, "Quantum Theory of Gravity. 1. The Canonical Theory," Phys. Rev. **160**, 1113 (1967).

[10] B. S. DeWitt, "Quantum Theory of Gravity. 2. The Manifestly Covariant Theory," Phys. Rev. **162** (1967) 1195.

[11] B. S. DeWitt, "Quantum Theory of Gravity. 3. Applications of the Covariant Theory," Phys. Rev. **162** (1967) 1239.

[12] W. Heisenberg, "Die Grenzen der Anwendbarkeit der bisherigen Quantentheorie" Z. Physik 110 (1938) 251.

[13] G. 't Hooft and M. J. G. Veltman, "One loop divergencies in the theory of gravitation," Annales Inst. Poincaré Phys. Theor. A **20** (1974) 69.

[14] S. Deser and P. van Nieuwenhuizen, "Nonrenormalizability of the Quantized Einstein-Maxwell System," Phys. Rev. Lett. **32** (1974) 245. S. Deser and P. van Nieuwenhuizen, "One Loop Divergences of Quantized Einstein-Maxwell Fields," Phys. Rev. D **10** (1974) 401.

[15] S. Deser, H. S. Tsao and P. van Nieuwenhuizen, "Nonrenormalizability of Einstein Yang-Mills Interactions at the One Loop Level," Phys. Lett. B **50** (1974) 491. S.

Deser, P. Van Nieuwenhuizen and H.S. Tsao, "One Loop Divergences of the Einstein Yang-Mills System", Phys. Rev. D10 (1974) 3337.

[16] S. Deser, P. Van Nieuwenhuizen, "Non-renormalizability of the quantized Dirac-Einstein system", Phys. Rev. D10 (1974) 411.

[17] A. O. Barvinsky and G. A. Vilkovisky, "Divergences and Anomalies for Coupled Gravitational and Majorana Spin 1/2 Fields," Nucl. Phys. B **191**, 237 (1981).

[18] A. Van Proeyen, "Gravitational Divergences of the Electromagnetic Interactions of Massive Vector Particles," Nucl. Phys. B **174** (1980) 189.

[19] E. Sezgin and P. van Nieuwenhuizen, "Renormalizability Properties of Antisymmetric Tensor Fields Coupled to Gravity," Phys. Rev. D **22** (1980) 301.

[20] M.H. Goroff and A. Sagnotti, "The Ultraviolet Behavior of Einstein Gravity", Nucl. Phys. **B** 266, 709 (1986);

[21] A. E. M. van de Ven, "Two loop quantum gravity", Nucl. Phys. **B** 378, 309 (1992).

[22] K. S. Stelle, "Renormalization Of Higher Derivative Quantum Gravity," Phys. Rev. D **16** (1977) 953.

[23] B. L. Voronov and I. V. Tyutin, "On Renormalization Of R**2 Gravitation. (In Russian)," Yad. Fiz. **39** (1984) 998.

[24] D.Z. Freedman, P. van Nieuwenhuizen and S. Ferrara, "Progress Toward A Theory Of Supergravity", Phys. Rev. D13 (1976) 32143218.

[25] C. W. Misner, "Feynman quantization of general relativity," Rev. Mod. Phys. **29** (1957) 497.

[26] S. W. Hawking, "Quantum Gravity and Path Integrals," Phys. Rev. D **18** (1978) 1747.

[27] S. W. Hawking, "Space-Time Foam," Nucl. Phys. B **144** (1978) 349.

[28] S. W. Hawking, "Gravitational Instantons," Phys. Lett. A **60** (1977) 81.

[29] G. W. Gibbons and S. W. Hawking, "Gravitational Multi - Instantons," Phys. Lett. B **78** (1978) 430.

[30] G. W. Gibbons and S. W. Hawking, "Classification of Gravitational Instanton Symmetries," Commun. Math. Phys. **66** (1979) 291.

[31] G. W. Gibbons and S. W. Hawking, "Action Integrals and Partition Functions in Quantum Gravity," Phys. Rev. D **15** (1977) 2752.

[32] J. B. Hartle and S. W. Hawking, "Wave Function of the Universe," Phys. Rev. D **28** (1983) 2960.

[33] H. W. Hamber, "Quantum Gravity on the Lattice," Gen. Rel. Grav. **41** (2009) 817 arXiv:0901.0964 [gr-qc].

[34] J. Ambjørn and J. Jurkiewicz, "Four-dimensional simplicial quantum gravity," Phys. Lett. B **278** (1992) 42.

[35] S. Catterall, J. B. Kogut and R. Renken, "Phase structure of four-dimensional simplicial quantum gravity," Phys. Lett. B **328** (1994) 277 [hep-lat/9401026].

[36] P. Bialas, Z. Burda, A. Krzywicki and B. Petersson, "Focusing on the fixed point of 4-D simplicial gravity," Nucl. Phys. B **472** (1996) 293 [hep-lat/9601024].

[37] B. V. de Bakker, "Further evidence that the transition of 4-D dynamical triangulation is first order," Phys. Lett. B **389** (1996) 238 [hep-lat/9603024].

[38] S. Weinberg, "Critical phenomena for field theorists", in the proceedings of the International School of Subnuclear Physics, Ettore Majorana Center for scientific culture, Erice, July 24-26, 1976.

[39] S. Weinberg, "Ultraviolet divergences in quantum theories of gravitation", in *General Relativity: An Einstein centenary survey*, ed. S. W. Hawking and W. Israel, pp.790–831; Cambridge University Press (1979).

[40] A. Ashtekar, "New hamiltonian formulation for general relativity", Phys. Rev. D36 (1987) 1587.

[41] C. Rovelli, L. Smolin, "Knot theory and quantum gravity", Phys. Rev. Lett. 61, (1988) 1155; "Loop space representation of quantum general relativity", Nucl. Phys. B331 (1990) 80.

[42] D. Oriti, "The Group field theory approach to quantum gravity," In *Oriti, D. (ed.): Approaches to quantum gravity* 310-331 [gr-qc/0607032].

[43] R. Gurau and V. Rivasseau, "The 1/N expansion of colored tensor models in arbitrary dimension," EuroPhys. Lett. **95** (2011) 50004 arXiv:1101.4182 [gr-qc].

[44] N. Straumann, "On Pauli's invention of non-abelian Kaluza-Klein Theory in 1953", proceddings of the IXth Marcel Grossmann Meeting, arXiv:gr-qc/0012054.

[45] C. N. Yang and R. L. Mills, "Conservation of Isotopic Spin and Isotopic Gauge Invariance," Phys. Rev. **96** (1954) 191.

[46] L. O'Raifeartaigh, "The dawning of gauge theory", Princeton University Press, (1997).

[47] R. Utiyama, "Invariant theoretical interpretation of interaction," Phys. Rev. **101**, 1597 (1956).

[48] D. J. Gross and F. Wilczek, "Ultraviolet Behavior of Nonabelian Gauge Theories," Phys. Rev. Lett. **30** (1973) 1343.

[49] H. D. Politzer, "Reliable Perturbative Results for Strong Interactions?," Phys. Rev. Lett. **30** (1973) 1346.

[50] S. Weinberg, "Gravitation and cosmology", John Wiley (1972).

[51] R.P. Feynman, F.B. Morinigo and W.G. Wagner, "Feynman lectures on gravitation", ed. B. Hatfield, Westview Press (2003).

[52] S. Deser, "Self interaction and gauge invariance", Gen. Rel. Grav. **1**, 9 (1970).

[53] C. de Rham, "Massive Gravity", Living Reviews in Relativity 17: 7, (2014), arXiv:1401.4173.

[54] K. Hinterbichler, "Theoretical Aspects of Massive Gravity", Reviews of Modern Physics 84 (2012) 671710, arXiv:1105.3735.

[55] B.P. Abbott et al. (LIGO Scientific Collaboration and Virgo Collaboration) "GW151226: Observation of Gravitational Waves from a 22-Solar-Mass Binary Black Hole Coalescence" Phys. Rev. Lett. 116, 241103 (2016).

[56] S. Boughn and T. Rothman, "Can gravitons be detected?," Found. Phys. **36**, 1801 (2006) [gr-qc/0601043].
"Aspects of graviton detection: Graviton emission and absorption by atomic hydrogen," Class. Quant. Grav. **23**, 5839 (2006) [gr-qc/0605052].

[57] L. M. Krauss and F. Wilczek, "Using Cosmology to Establish the Quantization of Gravity," Phys. Rev. D **89**, no. 4, 047501 (2014) arXiv:1309.5343 [hep-th]. "From *B*-Modes to Quantum Gravity and Unification of Forces," Int. J. Mod. Phys. D **23**, no. 12, 1441001 (2014) arXiv:1404.0634 [gr-qc].

[58] R. P. Woodard, "Perturbative Quantum Gravity Comes of Age," Int. J. Mod. Phys. D **23** (2014) no.09, 1430020 arXiv:1407.4748 [gr-qc].

[59] K. J. Barnes, Ph.D. thesis (1963).

[60] R. J. Rivers, Il Nuovo Cimento 34 (1964) 387.

[61] P. van Nieuwenhuizen, Nucl. Phys. B60 (1973) 478.
P. Van Nieuwenhuizen, "Supergravity", Phys. Rept. 68 (1981) 189.

[62] M. Asorey, J. L. Lopez and I. L. Shapiro, "Some remarks on high derivative quantum gravity," Int. J. Mod. Phys. A **12** (1997) 5711 arXiv:hep-th/9610006.

[63] J. S. Schwinger, "On gauge invariance and vacuum polarization," Phys. Rev. **82**, 664 (1951).

[64] B. S. DeWitt, "Quantum Field Theory in Curved Space-Time," Phys. Rept. **19** (1975) 295.

[65] B.S. DeWitt, "The Global Approach to Quantum Field Theory", Oxford University Press (2002).

[66] S.W. Hawking, "Zeta function regularization of path integrals in curved spacetime", Comm. Math. Phys. **55** (1977) 133.

[67] E. Elizalde, S. D. Odintsov, A. Romeo, A. A. Bytsenko and S. Zerbini, "Zeta regularization techniques with applications," Singapore, Singapore: World Scientific (1994)

[68] E. Elizalde, "Ten physical applications of spectral zeta functions," Lect. Notes Phys. M **35** (1995) 1.

[69] B. S. DeWitt, "Dynamical theory of groups and fields," Conf. Proc. C **630701** (1964) 585 [Les Houches Lect. Notes **13** (1964) 585].

[70] P. B. Gilkey, "Invariance theory, the heat equation and the Atiyah-Singer index theorem," http://pages.uoregon.edu/gilkey/dirPDF/InvarianceTheory1Ed.pdf CRC Press (1995).

[71] A.O. Barvinsky and G.A. Vilkovisky, "The Generalized Schwinger-Dewitt Technique in Gauge Theories and Quantum Gravity" Phys. Rept. **119**, 1-74, (1985).

[72] I.G. Avramidi, "Heat kernel and quantum gravity," Lect. Notes Phys. **M64** (2000) 1.

[73] M.E. Peskin and D.V. Schroeder, "An Introduction to Quantum Field Theory", Westview Press (1995).

[74] S.M. Christensen and M.J. Duff, "New gravitational index theorems and super theorems", Nucl. Phys. **B 154** 301 (1979).

[75] M. J. Duff, "Observations on Conformal Anomalies," Nucl. Phys. B **125** (1977) 334.
M. J. Duff, "Twenty years of the Weyl anomaly," Class. Quant. Grav. **11** (1994) 1387 hep-th/9308075.

[76] E. Witten, "(2+1)-Dimensional Gravity as an Exactly Soluble System," Nucl. Phys. B **311** (1988) 46.

[77] S. M. Christensen and M. J. Duff, "Quantizing Gravity With A Cosmological Constant," Nucl. Phys. B **170** (1980) 480.

[78] R. E. Kallosh, O. V. Tarasov and I. V. Tyutin, "One Loop Finiteness Of Quantum Gravity Off Mass Shell," Nucl. Phys. B **137**, 145 (1978).

[79] M. K. Chase, "Absence of Leading Divergences in Two Loop Quantum Gravity," Nucl. Phys. B **203** (1982) 434.

[80] N. Marcus and A. Sagnotti, "The Ultraviolet Behavior of $N = 4$ Yang-Mills and the Power Counting of Extended Superspace," Nucl. Phys. B **256** (1985) 77.

[81] A. O. Barvinsky and G. A. Vilkovisky, "The Effective Action In Quantum Field Theory: Two Loop Approximation," in Batalin, I.A. et al. (ed.): "Quantum Field

Theory and Quantum Statistics", essays in honor of the sixtieth birthday of E.S. Fradkin VOL. 1*, 245-275, Adam Hilger, Bristol (1987).

[82] Z. Bern, C. Cheung, H. H. Chi, S. Davies, L. Dixon and J. Nohle, "Evanescent Effects Can Alter Ultraviolet Divergences in Quantum Gravity without Physical Consequences," Phys. Rev. Lett. 115 (2015) no.21, 211301, arXiv:1507.06118 [hep-th].

[83] M. Y. Kalmykov and P. I. Pronin, "One loop effective action in gauge gravitational theory," Nuovo Cim. B **106** (1991) 1401.
M. Y. Kalmykov, P. I. Pronin and K. V. Stepanyantz, "Projective invariance and one loop effective action in affine metric gravity interacting with scalar field," Class. Quant. Grav. **11** (1994) 2645 [hep-th/9408032].
M. Y. Kalmykov and P. I. Pronin, "The One loop divergences and renormalizability of the minimal gauge theory of gravity," Gen. Rel. Grav. **27** (1995) 873 [hep-th/9412177].

[84] F. T. Brandt and D. G. C. McKeon, "Radiative Corrections and the Palatini Action," arXiv:1601.04944 [hep-th].

[85] K. Krasnov, "Pure Connection Action Principle for General Relativity," Phys. Rev. Lett. **106** (2011) 251103 arXiv:1103.4498 [gr-qc].
K. Krasnov, "Gravity as a diffeomorphism invariant gauge theory," Phys. Rev. D **84** (2011) 024034 arXiv:1101.4788 [hep-th].
K. Krasnov, "One-loop beta-function for an infinite-parameter family of gauge theories," JHEP **1503** (2015) 030 arXiv:1501.00849 [hep-th].

[86] A. D. Sakharov, Sov. Phys. Dokl. **12** (1968) 1040 [Dokl. Akad. Nauk Ser. Fiz. **177** (1967) 70] [Sov. Phys. Usp. **34** (1991) 394] [Gen. Rel. Grav. **32** (2000) 365].

[87] M. Visser, Mod. Phys. Lett. A **17** (2002) 977 [gr-qc/0204062].

[88] Stephen L. Adler, "Einstein Gravity as a Symmetry Breaking Effect in Quantum Field Theory " Rev. Mod. Phys. 54 (1982) 729, Rev.Mod.Phys. 55 (1983) 837.

[89] A.H. Chamseddine, A. Connes "The Spectral action principle" Commun.Math.Phys. 186 (1997) 731-750 arXiv: hep-th/9606001.

[90] T. Jacobson, Phys. Rev. Lett. 75, (1995) 1290. arXiv:gr-qc/0311082.

[91] T. Padmanabhan, Rept. Prog. Phys. 73 (2010) 046901, arXiv:0911.5004; J. Phys. Conf. Ser. 306 (2011) 012001, arXiv:1012.4476.

[92] K. Akama, "An Attempt at Pregeometry: Gravity With Composite Metric," Prog. Theor. Phys. **60**, 1900 (1978).

[93] R. Floreanini and R. Percacci, "Topological pregeometry," Mod. Phys. Lett. A **5**, 2247 (1990).

[94] C. Wetterich, "Gravity from spinors," Phys. Rev. D **70**, 105004 (2004) [hep-th/0307145].

[95] J. Alfaro, D. Espriu and D. Puigdomenech, "The emergence of geometry: a two-dimensional toy model," Phys. Rev. D **82** (2010) 045018 arXiv:1004.3664 [hep-th].

[96] R. Floreanini, R. Percacci and E. Spallucci, "Coleman-Weinberg effect in quantum gravity," Class. Quant. Grav. **8**, L193 (1991).

[97] R. Floreanini and R. Percacci, "Mean field quantum gravity," Phys. Rev. D **46**, 1566 (1992).

[98] R. Floreanini and R. Percacci, "Average effective potential for the conformal factor," Nucl. Phys. B **436**, 141 (1995) [hep-th/9305172].

[99] R. Floreanini and R. Percacci, "The Renormalization group flow of the Dilaton potential," Phys. Rev. D **52**, 896 (1995) [hep-th/9412181].

[100] R. Floreanini and R. Percacci, "Quantum mechanical breaking of local GL(4) invariance," Phys. Lett. B **379**, 87 (1996) [hep-th/9508157].

[101] R. Percacci, "The Higgs phenomenon in quantum gravity," Nucl. Phys. B **353** (1991) 271 arXiv:0712.3545 [hep-th].

[102] T. Dereli and R. W. Tucker, "Signature dynamics in general relativity," Class. Quant. Grav. **10** (1993) 365.

[103] A. Carlini and J. Greensite, "Why is space-time Lorentzian?," Phys. Rev. D **49** (1994) 866, arXiv:gr-qc/9308012.

[104] S. Mukohyama and J. P. Uzan, "From configuration to dynamics: Emergence of Lorentz signature in classical field theory," Phys. Rev. D **87** (2013) 6, 065020, arXiv:1301.1361 [hep-th];
J. Kehayias, S. Mukohyama and J. P. Uzan, "Emergent Lorentz Signature, Fermions, and the Standard Model," Phys. Rev. D **89** (2014) 10, 105017, arXiv:1403.0580 [hep-th].

[105] C. Barcelo, S. Liberati and M. Visser, "Analogue gravity," Living Rev. Rel. **8**, 12 (2005) [Living Rev. Rel. **14**, 3 (2011)] [gr-qc/0505065].

[106] G. E. Volovik, "Superfluid analogies of cosmological phenomena," Phys. Rept. **351**, 195 (2001) [gr-qc/0005091].

[107] T. Konopka, F. Markopoulou and S. Severini, "Quantum Graphity: A Model of emergent locality," Phys. Rev. D **77** (2008) 104029 arXiv:0801.0861 [hep-th].

[108] C. A. Trugenberger, "Quantum Gravity as an Information Network: Self-Organization of a 4D Universe," arXiv:1501.01408 [hep-th].

[109] D. N. Blaschke and H. Steinacker, "Curvature and Gravity Actions for Matrix Models II: The Case of general Poisson structure," Class. Quant. Grav. **27**, 235019 (2010) arXiv:1007.2729 [hep-th].

[110] S. Carlip, "Challenges for Emergent Gravity," Stud. Hist. Philos. Mod. Phys. **46**, 200 (2014) arXiv:1207.2504 [gr-qc].

[111] G. Stephenson, Nuovo Cimento 9, 263 (1958).

[112] P.W. Higgs, Nuovo Cimento 11, 816 (1959).

[113] C.W. Kilmister, D.L. Newman, Proc. Cambridge Phil. Soc. (Math. Phys. Sci.) 57, 851 (1961).

[114] C.N. Yang, "Integral formalism for gauge fields", Phys. Rev. Lett. 33, 445447 (1974)

[115] J. Julve and M. Tonin, "Quantum Gravity with Higher Derivative Terms," Nuovo Cim. B **46** (1978) 137.

[116] E.S. Fradkin, A.A. Tseytlin, Phys. Lett. **104 B**, 377 (1981); Nucl. Phys. **B 201**, 469 (1982).

[117] I. G. Avramidi and A. O. Barvinsky, "Asymptotic Freedom In Higher Derivative Quantum Gravity," Phys. Lett. B **159** (1985) 269.

[118] A. Salam and J. A. Strathdee, "Remarks On High-Energy Stability And Renormalizability Of Gravity Theory," Phys. Rev. D **18** (1978) 4480.

[119] A. Bonanno and M. Reuter, "Modulated Ground State of Gravity Theories with Stabilized Conformal Factor," Phys. Rev. D **87**, no. 8, 084019 (2013) arXiv:1302.2928 [hep-th].

[120] T. Biswas, E. Gerwick, T. Koivisto and A. Mazumdar, "Towards singularity and ghost free theories of gravity," Phys. Rev. Lett. **108** (2012) 031101 arXiv:1110.5249 [gr-qc].

[121] L. Modesto, "Super-renormalizable Quantum Gravity," Phys. Rev. D **86** (2012) 044005 arXiv:1107.2403 [hep-th].

[122] E. Tomboulis, "1/N Expansion And Renormalization In Quantum Gravity," Phys. Lett. B **70** (1977) 361.
E. Tomboulis, "Renormalizability And Asymptotic Freedom In Quantum Gravity," Phys. Lett. B **97** (1980) 77.
E. T. Tomboulis, "Unitarity in Higher Derivative Quantum Gravity," Phys. Rev. Lett. **52** (1984) 1173.

[123] P. Horava, "Quantum Gravity at a Lifshitz Point," Phys. Rev. D **79**, 084008 (2009) arXiv:0901.3775 [hep-th].

[124] D. Blas, O. Pujolas and S. Sibiryakov, "Models of non-relativistic quantum gravity: The Good, the bad and the healthy," JHEP **1104** (2011) 018 arXiv:1007.3503 [hep-th].

[125] A. Contillo, S. Rechenberger and F. Saueressig, "Renormalization group flow of Hoava-Lifshitz gravity at low energies," JHEP **1312** (2013) 017 arXiv:1309.7273 [hep-th].

[126] D. Benedetti and F. Guarnieri, "One-loop renormalization in a toy model of Hoava-Lifshitz gravity," JHEP **1403** (2014) 078 arXiv:1311.6253 [hep-th].

[127] G. D'Odorico, F. Saueressig and M. Schutten, "Asymptotic Freedom in Hoava-Lifshitz Gravity," Phys. Rev. Lett. **113**, no. 17, 171101 (2014) arXiv:1406.4366 [gr-qc].

[128] G. D'Odorico, J. W. Goossens and F. Saueressig, "Covariant computation of effective actions in Horava-Lifshitz gravity," JHEP **1510** (2015) 126 arXiv:1508.00590 [hep-th].

[129] D. Benedetti and J. Henson, "Spectral geometry as a probe of quantum spacetime," Phys. Rev. D **80** (2009) 124036 arXiv:0911.0401 [hep-th].

[130] J. Ambjørn, A. Görlich, S. Jordan, J. Jurkiewicz and R. Loll, "CDT meets Horava-Lifshitz gravity," Phys. Lett. B 690 (2010) 413, arXiv:1002.3298.

[131] C. Anderson, S. J. Carlip, J. H. Cooperman, P. Horava, R. K. Kommu and P. R. Zulkowski, "Quantizing Horava-Lifshitz Gravity via Causal Dynamical Triangulations," Phys. Rev. D 85 (2012) 044027, arXiv:1111.6634.

[132] J. Ambjørn, A. Görlich, J. Jurkiewicz, A. Kreienbuhl and R. Loll, "Renormalization Group Flow in CDT," Class.Quant.Grav. 31 (2014) 165003, arXiv:1405.4585.

[133] R. Iengo, J. G. Russo and M. Serone, "Renormalization group in Lifshitz-type theories," JHEP 0911 (2009) 020, arXiv:0906.3477.

[134] M. T. Grisaru, P. van Nieuwenhuizen and J. A. M. Vermaseren, "One Loop Renormalizability of Pure Supergravity and of Maxwell-Einstein Theory in Extended Supergravity," Phys. Rev. Lett. **37** (1976) 1662.

[135] M. T. Grisaru, "Two Loop Renormalizability of Supergravity," Phys. Lett. B **66** (1977) 75.

[136] S. Deser, J. H. Kay and K. S. Stelle, "Renormalizability Properties of Supergravity," Phys. Rev. Lett. **38** (1977) 527 arXiv:1506.03757 [hep-th].

[137] Z. Bern, J. J. Carrasco, L. J. Dixon, H. Johansson, D. A. Kosower and R. Roiban, "Three-Loop Superfiniteness of N=8 Supergravity," Phys. Rev. Lett. **98** (2007) 161303 arXiv:hep-th/0702112.

[138] Z. Bern, J. J. Carrasco, L. J. Dixon, H. Johansson and R. Roiban, "The Ultraviolet Behavior of N=8 Supergravity at Four Loops," Phys. Rev. Lett. **103** (2009) 081301 arXiv:0905.2326 [hep-th].

[139] H. Elvang, D. Z. Freedman and M. Kiermaier, "A simple approach to counterterms in N=8 supergravity," JHEP **1011** (2010) 016 arXiv:1003.5018 [hep-th].
N. Beisert, H. Elvang, D. Z. Freedman, M. Kiermaier, A. Morales and S. Stieberger, "E7(7) constraints on counterterms in N=8 supergravity," Phys. Lett. B **694** (2010) 265 arXiv:1009.1643 [hep-th].

[140] Z. Bern, S. Davies and T. Dennen, "Enhanced ultraviolet cancellations in $\mathcal{N} = 5$ supergravity at four loops," Phys. Rev. D **90** (2014) 10, 105011 arXiv:1409.3089 [hep-th].

[141] S. Weinberg "What is quantum field theory, and what did we think it is?" in *Boston 1996, Conceptual foundations of quantum field theory* 241-251 arXiv: hep-th/9702027.

[142] S. Weinberg, "Phenomenological Lagrangians," Physica A **96** (1979) 327.

[143] J. Gasser and H. Leutwyler, "Chiral Perturbation Theory to One Loop," Annals Phys. **158** (1984) 142.

[144] J. F. Donoghue, "Leading quantum correction to the Newtonian potential," Phys. Rev. Lett. **72** (1994) 2996; arXiv:gr-qc/9310024];
J. F. Donoghue, "General Relativity As An Effective Field Theory: The Leading Quantum Corrections," Phys. Rev. D **50** (1994) 3874; arXiv:gr-qc/9405057].

[145] Cliff P. Burgess, "Quantum gravity in everyday life: General Relativity as an effective field theory", Living Rev. in Rel. 7, (2004), 5.

[146] D. Espriu and D. Puigdomenech, "Gravity as an effective theory," Acta Phys. Polon. B **40** (2009) 3409 arXiv:0910.4110 [hep-th].

[147] K. S. Stelle, "Classical Gravity with Higher Derivatives," Gen. Rel. Grav. **9** (1978) 353.

[148] D. Anselmi, "Absence of higher derivatives in the renormalization of propagators in quantum field theories with infinitely many couplings," Class. Quant. Grav. **20** (2003) 2355 hep-th/0212013.

[149] B. M. Barker, S. N. Gupta and R. D. Haracz, "One-Graviton Exchange Interaction of Elementary Particles," Phys. Rev. **149** (1966) 1027.

[150] M. J. Duff, "Quantum corrections to the Schwarzschild solution," Phys. Rev. D **9** (1974) 1837.

[151] H.W. Hamber, S. Liu, "On the quantum corrections to the Newtonian potential", Phys. Lett. B357, 51 (1995) arXiv:hep-th/9505182.

[152] A. Akhundov, S. Bellucci, A. Shiekh, "Gravitational interaction to one loop in effective quantum gravity" Phys. Lett. B395, 16 (1997).

[153] N.E.J. Bjerrum-Bohr (2002) "Leading quantum gravitational corrections to scalar QED", Phys. Rev. D66, 084023 arXiv:hep-th/0206236.

[154] I. B. Khriplovich and G. G. Kirilin, "Quantum power correction to the Newton law". J. Exp. Theor. Phys. **95** (2002) 981 [Zh. Eksp. Teor. Fiz. **95** (2002) 1139]; arXiv:gr-qc/0207118.

[155] N. E. J. Bjerrum-Bohr, J. F. Donoghue and B. R. Holstein, "Quantum gravitational corrections to the nonrelativistic scattering potential of two masses," Phys. Rev. D **67** (2003) 084033 [Erratum-ibid. D **71** (2005) 069903];arXiv:hep-th/0211072.

[156] I.B. Khriplovich, G.G. Kirilin, "Quantum long range interactions in general relativity", J. Exp. Theor. Phys. 98 (2004) 1063-1072 arXiv:gr-qc/0402018.

[157] Y. Iwasaki, "Quantum theory of gravitation vs. classical theory. - fourth-order potential," Prog. Theor. Phys. **46** (1971) 1587.

[158] N. E. J. Bjerrum-Bohr, J. F. Donoghue and B. R. Holstein, "Quantum corrections to the Schwarzschild and Kerr metrics," Phys. Rev. D **68** (2003) 084005 [Erratum-ibid. D **71** (2005) 069904]; arXiv:hep-th/0211071].

[159] J. F. Donoghue, B. R. Holstein, B. Garbrecht and T. Konstandin, "Quantum corrections to the Reissner-Nordstrom and Kerr-Newman metrics," Phys. Lett. B **529** (2002) 132, arXiv:hep-th/0112237.

[160] B. R. Holstein and J. F. Donoghue, "Classical physics and quantum loops," Phys. Rev. Lett. **93** (2004) 201602 arXiv:hep-th/0405239.

[161] N. E. J. Bjerrum-Bohr, "Leading quantum gravitational corrections to scalar QED," Phys. Rev. D **66** (2002) 084023, arXiv:hep-th/0206236.

[162] S. Faller, "Effective Field Theory of Gravity: Leading Quantum Gravitational Corrections to Newtons and Coulombs Law," Phys. Rev. D **77** (2008) 124039 arXiv:0708.1701 [hep-th].

[163] N. E. J. Bjerrum-Bohr, J. F. Donoghue and P. Vanhove, "On-shell Techniques and Universal Results in Quantum Gravity," JHEP **1402** (2014) 111 arXiv:1309.0804 [hep-th].

[164] A. Satz, A. Codello and F. D. Mazzitelli, "Low energy Quantum Gravity from the Effective Average Action," Phys. Rev. D **82** (2010) 084011 arXiv:1006.3808 [hep-th].

[165] C. F. Steinwachs and A. Y. Kamenshchik, Phys. Rev. D **84** (2011) 024026 doi:10.1103/PhysRevD.84.024026 arXiv:1101.5047 [gr-qc].

[166] A. Codello, R. Percacci, L. Rachwal and A. Tonero, "Computing the Effective Action with the Functional Renormalization Group," Eur. Phys. J. C76 (2016) no.4, 226 arXiv:1505.03119 [hep-th].

[167] J.W. York, J. Math. Phys. 14, 456 (1973).

[168] P. Candelas and D. J. Raine, "Feynman Propagator in Curved Space-Time," Phys. Rev. D **15** (1977) 1494.

[169] M. Visser, "How to Wick rotate generic curved spacetime" http://www.gravityresearchfoundation.org/pdf/awarded/1991/vissar.pdf.

[170] S.W. Hawking and G.F.R. Ellis, "The Large Scale Structure of Space-Time", Cambridge University Press (1973).

[171] A. Jaffe and G. Ritter, "Quantum field theory on curved backgrounds. I. The Euclidean functional integral," Commun. Math. Phys. **270**, 545 (2007) [hep-th/0609003].
A. Jaffe and G. Ritter, "Reflection Positivity and Monotonicity," J. Math. Phys. **49** (2008) 052301 arXiv:0705.0712 [math-ph].

[172] J. Ambjørn, J. Jurkiewicz and R. Loll, Phys. Rev. Lett. **85** 924 (2000) arXiv:hep-th/0002050]; Nucl. Phys. B **610** 347 (2001) arXiv:hep-th/0105267]; Phys. Rev. Lett. **93** 131301 (2004) arXiv:hep-th/0404156]; Phys. Rev. Lett. **95** 171301 (2005) arXiv:hep-th/0505113]; Phys. Rev. **D72** 064014 (2005) arXiv:hep-th/0505154].

[173] A. Lichnerowicz, "Propagateurs et commutateurs en relativité générale", Inst. Hautes Etude Sci. Publ. Math., 10 (1961).

[174] I.L. Buchbinder, S.D. Odintsov and I.L. Shapiro, "Effective action in quantum gravity", IOPP Publishing, Bristol (1992).

[175] J. Z. Simon, "The Stability of flat space, semiclassical gravity, and higher derivatives," Phys. Rev. D **43** (1991) 3308.

[176] G. W. Gibbons, S. W. Hawking and M. J. Perry, "Path Integrals and the Indefiniteness of the Gravitational Action," Nucl. Phys. B **138** (1978) 141.

[177] P. O. Mazur and E. Mottola, "The Gravitational Measure, Solution of the Conformal Factor Problem and Stability of the Ground State of Quantum Gravity," Nucl. Phys. B **341** (1990) 187.

[178] K. Schleich, "Conformal Rotation in Perturbative Gravity," Phys. Rev. D **36** (1987) 2342.

[179] Z. Bern, E. Mottola and S. K. Blau, "General covariance of the path integral for quantum gravity," Phys. Rev. D **43** (1991) 1212.

[180] E. Mottola, "Functional integration over geometries," J. Math. Phys. **36**, 2470 (1995) [hep-th/9502109].

[181] M. R. Gaberdiel, D. Grumiller and D. Vassilevich, JHEP 1011 (2010) 094 arXiv:1007.5189 [hep-th].

[182] H. -b. Zhang and X. Zhang, Class. Quant. Grav. 29 (2012) 145013 arXiv:1205.3681 [hep-th].

[183] J. Alexandre, N. Houston, N.E. Mavromatos, "Dynamical Supergravity Breaking via the Super-Higgs Effect Revisited", Phys. Rev. D88 (2013), arXiv:1310.4122 [hep-th].

[184] M. A. Rubin and C. R. Ordonez, "Symmetric Tensor Eigen Spectrum of the Laplacian on n Spheres," J. Math. Phys. **26** (1985) 65.

[185] R. Camporesi and A. Higuchi, "Spectral functions and zeta functions in hyperbolic spaces," J. Math. Phys. **35** (1994) 4217.

[186] A. H. Chamseddine and A. Connes, "The spectral action principle," Commun. Math. Phys. **186** (1997) 731 arXiv:hep-th/9606001].

[187] D. F. Litim, "Optimized renormalization group flows," Phys. Rev. D **64** (2001) 105007 [hep-th/0103195].
       "Critical exponents from optimised renormalisation group flows," Nucl. Phys. B **631** (2002) 128 arXiv:hep-th/0203006].

[188] C. Wetterich, "Average Action and the Renormalization Group Equations", Nucl. Phys. **B352** 529 (1991).

[189] C. Wetterich, "Exact Evolution Equation For The Effective Potential," Phys. Lett. **B 301** (1993) 90.

[190] T. R. Morris, "Derivative expansion of the exact renormalization group," Phys. Lett. B **329** (1994) 241, arXiv:hep-ph/9403340].

[191] T. R. Morris, "On truncations of the exact renormalization group," Phys. Lett. B **334** (1994) 355, arXiv:hep-th/9405190].

[192] K. G. Wilson and M. E. Fisher, "Critical exponents in 3.99 dimensions," Phys. Rev. Lett. **28** (1972) 240.

[193] T.R. Morris, Prog. Theor. Phys. Suppl. **131** 395-414 (1998) arXiv:hep-th/9802039].

[194] C. Bagnuls and C. Bervillier, "Exact renormalization group equations: An introductory review," Phys. Rept. **348** (2001) 91 arXiv:hep-th/0002034].

[195] J. Berges, N. Tetradis and C. Wetterich, "Non-perturbative renormalization flow in quantum field theory and statistical physics," Phys. Rept. **363** (2002) 223 arXiv:hep-ph/0005122].

[196] B. Delamotte, "An introduction to the nonperturbative renormalization group", Lect. Notes Phys. **852** (2012) 49–132 arXiv:cond-mat/0702365.

[197] M. Reuter and C. Wetterich, "Effective average action for gauge theories and exact evolution equations," Nucl. Phys. B **417**, 181 (1994).

[198] H. Gies, "Running coupling in Yang-Mills theory: A flow equation study," Phys. Rev. D **66** (2002) 025006 arXiv:hep-th/0202207].

[199] J. M. Pawlowski, "Aspects of the functional renormalisation group," Annals Phys. **322** (2007) 2831–2915 arXiv:hep-th/0512261.

[200] S. Arnone, T.R. Morris, O.J. Rosten Eur. Phys. J. **C50** 467-504 (2007) e-Print: hep-th/0507154; T.R. Morris, O.J. Rosten, J. Phys. **A39** 11657-11681 (2006) e-Print: hep-th/0606189.

[201] A. Codello, "Renormalization group flow equations for the proper vertices of the background effective average action," Phys. Rev. D **91**, no. 6, 065032 (2015) arXiv:1304.2059 [hep-th].

[202] T.A. Ryttov and F. Sannino, Phys. Rev. D 78 (2008) 065001, arXiv:0711.3745.

[203] R.I. Nepomechie, "Remarks on Quantized Yang-Mills Theory in Twentysix-dimensions" Phys. Lett. B128 (1983) 177.

[204] R. Camporesi and A. Higuchi, "On the Eigen functions of the Dirac operator on spheres and real hyperbolic spaces," J. Geom. Phys. **20** (1996) 1 arXiv:gr-qc/9505009.

[205] G. de Berredo-Peixoto, D. D. Pereira and I. L. Shapiro, "Universality and ambiguity in fermionic effective actions," Phys. Rev. D **85** (2012) 064025 arXiv:1201.2649 [hep-th].

[206] P. Doná and R. Percacci, "Functional renormalization with fermions and tetrads," Phys. Rev. D **87** (2013) 4, 045002 arXiv:1209.3649 [hep-th].

[207] M. Reuter, "Nonperturbative evolution equation for quantum gravity" Phys. Rev. **D57**, 971 (1998) arXiv:hep-th/9605030.

[208] D. Dou and R. Percacci, "The running gravitational couplings," Class. Quant. Grav. **15** (1998) 3449; arXiv:hep-th/9707239].

[209] K. G. Wilson and J. B. Kogut, "The Renormalization group and the epsilon expansion," Phys. Rept. **12** (1974) 75; K.G. Wilson, Rev. Mod. Phys. **47** 773 - 840 (1975).

[210] G. Parisi, "The Theory of Nonrenormalizable Interactions. 1. The Large N Expansion," Nucl. Phys. B **100** (1975) 368.

[211] K. Gawedzki, A. Kupiainen (1985a). "Renormalizing The Nonrenormalizable", Phys. Rev. Lett. **55** 363-365 (1985); Nucl. Phys. **B262** 33 (1985).

[212] C. de Calan, P.A. Faria da Veiga, J. Magnen, R. Seneor, "Constructing the three-dimensional Gross-Neveu model with a large number of flavor components". Phys. Rev. Lett. **66** 3233-3236 (1991).

[213] D. Becker and M. Reuter "En route to background independence: broken split-symmetry and how to restore it with bi-metric average actions", Annals Phys. 350 (2014) 225-301, arXiv:1404.4537 [hep-th].

[214] N. Ohta, R. Percacci and A. D. Pereira, "Gauges and functional measures in quantum gravity I: Einstein theory," JHEP **1606** (2016) 115, arXiv:1605.00454 [hep-th].

[215] G. de Berredo-Peixoto and I. L. Shapiro, "Conformal quantum gravity with the Gauss-Bonnet term," Phys. Rev. D **70** (2004) 044024 [hep-th/0307030].

[216] G. de Berredo-Peixoto and I. L. Shapiro, "Higher derivative quantum gravity with Gauss-Bonnet term," Phys. Rev. D **71** (2005) 064005 [hep-th/0412249].

[217] A. Codello and R. Percacci, "Fixed points of higher derivative gravity," Phys. Rev. Lett. **97** (2006) 221301 [hep-th/0607128].

[218] M. Niedermaier, "Gravitational Fixed Points from Perturbation Theory," Phys. Rev. Lett. **103** (2009) 101303;
"Gravitational fixed points and asymptotic safety from perturbation theory," Nucl. Phys. B **833** (2010) 226.

[219] N. Ohta and R. Percacci, "Higher Derivative Gravity and Asymptotic Safety in Diverse Dimensions," Class. Quant. Grav. **31** (2014) 015024 arXiv:1308.3398 [hep-th].

[220] N. H. Barth and S. M. Christensen, "Quantizing Fourth Order Gravity Theories. 1. The Functional Integral," Phys. Rev. D **28** (1983) 1876.

[221] I. G. Avramidi, "Covariant methods for the calculation of the effective action in quantum field theory and investigation of higher derivative quantum gravity," hep-th/9510140.

[222] A. O. Barvinsky and G. A. Vilkovisky, "The Generalized Schwinger-Dewitt Technique in Gauge Theories and Quantum Gravity," Phys. Rept. **119** (1985) 1.

[223] H. W. Lee, P. Y. Pac and H. K. Shin, "New Algorithm for Asymptotic Expansions of the Heat Kernel," Phys. Rev. D **35** (1987) 2440.

[224] V. P. Gusynin, "Seeley-gilkey Coefficients For The Fourth Order Operators On A Riemannian Manifold," Nucl. Phys. B **333** (1990) 296.

[225] V. P. Gusynin and V. V. Kornyak, "Complete Computation of DeWitt-Seeley-Gilkey Coefficient E_4 for Nonminimal Operator on Curved Manifolds," arXiv:math/9909145.

[226] K. Groh, S. Rechenberger, F. Saueressig and O. Zanusso, "Higher Derivative Gravity from the Universal Renormalization Group Machine," PoS EPS **-HEP2011** (2011) 124 arXiv:1111.1743 [hep-th].

[227] Bergshoeff, E. A., Hohm, O. and Townsend, P. K., "Massive Gravity in Three Dimensions", Phys. Rev. Lett., 102, 201301 (2009). arXiv:0901.1766 [hep-th].

[228] N. Ohta, "Beta function and asymptotic safety in three-dimensional higher derivative gravity", Class. Quantum Grav. 29 205012 (2012) arXiv:1205.0476 [hep-th].

[229] S. Deser, R. Jackiw and S. Templeton, "Topologically massive gauge theories," Annals Phys. **140** (1982) 372. Erratum: ibid. **185** (1988) 406.

[230] S. Deser and Z. Yang, "Is Topologically Massive Gravity Renormalizable?," Class. Quant. Grav. **7** (1990) 1603.

[231] B. Keszthelyi and G. Kleppe, "Renormalizability of D = 3 topologically massive gravity," Phys. Lett. B **281** (1992) 33.

[232] I. Oda, "Renormalizability of Topologically Massive Gravity," arXiv:0905.1536 [hep-th].

[233] R. Percacci and E. Sezgin, "One Loop Beta Functions in Topologically Massive Gravity," Class. Quant. Grav. **27** (2010) 155009 arXiv:1002.2640 [hep-th].

[234] R. Percacci, M. J. Perry, C. N. Pope and E. Sezgin, "Beta Functions of Topologically Massive Supergravity," JHEP **1403** (2014) 083 arXiv:1302.0868 [hep-th].

[235] E. Witten, "Analytic Continuation Of Chern-Simons Theory," AMS/IP Stud. Adv. Math. **50** (2011) 347 arXiv:1001.2933 [hep-th].

[236] R. D. Pisarski and S. Rao, "Topologically Massive Chromodynamics in the Perturbative Regime," Phys. Rev. D **32** (1985) 2081.

[237] E. Witten, "Quantum Field Theory and the Jones Polynomial," Commun. Math. Phys. **121** (1989) 351.

[238] M. A. Shifman, "Four-dimension aspect of the perturbative renormalization in three-dimensional Chern-Simons theory," Nucl. Phys. B **352** (1991) 87.

[239] M. Reuter, "Effective average action of Chern-Simons field theory," Phys. Rev. D **53** (1996) 4430 [hep-th/9511128].

[240] R. Percacci, "On the Topological Mass in Three-dimensional Gravity," Annals Phys. **177** (1987) 27.

[241] R. K. Gupta and A. Sen, "Consistent Truncation to Three Dimensional (Super)-gravity," JHEP **0803** (2008) 015 arXiv:0710.4177 [hep-th].

[242] P. Donà, A. Eichhorn and R. Percacci, "Matter matters in asymptotically safe quantum gravity," Phys. Rev. D **89**, no. 8, 084035 (2014) arXiv:1311.2898 [hep-th].

[243] A. Codello, G. D'Odorico and C. Pagani, "Consistent closure of renormalization group flow equations in quantum gravity," Phys. Rev. D **89** (2014) 8, 081701 arXiv:1304.4777 [gr-qc].

[244] G. Cognola, E. Elizalde, S. Nojiri, S. D. Odintsov and S. Zerbini, "One-loop $f(R)$ gravity in de Sitter universe," JCAP **0502** (2005) 010; arXiv:hep-th/0501096].

[245] A. Codello, R. Percacci and C. Rahmede, "Ultraviolet properties of f(R)-gravity," Int. J. Mod. Phys. A **23** (2008) 143 arXiv:0705.1769 [hep-th].

[246] A. Codello, R. Percacci and C. Rahmede, "Investigating the ultraviolet properties of gravity with a Wilsonian renormalization group equation", Ann. Phys. 324, 414-469 (2009) arXiv:0805.2909[hep-th].

[247] K. Falls, D. F. Litim, K. Nikolakopoulos and C. Rahmede, "A bootstrap towards asymptotic safety" arXiv:1301.4191 [hep-th];
K. Falls, D. F. Litim, K. Nikolakopoulos and C. Rahmede, "Further evidence for asymptotic safety of quantum gravity," Phys. Rev. D93 (2016) no.10, 104022 arXiv:1410.4815 [hep-th].

[248] L. Smolin, "A fixed point for quantum gravity", Nucl. Phys. B 208, 439-466 (1982).

[249] R. Percacci, "Further evidence for a gravitational fixed point," Phys. Rev. D **73** (2006) 041501 [hep-th/0511177].

[250] D. F. Litim and F. Sannino, "Asymptotic safety guaranteed," JHEP **1412** (2014) 178 arXiv:1406.2337 [hep-th].

[251] P. van Nieuwenhuizen, "Classical Gauge Fixing in Quantum Field Theory," Phys. Rev. D **24** (1981) 3315.

[252] R. P. Woodard, "The Vierbein Is Irrelevant in Perturbation Theory," Phys. Lett. B **148** (1984) 440.

[253] V. Silveira and A. Zee, "Scalar Phantoms," Phys. Lett. B **161** (1985) 136.

[254] J. McDonald, "Gauge singlet scalars as cold dark matter," Phys. Rev. D **50** (1994) 3637 [hep-ph/0702143 [HEP-PH].

[255] C. P. Burgess, M. Pospelov and T. ter Veldhuis, "The Minimal model of nonbaryonic dark matter: A Singlet scalar," Nucl. Phys. B **619** (2001) 709 [hep-ph/0011335].

[256] J. M. Cline, K. Kainulainen, P. Scott and C. Weniger, "Update on scalar singlet dark matter," Phys. Rev. D **88** (2013) 055025 [Phys. Rev. D **92** (2015) 3, 039906] arXiv:1306.4710 [hep-ph].

[257] R. D. Peccei and H. R. Quinn, "CP Conservation in the Presence of Instantons," Phys. Rev. Lett. **38** (1977) 1440.

[258] S. Weinberg, "A New Light Boson?," Phys. Rev. Lett. **40** (1978) 223.

[259] F. Wilczek, "Problem of Strong p and t Invariance in the Presence of Instantons," Phys. Rev. Lett. **40** (1978) 279.

[260] R. Essig *et al.*, "Working Group Report: New Light Weakly Coupled Particles," arXiv:1311.0029 [hep-ph].

[261] S. Bertolini, L. Di Luzio and M. Malinsky, "Intermediate mass scales in the non-supersymmetric SO(10) grand unification: A Reappraisal," Phys. Rev. D **80** (2009) 015013 arXiv:0903.4049 [hep-ph].

[262] F. Sannino, "Conformal Dynamics for TeV Physics and Cosmology," Acta Phys. Polon. B **40** (2009) 3533 arXiv:0911.0931 [hep-ph].

[263] P. Fischer and D. F. Litim, "Fixed points of quantum gravity in extra dimensions," Phys. Lett. B **638** (2006) 497 arXiv:hep-th/0602203].

[264] D. F. Litim and T. Plehn, "Signatures of gravitational fixed points at the LHC," Phys. Rev. Lett. **100** (2008) 131301 arXiv:0707.3983 [hep-ph];

[265] D. F. Litim and T. Plehn, "Virtual Gravitons at the LHC," arXiv:0710.3096 [hep-ph];

[266] E. Gerwick, D. Litim and T. Plehn, "Asymptotic safety and Kaluza-Klein gravitons at the LHC," Phys. Rev. D **83** (2011) 084048 arXiv:1101.5548 [hep-ph].

[267] B. Koch, "Black Hole Resonances or no Black Holes due to Large Extra Dimensions with Gravitational Fixed Point?," arXiv:0707.4644 [hep-ph].

[268] B. Dobrich and A. Eichhorn, "Can we see quantum gravity? Photons in the asymptotic-safety scenario," JHEP **1206** (2012) 156 arXiv:1203.6366 [gr-qc].

[269] G. Aad *et al.* [ATLAS Collaboration], "Search for dark matter candidates and large extra dimensions in events with a jet and missing transverse momentum with the ATLAS detector," JHEP **1304** (2013) 075 arXiv:1210.4491 [hep-ex].

[270] G. Aad *et al.* [ATLAS Collaboration], "Search for Extra Dimensions in diphoton events using proton-proton collisions recorded at $\sqrt{s} = 7$ TeV with the ATLAS detector at the LHC," New J. Phys. **15** (2013) 043007 arXiv:1210.8389 [hep-ex].

[271] S. M. Christensen and M. J. Duff, "Quantum Gravity In Two + Epsilon Dimensions," Phys. Lett. B **79** (1978) 213.

[272] R. Gastmans, R. Kallosh and C. Truffin, "Quantum Gravity Near Two-Dimensions," Nucl. Phys. B **133** (1978) 417.

[273] I. Jack and D. R. T. Jones, "The Epsilon expansion of two-dimensional quantum gravity," Nucl. Phys. B **358** (1991) 695.

[274] H. Kawai and M. Ninomiya, "Renormalization Group and Quantum Gravity," Nucl. Phys. B **336** (1990) 115;

[275] H. Kawai, Y. Kitazawa and M. Ninomiya, "Ultraviolet stable fixed point and scaling relations in (2+epsilon)-dimensional quantum gravity," Nucl. Phys. B **404** (1993) 684 arXiv:hep-th/9303123].

[276] T. Aida, Y. Kitazawa, J. Nishimura and A. Tsuchiya, "Two loop renormalization in quantum gravity near two-dimensions," Nucl. Phys. B **444** (1995) 353 arXiv:hep-th/9501056].

[277] L.N. Granda, Sergei D. Odintsov "Exact renormalization group for O(4) gauged supergravity", Phys. Lett. B409 206-212 (1997) arXiv:hep-th/9706062.

[278] A.A. Bytsenko, L.N. Granda, Sergei D. Odintsov (1997) "Exact renormalization group and running Newtonian coupling in higher derivative gravity". JETP Lett. 65, 600-604 arXiv:hep-th/9705008.

[279] L.N. Granda, Sergei D. Odintsov (1998) "Effective average action and nonperturbative renormalization group equation in higher derivative quantum gravity", Grav. Cosmol. 4, 85-95 arXiv:gr-qc/9801026.

[280] W. Souma, Prog. Theor. Phys. "Nontrivial ultraviolet fixed point in quantum gravity", **102**, 181 (1999); arXiv:hep-th/9907027].

[281] O. Lauscher and M. Reuter, "Ultraviolet fixed point and generalized flow equation of quantum gravity", Phys. Rev. **D65**, 025013 (2002); arXiv:hep-th/0108040].

[282] O. Lauscher and M. Reuter, "Is quantum Einstein gravity nonperturbatively renormalizable?" Class. Quant. Grav. **19**, 483 (2002); arXiv:hep-th/0110021];
O. Lauscher and M. Reuter, "Towards nonperturbative renormalizability of quantum Einstein gravity", Int. J. Mod. Phys. **A 17**, 993 (2002); arXiv:hep-th/0112089];
M. Reuter and F. Saueressig, "Renormalization group flow of quantum gravity in the Einstein–Hilbert truncation", Phys. Rev. **D65**, 065016 (2002). arXiv:hep-th/0110054].

[283] D. F. Litim, "Fixed points of quantum gravity," Phys. Rev. Lett. **92** (2004) 201301 [hep-th/0312114].

[284] R. Floreanini and R. Percacci, "The Heat kernel and the average effective potential," Phys. Lett. B **356** (1995) 205 [hep-th/9505172].

[285] S. B. Liao, "On connection between momentum cutoff and the proper time regularizations," Phys. Rev. D **53** (1996) 2020 [hep-th/9501124].
S. B. Liao, "Operator cutoff regularization and renormalization group in Yang-Mills theory," Phys. Rev. D **56** (1997) 5008 [hep-th/9511046].

[286] D. Zappala, "Perturbative and non-perturbative aspects of the proper time renormalization group," Phys. Rev. D **66** (2002) 105020 arXiv:hep-th/0202167];
M. Mazza and D. Zappala, "Proper time regulator and renormalization group flow," Phys. Rev. D **64** (2001) 105013 arXiv:hep-th/0106230].

[287] A. Bonanno and M. Reuter, "Proper time flow equation for gravity," JHEP **0502** (2005) 035 arXiv:hep-th/0410191].

[288] O. Lauscher and M. Reuter, "Flow equation of quantum Einstein gravity in a higher derivative truncation", Phys. Rev. **D 66**, 025026 (2002) arXiv:hep-th/0205062].

[289] M. Reuter and F. Saueressig, "A Class of nonlocal truncations in quantum Einstein gravity and its renormalization group behavior" Phys. Rev. D66, 125001 arXiv:hep-th/0206145;
M. Reuter, F. Saueressig, "Nonlocal quantum gravity and the size of the universe". Fortsch. Phys. 52, 650-654 (2004) arXiv:hep-th/0311056.

[290] P. Forgacs and M. Niedermaier, "A fixed point for truncated quantum Einstein gravity," arXiv:hep-th/0207028; M. Niedermaier, *Nucl. Phys.* **B 673**, 131-169 (2003) arXiv:hep-th/0304117].

[291] R. Percacci and D. Perini, Phys. Rev. **D67**, 081503(R) (2003) arXiv:hep-th/0207033].

[292] R. Percacci and D. Perini, Phys. Rev. **D68**, 044018 (2003); arXiv:hep-th/0304222].

[293] O. Lauscher and M. Reuter "Fractal spacetime structure in asymptotically safe gravity", JHEP 0510, 050 (2005) arXiv:hep-th/0508202;
M. Reuter and J.-M. Schwindt "Scale-dependent metric and causal structures in Quantum Einstein Gravity", JHEP 0701, 049 (2007a) arXiv:hep-th/0611294.

[294] R. Percacci and D. Perini, "On the ultraviolet behaviour of Newton's constant", (archive version titled "Should we expect a fixed point for Newton's constant?") Class. Quant. Grav. **21** (2004) 5035 [hep-th/0401071];
R. Percacci, "The Renormalization group, systems of units and the hierarchy problem," J. Phys. A **40** (2007) 4895 [hep-th/0409199].

[295] M. Reuter and J.-M. Schwindt "A Minimal length from the cutoff modes in asymptotically safe quantum gravity", JHEP 0601, 070 (2006) arXiv:hep-th/0511021.

[296] A. Bonanno and M. Reuter, "Renormalization group improved black hole spacetimes," Phys. Rev. D **62** (2000) 043008 [hep-th/0002196];
A. Bonanno and M. Reuter, "Spacetime structure of an evaporating black hole in quantum gravity," Phys. Rev. D **73** (2006) 083005 [hep-th/0602159].

[297] A. Bonanno and M. Reuter, "Cosmology of the Planck era from a renormalization group for quantum gravity," Phys. Rev. D **65** (2002) 043508 [hep-th/0106133];
A. Bonanno and M. Reuter, "Cosmological perturbations in renormalization group derived cosmologies," Int. J. Mod. Phys. D **13** (2004) 107 [astro-ph/0210472].

[298] A. Bonanno and M. Reuter, "Cosmology with selfadjusting vacuum energy density from a renormalization group fixed point," Phys. Lett. B **527** (2002) 9 [astro-ph/0106468];
E. Bentivegna, A. Bonanno and M. Reuter, "Confronting the IR fixed point cosmology with high redshift supernova data," JCAP **0401** (2004) 001 [astro-ph/0303150].

[299] M. Reuter and H. Weyer, "Quantum gravity at astrophysical distances?," JCAP **0412** (2004) 001 [hep-th/0410119].

[300] M. Reuter and H. Weyer, "Running Newton constant, improved gravitational actions, and galaxy rotation curves," Phys. Rev. D **70** (2004) 124028 [hep-th/0410117].

[301] M. Reuter and H. Weyer, "Renormalization group improved gravitational actions: A Brans-Dicke approach," Phys. Rev. D **69** (2004) 104022 [hep-th/0311196].

[302] M. Reuter and F. Saueressig, "From big bang to asymptotic de Sitter: Complete cosmologies in a quantum gravity framework," JCAP **0509** (2005) 012 [hep-th/0507167].

[303] Max Niedermaier and Martin Reuter, "The Asymptotic Safety Scenario in Quantum Gravity", Living Rev. Relativity 9, (2006), 5.

[304] M. Niedermaier, "The asymptotic safety scenario in quantum gravity: An introduction," Class. Quant. Grav. **24** (2007) R171 arXiv:gr-qc/0610018].

[305] R. Percacci, "Asymptotic Safety", in "Approaches to Quantum Gravity: Towards a New Understanding of Space, Time and Matter" ed. D. Oriti, Cambridge University Press; e-Print: arXiv:0709.3851 [hep-th].

[306] D. Benedetti, P.F. Machado and F. Saueressig "Asymptotic safety in higher-derivative gravity", Mod. Phys. Lett. A24, 2233-2241 (2009) arXiv:0901.2984 [hep-th].

[307] D. Benedetti, P.F. Machado and F. Saueressig, "Taming perturbative divergences in asymptotically safe gravity", Nucl. Phys. B824, 168-191 (2010). arXiv:0902.4630 [hep-th].

[308] S. Rechenberger and F. Saueressig, "The R^2 phase diagram of QEG and its spectral dimension", Phys.Rev. D86 (2012) 024018 arXiv:1206.0657 [hep-th].

[309] P. F. Machado and F. Saueressig, "On the renormalization group flow of f(R)-gravity," arXiv:0712.0445 [hep-th].

[310] G. Narain and R. Percacci (2009b) Renormalization group flow in scalar-tensor theories I. Class. and Quantum Grav. 27, 075001 (2010) arXiv:0911.0386 [hep-th].

[311] G. Narain and C. Rahmede (2009) Renormalization group flow in scalar-tensor theories II. Class. and Quantum Grav. 27, 075002 (2010) arXiv:0911.0394 [hep-th].

[312] U. Harst and M. Reuter, "QED coupled to QEG," JHEP **1105** (2011) 119 arXiv:1101.6007 [hep-th].

[313] S. P. Robinson and F. Wilczek, Phys. Rev. Lett. **96** (2006) 231601 [hep-th/0509050].

[314] A. R. Pietrykowski, Phys. Rev. Lett. **98** (2007) 061801 [hep-th/0606208].

[315] D. Ebert, J. Plefka and A. Rodigast, "Absence of gravitational contributions to the running Yang-Mills coupling," Phys. Lett. B **660** (2008) 579 arXiv:0710.1002 [hep-th].

[316] D. J. Toms, "Quantum gravity and charge renormalization," Phys. Rev. D **76** (2007) 045015 arXiv:0708.2990 [hep-th];
D. J. Toms, "Cosmological constant and quantum gravitational corrections to the running fine structure constant," Phys. Rev. Lett. **101** (2008) 131301 arXiv:0809.3897 [hep-th].

[317] Y. Tang and Y. L. Wu, "Gravitational Contributions to the Running of Gauge Couplings," Commun. Theor. Phys. **54** (2010) 1040 arXiv:0807.0331 [hep-ph].

[318] J. E. Daum, U. Harst and M. Reuter, "Running Gauge Coupling in Asymptotically Safe Quantum Gravity," JHEP **1001** (2010) 084 arXiv:0910.4938 [hep-th].

[319] M. M. Anber, J. F. Donoghue and M. El-Houssieny, "Running couplings and operator mixing in the gravitational corrections to coupling constants," Phys. Rev. D **83** (2011) 124003 arXiv:1011.3229 [hep-th].

[320] S. Folkerts, D. F. Litim and J. M. Pawlowski, "Asymptotic freedom of Yang-Mills theory with gravity," Phys. Lett. B **709** (2012) 234 arXiv:1101.5552 [hep-th].

[321] A. Nink and M. Reuter, "On the physical mechanism underlying Asymptotic Safety," JHEP **1301** (2013) 062 arXiv:1208.0031 [hep-th].

[322] U. Harst and M. Reuter, "The 'Tetrad only' theory space: Nonperturbative renormalization flow and Asymptotic Safety," JHEP **1205** (2012) 005 arXiv:1203.2158 [hep-th].

[323] J.-E. Daum and M. Reuter, "Renormalization Group Flow of the Holst Action," Phys. Lett. B **710** (2012) 215 arXiv:1012.4280 [hep-th];
J. E. Daum and M. Reuter, "Running Immirzi Parameter and Asymptotic Safety," PoS CNCFG **2010** (2010) 003 arXiv:1111.1000 [hep-th];
J. E. Daum and M. Reuter, "Einstein-Cartan gravity, Asymptotic Safety, and the running Immirzi parameter," Annals Phys. **334** (2013) 351 arXiv:1301.5135 [hep-th].

[324] U. Harst and M. Reuter, "A new functional flow equation for EinsteinCartan quantum gravity," Annals Phys. **354** (2015) 637 arXiv:1410.7003 [hep-th].

[325] D. Benedetti and S. Speziale, "Perturbative quantum gravity with the Immirzi parameter," JHEP **1106** (2011) 107 arXiv:1104.4028 [hep-th].

[326] C. Pagani and R. Percacci, "Quantum gravity with torsion and non-metricity," arXiv:1506.02882 [gr-qc].

[327] K. Groh and F. Saueressig, "Ghost wave-function renormalization in Asymptotically Safe Quantum Gravity," J. Phys. A **43** (2010) 365403 arXiv:1001.5032 [hep-th].

[328] A. Eichhorn and H. Gies, "Ghost anomalous dimension in asymptotically safe quantum gravity," Phys. Rev. D **81** (2010) 104010 arXiv:1001.5033 [hep-th].

[329] A. Eichhorn, H. Gies and M. M. Scherer, "Asymptotically free scalar curvature-ghost coupling in Quantum Einstein Gravity," Phys. Rev. D **80** (2009) 104003 arXiv:0907.1828 [hep-th].

[330] E. Manrique, S. Rechenberger and F. Saueressig, "Asymptotically Safe Lorentzian Gravity," Phys. Rev. Lett. **106** (2011) 251302 arXiv:1102.5012 [hep-th].

[331] M. Reuter and H. Weyer, "Background Independence and Asymptotic Safety in Conformally Reduced Gravity," Phys. Rev. D **79** (2009) 105005 arXiv:0801.3287 [hep-th].

[332] M. Reuter and H. Weyer, "Conformal sector of Quantum Einstein Gravity in the local potential approximation: Non-Gaussian fixed point and a phase of unbroken diffeomorphism invariance," Phys. Rev. D **80** (2009) 025001 arXiv:0804.1475 [hep-th].

[333] P.F. Machado and R. Percacci, "Conformally reduced quantum gravity revisited", Phys. Rev. D80, 024020 (2009) arXiv:0904.2510 [hep-th].

[334] I. Antoniadis and E. Mottola, Phys. Rev. D45 2013 (1992).

[335] A. Bonanno and F. Guarnieri, "Universality and symmetry breaking in conformally reduced quantum gravity", Phys. Rev. D 86, 105027 (2012) arXiv:1206.6531 [hep-th].

[336] K. Falls, "Asymptotic safety and the cosmological constant," JHEP 1601 (2016) 069 arXiv:1408.0276 [hep-th].

[337] R. Percacci, "Renormalization group flow of Weyl invariant dilaton gravity," New J. Phys. **13** (2011) 125013 arXiv:1110.6758 [hep-th].

[338] C. Pagani and R. Percacci, "Quantization and fixed points of non-integrable Weyl theory," Class. Quant. Grav. **31** (2014) 115005 arXiv:1312.7767 [hep-th].

[339] A. Codello, G. D'Odorico, C. Pagani and R. Percacci, "The Renormalization Group and Weyl-invariance," Class. Quant. Grav. **30** (2013) 115015 arXiv:1210.3284 [hep-th].

[340] A. Codello, G. D'Odorico and C. Pagani, "A functional RG equation for the c-function," JHEP **1407** (2014) 040 arXiv:1312.7097 [hep-th].

[341] A. Codello and G. DOdorico, "Scaling and Renormalization in two dimensional Quantum Gravity," Phys. Rev. D **92** (2015) 2, 024026 arXiv:1412.6837 [gr-qc].

[342] A. Codello, G. DOdorico and C. Pagani, "Functional and Local Renormalization Groups," Phys. Rev. D **91** (2015) 12, 125016 arXiv:1502.02439 [hep-th].

[343] D. L. Jafferis, I. R. Klebanov, S. S. Pufu and B. R. Safdi, "Towards the F-Theorem: N=2 Field Theories on the Three-Sphere," JHEP **1106** (2011) 102 arXiv:1103.1181 [hep-th].

[344] D. Becker and M. Reuter, "Towards a *C*-function in 4D quantum gravity," JHEP **1503** (2015) 065 arXiv:1412.0468 [hep-th].

[345] J. E. Daum and M. Reuter, "Effective Potential of the Conformal Factor: Gravitational Average Action and Dynamical Triangulations," Adv. Sci. Lett. **2** (2009) 255 arXiv:0806.3907 [hep-th].

[346] Martin Reuter and Frank Saueressig, "Fractal space-times under the microscope: a renormalization group view of Monte Carlo data", JHEP 1112, 012 (2011) arXiv:1110.5224 [hep-th].

[347] M. Reuter and F. Saueressig, "Quantum Einstein Gravity," New J. Phys. **14** (2012) 055022 arXiv:1202.2274 [hep-th].

[348] G. Calcagni, A. Eichhorn and F. Saueressig, "Probing the quantum nature of spacetime by diffusion," Phys. Rev. D **87** (2013) 12, 124028 arXiv:1304.7247 [hep-th].

[349] N. Alkofer, F. Saueressig and O. Zanusso, "Spectral dimensions from the spectral action," Phys. Rev. D **91** (2015) 2, 025025 arXiv:1410.7999 [hep-th].

[350] Y. F. Cai and D. A. Easson, "Asymptotically safe gravity as a scalar-tensor theory and its cosmological implications," Phys. Rev. D **84** (2011) 103502 arXiv:1107.5815 [hep-th].

[351] B. Koch and I. Ramirez, "Exact renormalization group with optimal scale and its application to cosmology," Class. Quant. Grav. **28** (2011) 055008 arXiv:1010.2799 [gr-qc].

[352] S. Domazet and H. Stefancic, "Renormalization group scale-setting in astrophysical systems," Phys. Lett. B **703** (2011) 1 arXiv:1010.3585 [gr-qc].

[353] A. Bonanno and M. Reuter, "Entropy signature of the running cosmological constant," JCAP **0708** (2007) 024 arXiv:0706.0174 [hep-th].

[354] M. Hindmarsh, D. Litim and C. Rahmede, "Asymptotically Safe Cosmology," JCAP **1107** (2011) 019 arXiv:1101.5401 [gr-qc].

[355] S. Weinberg, "Asymptotically Safe Inflation," Phys. Rev. D **81** (2010) 083535 arXiv:0911.3165 [hep-th].

[356] S.-H. H. Tye and J. Xu, "Comment on Asymptotically Safe Inflation," Phys. Rev. D **82** (2010) 127302 arXiv:1008.4787 [hep-th].

[357] A. Bonanno, "An effective action for asymptotically safe gravity," Phys. Rev. D **85** (2012) 081503 arXiv:1203.1962 [hep-th].

[358] M. Hindmarsh and I. D. Saltas, "f(R) Gravity from the renormalisation group," Phys. Rev. D **86** (2012) 064029 arXiv:1203.3957 [gr-qc].

[359] E. J. Copeland, C. Rahmede and I. D. Saltas, "Asymptotically Safe Starobinsky Inflation," Phys. Rev. D **91** (2015) 10, 103530 arXiv:1311.0881 [gr-qc].

[360] B. F. L. Ward, "Planck Scale Cosmology in Resummed Quantum Gravity," Mod. Phys. Lett. A **23** (2008) 3299 arXiv:0808.3124 [gr-qc].

[361] A. Bonanno, A. Contillo and R. Percacci, "Inflationary solutions in asymptotically safe f(R) theories," Class. Quant. Grav. **28** (2011) 145026 arXiv:1006.0192 [gr-qc].

[362] A. Contillo, M. Hindmarsh and C. Rahmede, "Renormalisation group improvement of scalar field inflation," Phys. Rev. D **85** (2012) 043501 arXiv:1108.0422 [gr-qc].

[363] S. E. Hong, Y. J. Lee and H. Zoe, "The Possibility of Inflation in Asymptotically Safe Gravity," Int. J. Mod. Phys. D **21** (2012) 1250062 arXiv:1108.5886 [gr-qc].

[364] Y. F. Cai, Y. C. Chang, P. Chen, D. A. Easson and T. Qiu, "Planck constraints on Higgs modulated reheating of renormalization group improved inflation", Phys. Rev. D **88** (2013) 083508 arXiv:1304.6938 [hep-th].

[365] T. Henz, J. M. Pawlowski, A. Rodigast and C. Wetterich, "Dilaton Quantum Gravity," Phys. Lett. B **727** (2013) 298 arXiv:1304.7743 [hep-th].

[366] Z. Z. Xianyu and H. J. He, "Asymptotically Safe Higgs Inflation," JCAP **1410** (2014) 083 arXiv:1407.6993 [astro-ph.CO].

[367] Y. F. Cai and D. A. Easson, "Black holes in an asymptotically safe gravity theory with higher derivatives," JCAP **1009** (2010) 002 arXiv:1007.1317 [hep-th].

[368] R. Casadio, S. D. H. Hsu and B. Mirza, "Asymptotic Safety, Singularities, and Gravitational Collapse," Phys. Lett. B **695** (2011) 317 arXiv:1008.2768 [gr-qc].

[369] G. Kofinas and V. Zarikas, "Avoidance of singularities in asymptotically safe Quantum Einstein Gravity," JCAP 1510 (2015) no.10, 069 arXiv:1506.02965 [hep-th].

[370] M. Reuter and E. Tuiran, "Quantum Gravity Effects in the Kerr Spacetime," Phys. Rev. D **83** (2011) 044041 arXiv:1009.3528 [hep-th].

[371] B. Koch and F. Saueressig, "Structural aspects of asymptotically safe black holes," Class. Quant. Grav. **31** (2014) 015006 arXiv:1306.1546 [hep-th].

[372] B. Koch, C. Contreras, P. Rioseco and F. Saueressig, "Black holes and running couplings: A comparison of two complementary approaches," arXiv:1311.1121 [hep-th].

[373] A. Shomer, "A Pedagogical explanation for the non-renormalizability of gravity," arXiv:0709.3555 [hep-th]. ·

[374] K. Falls, D. F. Litim and A. Raghuraman, "Black Holes and Asymptotically Safe Gravity," Int. J. Mod. Phys. A **27** (2012) 1250019 arXiv:1002.0260 [hep-th].

[375] K. Falls and D. F. Litim, "Black hole thermodynamics under the microscope," Phys. Rev. D **89** (2014) 084002 arXiv:1212.1821 [gr-qc].

[376] S. Basu and D. Mattingly, "Asymptotic Safety, Asymptotic Darkness, and the hoop conjecture in the extreme UV," Phys. Rev. D **82** (2010) 124017 arXiv:1006.0718 [hep-th].

[377] B. F. L. Ward, "Massive elementary particles and black holes," JCAP **0402** (2004) 011 [hep-ph/0312188].

[378] J. Hewett and T. Rizzo, "Collider Signals of Gravitational Fixed Points," JHEP **0712** (2007) 009 arXiv:0707.3182 [hep-ph].

[379] M. Shaposhnikov and C. Wetterich, "Asymptotic safety of gravity and the Higgs boson mass," Phys. Lett. B **683** (2010) 196 arXiv:0912.0208 [hep-th].

[380] F. Bezrukov, M. Y. Kalmykov, B. A. Kniehl and M. Shaposhnikov, "Higgs Boson Mass and New Physics," JHEP **1210** (2012) 140 arXiv:1205.2893 [hep-ph].

[381] D. F. Litim and J. M. Pawlowski, "Renormalization group flows for gauge theories in axial gauges," JHEP **0209** (2002) 049 [hep-th/0203005].

[382] M. M. Anber and J. F. Donoghue, "On the running of the gravitational constant," Phys. Rev. D **85** (2012) 104016 arXiv:1111.2875 [hep-th].

[383] R. Percacci and G. P. Vacca, "Search of scaling solutions in scalar-tensor gravity," Eur. Phys. J. C **75** (2015) 5, 188 arXiv:1501.00888 [hep-th].

[384] D. Benedetti, "Asymptotic safety goes on shell," New J. Phys. **14** (2012) 015005 arXiv:1107.3110 [hep-th].

[385] A. Eichhorn, "On unimodular quantum gravity," Class. Quant. Grav. **30** (2013) 115016 arXiv:1301.0879 [gr-qc].

[386] A. Nink, "Field Parametrization Dependence in Asymptotically Safe Quantum Gravity," Phys. Rev. D **91** (2015) 4, 044030 arXiv:1410.7816 [hep-th].

[387] M. Demmel and A. Nink, "Connections and geodesics in the space of metrics," Phys. Rev. D92 (2015) no.10, 104013 arXiv:1506.03809 [gr-qc].

[388] P. Labus, R. Percacci and G. P. Vacca, "Asymptotic safety in $O(N)$ scalar models coupled to gravity," Phys. Lett. B753 (2016) 274-281, arXiv:1505.05393 [hep-th].

[389] K. Falls, "On the renormalisation of Newton's constant," Phys. Rev. D92 (2015) no.12, 124057 arXiv:1501.05331 [hep-th].

[390] K. Falls, "Critical scaling in quantum gravity from the renormalisation group," arXiv:1503.06233 [hep-th].

[391] H. W. Hamber, "On the gravitational scaling dimensions," Phys. Rev. D **61** (2000) 124008 [hep-th/9912246].

[392] H. Gies, B. Knorr and S. Lippoldt, "Generalized Parametrization Dependence in Quantum Gravity," Phys. Rev. D92 (2015) no.8, 084020 arXiv:1507.08859 [hep-th].

[393] D. Benedetti, F. Caravelli, "The local potential approximation in quantum gravity" JHEP 1206 (2012) 017, Erratum-ibid. 1210 (2012) 157 arXiv:1204.3541 [hep-th].

[394] D. Benedetti, "On the number of relevant operators in asymptotically safe gravity," EuroPhys. Lett. **102** (2013) 20007 arXiv:1301.4422 [hep-th].

[395] J. A. Dietz and T. R. Morris, "Asymptotic safety in the f(R) approximation," JHEP **1301** (2013) 108 arXiv:1211.0955 [hep-th].

[396] J. A. Dietz and T. R. Morris, "Redundant operators in the exact renormalisation group and in the f(R) approximation to asymptotic safety," JHEP **1307** (2013) 064, arXiv:1306.1223 [hep-th].

[397] M. Demmel, F. Saueressig, O. Zanusso, "Fixed-functionals of three-dimensional Quantum Einstein Gravity," JHEP 1211 (2012) 131, arXiv:1208.2038 [hep-th].

[398] M. Demmel, F. Saueressig and O. Zanusso, "RG flows of Quantum Einstein Gravity on maximally symmetric spaces," JHEP **1406** (2014) 026 arXiv:1401.5495 [hep-th].

[399] M. Demmel, F. Saueressig, O. Zanusso, "RG flows of Quantum Einstein Gravity in the linear-geometric approximation," Annals Phys. 359 (2015) 141–165 arXiv:1412.7207 [hep-th];

[400] M. Demmel, F. Saueressig and O. Zanusso, "A proper fixed functional for four-dimensional Quantum Einstein Gravity," JHEP 1508 (2015) 113 arXiv:1504.07656 [hep-th].

[401] J. Borchardt and B. Knorr, "Global solutions of functional fixed point equations via pseudospectral methods," Phys. Rev. D **91** (2015) 10, 105011 arXiv:1502.07511 [hep-th].

[402] N. Ohta, R. Percacci and G. P. Vacca, "Flow equation for $f(R)$ gravity and some of its exact solutions," Phys. Rev. D **92** (2015) 6, 061501 arXiv:1507.00968 [hep-th]; "Renormalization Group Equation and scaling solutions for f(R) gravity in exponential parametrization," Eur. Phys. J. C76 (2016) no.2, 46 arXiv:1511.09393 [hep-th].

[403] L. F. Abbott, M. T. Grisaru and R. K. Schaefer, "The Background Field Method and the S Matrix," Nucl. Phys. B **229** (1983) 372.

[404] M. Safari, "Splitting Ward identity," Eur. Phys. J. C76 (2016) no.4, 201 arXiv:1508.06244 [hep-th].

[405] I. Donkin and J. Pawlowski, "The phase diagram of quantum gravity from diffeomorphism invariant RG flows" arXiv:1203.4207 [hep-th].

[406] Elisa Manrique and Martin Reuter (2009b) Bimetric Truncations for Quantum Einstein Gravity and Asymptotic Safety. Annals Phys. 325 785-815 (2010) arXiv:0907.2617 [hep-th].

[407] Elisa Manrique, Martin Reuter and Frank Saueressig (2010a) Matter induced bimetric actions for gravity. Ann. Phys. 326, 440-462 (2011) arXiv:1003.5129 [hep-th].

[408] Elisa Manrique, Martin Reuter and Frank Saueressig (2010b) Bimetric Renormalization Group Flows in Quantum Einstein Gravity. Ann. Phys. 326, 463-485 (2011) arXiv:1006.0099 [hep-th].

[409] N. Christiansen, D. Litim, J. Pawlowski and A. Rodigast, "Fixed points and infrared completion of quantum gravity", Phys. Lett. B728 (2014) 114-117 arXiv:1209.4038 [hep-th].

[410] N. Christiansen, B. Knorr, J. M. Pawlowski and A. Rodigast, "Global Flows in Quantum Gravity," Phys. Rev. D93 (2016) no.4, 044036 arXiv:1403.1232 [hep-th].

[411] N. Christiansen, B. Knorr, J. Meibohm, J. M. Pawlowski and M. Reichert, "Local Quantum Gravity," Phys. Rev. D92 (2015) no.12, 121501 arXiv:1506.07016 [hep-th].

[412] J. A. Dietz and T. R. Morris, "Background independent exact renormalization group for conformally reduced gravity," JHEP **1504** (2015) 118 arXiv:1502.07396 [hep-th].

[413] A. Codello, "Polyakov Effective Action from Functional Renormalization Group Equation," Annals Phys. **325** (2010) 1727 arXiv:1004.2171 [hep-th].

[414] A. Codello, "Large N Quantum Gravity," New J. Phys. **14** (2012) 015009 arXiv:1108.1908 [gr-qc].

[415] A. Codello and R. K. Jain, "Covariant Effective Field Theory of Gravity I: Formalism and Curvature expansion," arXiv:1507.06308 [gr-qc].

[416] A. Codello and R. K. Jain, "Covariant Effective Field Theory of Gravity II: Cosmological Implications," arXiv:1507.07829 [astro-ph.CO].

[417] A. Codello and A. Tonero, "A renormalization group improved computation of correlation functions in theories with non-trivial phase diagram," Phys. Rev. D94 (2016) no.2, 025015 arXiv:1504.00225 [hep-th].

[418] R. Percacci and G. P. Vacca, "Asymptotic Safety, Emergence and Minimal Length," Class. Quant. Grav. **27** (2010) 245026 arXiv:1008.3621 [hep-th].

[419] C. J. Isham, "Prima facie questions in quantum gravity," Lect. Notes Phys. **434** (1994) 1 [gr-qc/9310031].

Subject Index

Author Index

Printed in the United States
By Bookmasters